T0344497

A PHILOSOPHICAL APPROACH TO MOND

Dark matter is a fundamental component of the standard cosmological model, but in spite of four decades of increasingly sensitive searches, no-one has yet detected a single dark matter particle in the laboratory. An alternative cosmological paradigm exists: MOND (MOdified Newtonian Dynamics). Observations explained in the standard model by postulating dark matter are explained in MOND by proposing a modification of Newton's laws of motion. Both MOND and the standard model have had successes and failures – but only MOND has repeatedly predicted observational facts in advance of their discovery. In this volume, David Merritt outlines why such predictions are considered by many philosophers of science to be the 'gold standard' when it comes to judging a theory's validity. In a world where the standard model receives most attention, the author applies criteria from the philosophy of science to assess, in a systematic way, the viability of this alternative cosmological paradigm.

DAVID MERRITT was a professor at the Rochester Institute of Technology, and before that at Rutgers University, whose research interests include galaxy dynamics and evolution, supermassive black holes, and computational astrophysics. He is a former Chair of the Division on Dynamical Astronomy of the American Astronomical Society and is a founding member of the Center for Computational Relativity and Gravitation at RIT. He is author of the graduate textbook *Dynamics and Evolution of Galactic Nuclei*.

A PHILOSOPHICAL APPROACH TO MOND

Assessing the Milgromian Research Program in Cosmology

DAVID MERRITT

Rochester Institute of Technology, New York

CAMBRIDGE
UNIVERSITY PRESS

CAMBRIDGE
UNIVERSITY PRESS

Shaftesbury Road, Cambridge CB2 8EA, United Kingdom

One Liberty Plaza, 20th Floor, New York, NY 10006, USA

477 Williamstown Road, Port Melbourne, VIC 3207, Australia

314–321, 3rd Floor, Plot 3, Splendor Forum, Jasola District Centre, New Delhi – 110025, India

103 Penang Road, #05–06/07, Visioncrest Commercial, Singapore 238467

Cambridge University Press is part of Cambridge University Press & Assessment,
a department of the University of Cambridge.

We share the University's mission to contribute to society through the pursuit of
education, learning and research at the highest international levels of excellence.

www.cambridge.org
Information on this title: www.cambridge.org/9781108492690

DOI: 10.1017/9781108610926

First published 2020

A catalogue record for this publication is available from the British Library

ISBN   978-1-108-49269-0   Hardback

There are different sorts of conflicts between theories. One familiar kind of conflict is that in which two or more theorists offer rival solutions of the same problem. . . . often, naturally, the issue is a fairly confused one, in which each of the solutions proffered is in part right, in part wrong and in part just incomplete or nebulous. There is nothing to regret in the existence of disagreements of this sort. Even if, in the end, all the rival theories but one are totally demolished, still their contest has helped to test and develop the power of the arguments in favour of the survivor.

(Ryle (1954))

Horatio: O day and night, but this is wondrous strange!
Hamlet: And therefore as a stranger give it welcome.
   There are more things in heaven and earth, Horatio,
   Than are dreamt of in your philosophy.

(Hamlet, Act I, scene v)

Contents

Preface

Some species of ant, bee and termite are 'eusocial': they live in colonies of overlapping generations in which all the offspring are produced from one or a few individuals (the queen bee, for instance) while the other, non-reproducing members of the colony devote their lives to selfless behavior, protecting the colony and collectively rearing the young. Explanations for the origin of eusocial behavior start from the observation that eusocial insects share a large fraction of their DNA with the other members of the colony – in the case of honeybees, the degree of relatedness is 75%. Worker bees can pass on more of their genetic material by helping their 'sisters' than by having offspring of their own, and natural selection responds to this state of affairs by endowing them with the motivation to act altruistically toward the other bees in the colony.

In May 1976, the biologist Richard Alexander gave a lecture at Northern Arizona University on eusociality in which he tried to explain why it had never evolved in vertebrates. As a thought experiment, he speculated on what a eusocial mammal might be like. The need to accommodate a large and growing colony would favor subterranean rodents. He predicted that the ideal niche would be tropical and that the burrowing rodents would prefer to live in heavy clay soil that is inaccessible to most predators, and to feed on large tubers. And of course there would be one 'queen' rodent that gave birth to all of the offspring, who would behave altruistically toward each other. After the lecture, Alexander was surprised to be told by a member of the audience that he had just given a perfect description of the naked mole-rat, a species native to East Africa and which had just begun to be studied by biologists (Sherman et al., 1991, p. vii–viii).

Scientists tend to be very impressed by episodes like this. It is hard to believe that the theory on which Alexander's prediction was based – in this case, Darwin's theory of evolution by natural selection – could be very wrong if it correctly predicts something as *a priori* unlikely as a naked mole-rat.

There is a research tradition in cosmology that has been repeatedly successful in just this way, correctly predicting facts and relations – some quite surprising – in advance of their discovery. I am not referring here to the standard model of cosmology, the one that you will find in the textbooks. The theory I have in mind goes under a number of names; the most common is 'MOND,' which stands for 'MOdified Newtonian Dynamics,' but many researchers prefer the name 'Milgromian dynamics,' since the theory was originated by the astrophysicist Mordehai Milgrom.

Milgrom published the foundational postulates of his new theory in a set of three papers in 1983. At the time, the standard cosmological model was facing a major crisis – or, in the language of philosopher Karl Popper, a 'falsifying instance.' Observations of spiral galaxies like the Milky Way had revealed that the motion of stars and gas in their outskirts always fails to match with the predictions of Newtonian gravity: orbital speeds are always greater than predicted, sometimes much greater, and they are never observed to fall off with distance as Newton's equations generically predict. Anyone who was doing research in astrophysics at that time (as I was) will remember how quickly the community closed ranks and agreed on a consensus explanation for the anomaly: galaxies, it was postulated, are surrounded by 'dark matter,' which does not interact with radiation but which does generate gravitational force. Standard-model cosmologists still universally *assume* that the dark matter in every galaxy has whatever spatial distribution is needed to reconcile the observed motions with Newton's laws.

Milgrom was one of the scientists who expressed reservations about the dark matter hypothesis. What he found most impressive was not the anomalously high orbital speeds in the outskirts of galaxies, but rather the fact that every galactic rotation curve (the plot of orbital speed versus distance from the center) is 'asymptotically flat': it tends toward a constant value, different in each galaxy, and remains at this value as far out as observations permit. This is extremely hard to understand under the dark matter hypothesis, since every galaxy has a different history of formation and evolution and the dark matter would need to repeatedly redistribute itself to maintain that flat rotation curve. Milgrom proposed a different, and quite bold, response to the rotation-curve anomaly: a modification to Newton's laws. Many such modifications could do the trick, but Milgrom singled out one: he proposed that Newton's relation between gravitational force and acceleration should be modified in regions where the gravitational acceleration (that is, the gravitational force per unit of mass) falls below a certain, universal value. Milgrom labelled this new constant a_0 and estimated a value of order 10^{-10} m s^{-2} – so small that the proposed modification to the laws of motion would be practically undetectable anywhere in the solar system; it would only become important in regions of very small gravitational force like the outskirts of galaxies.

On its face, Milgrom's explanation for the rotation-curve anomaly is neither more nor less ad hoc than the dark matter postulate. Both are examples of what philosophers of science call 'auxiliary hypotheses': assumptions that are added to a theory in order to (in this case) reconcile it with falsifying data. There are likely to be an infinite number of auxiliary hypotheses that can account for any given anomaly. How do we decide which (if any) is correct?

This state of affairs has arisen many times in science, and philosophers of science have come up with a set of criteria. Clearly it is not enough for an auxiliary hypothesis to explain facts that are already known; as philosopher Elie Zahar (1973, p. 103) put it, theories can always be "cleverly engineered to yield the known facts." To be acceptable, an auxiliary hypothesis should also predict some *new* facts: the more unlikely the predictions (in the light of the pre-existing theory), the better. (Remember the naked mole-rat.) And ideally, at least some of those novel predictions should be confirmed by observation or experiment – this gives us confidence that we are moving in the direction of the *correct* theory, which, after all, *only* makes correct predictions.

How well does the galactic dark matter hypothesis meet these requirements? In explaining the rotation curve of the Milky Way galaxy, that hypothesis does make a novel prediction: that there should be dark matter near the Sun with a density (mass per unit volume) that is approximately known. The particles that make up the dark matter (if particles they be) are passing continuously through every laboratory on the Earth and could be detected. Attempts to verify this prediction (so-called 'direct detection' experiments) got underway shortly after the dark matter hypothesis was agreed upon, in the early 1980s, and they have continued unabated since then; the detectors currently in use are about ten million times more sensitive than those of the 1980s. But all attempts to detect the dark matter particles have failed: no one has ever observed anything that might reasonably be interpreted as the signal of a dark matter particle passing through their detector.

The situation is very different for Milgrom's hypothesis. Already in his first papers from 1983, Milgrom wrote down a number of novel predictions that follow from his postulates. For instance, he showed that his modification to Newton's laws predicts not only that rotation curves should be asymptotically flat (that result was built into the postulates, just as it is built into the dark matter hypothesis), but also that there should be a universal relation between the orbital speed in the outer parts of a galaxy and the galaxy's total (not dark!) mass. No one had even thought to look for such a relation before Milgrom predicted it; no doubt because – according to the standard model – it is the dark matter, not the ordinary matter, that sets the rotation speed. But Milgrom's prediction has been beautifully confirmed – a splendid example of a verified, novel prediction.[1]

Note that this prediction of Milgrom's is *refutable*: it could, in principle, have been found to be incorrect. By contrast, the standard-model prediction that dark matter particles are passing through an Earth-bound laboratory is *not* refutable, since nothing whatsoever is known about the properties of the putative dark

---

[1] This is the 'baryonic Tully–Fisher relation'. See Figure 4.1 and the discussion in Chapter 4.

particles. A failure to detect them might simply mean that their cross section for interaction with normal matter is very small (and that is, in fact, the explanation that standard-model cosmologists currently promote). On these grounds, as well, Milgrom's hypothesis 'wins': it is epistemically the preferred explanation.

Of course, that is not the same as saying that Milgrom's hypothesis is *correct*. I will not, in fact, be arguing that – although I know of nothing that would preclude such a conclusion. My goal is more modest: to assess the degree to which the Milgromian research program is progressive.

The terms 'research program' and 'progressive' will be familiar to philosophers of science but not to most scientists – at least, not with the specific meanings that philosophers attach to them. Both terms are due to Imre Lakatos, a student and colleague of Karl Popper. Lakatos recognized (as did Popper, and Thomas Kuhn) that scientific theories evolve, and they do so in characteristic ways. Typically there is a fixed set of assumptions, which Lakatos called the 'hard core' (and which Kuhn, at least sometimes, referred to as a 'paradigm'); for instance, in the standard cosmological model, the hard core contains the assumption that the general theory of relativity is correct. When a prediction of a theory is shown to be incorrect – when the theory is 'falsified', to use Popper's term – scientists, Lakatos said, rarely modify the hard core; instead they are likely to add an auxiliary hypothesis that targets the anomaly and 'explains' it, leaving the hard core intact.

Since theories change over time, Lakatos argued that the proper unit of appraisal is not a single theory, but rather the evolving *set* of theories that share the same hard core postulates over time – what Lakatos called a 'research program.' To the extent that this is correct, the central question for epistemologists of science is no longer "Has this theory been falsified?" ("All theories", said Lakatos, "are born refuted and die refuted") but rather "Is this research program progressing or degenerating?" Based on his analysis of the historical record, and being guided whenever possible by Popper's epistemic insights, Lakatos identified two conditions that characterize theory change in successful research programs. First, Lakatos found that changes to a theory should not be ad hoc: they should enlarge its scope and create the potential for new predictions – some of which, ideally, should be confirmed. Indeed, Lakatos argued that the *only* experiments or observations of any evidentiary value were those that targeted *novel* predictions ("the only relevant evidence is the evidence anticipated by a theory"). Second, Lakatos noted that successful research programs tend to develop autonomously, and not simply in response to anomalies. "A research programme is said to be *progressing* as long as its theoretical growth anticipates its empirical growth, that is, as long as it keeps predicting novel facts with some success," he wrote. Whereas a stagnating, or "degenerating," research program is one that "gives only *post hoc* explanations either of chance discoveries or of facts anticipated by, and discovered in, a rival programme" (Lakatos, 1971, p. 112).

Lakatos's criteria should give pause to anyone familiar with the history of the standard cosmological model since about 1980. The hypotheses in that model relating to dark matter and dark energy both were added to the theory in response to "chance discoveries." Furthermore, an enormous amount of effort has been – and continues to be – expended by standard-model cosmologists in attempts to find explanations for "facts anticipated by, and discovered in, a rival programme": that is, the Milgromian research program. It is fair to say that standard-model cosmologists have not yet succeeded in convincingly explaining even one of the many novel facts that were first predicted by Milgrom and later confirmed by observation.

At this point, many readers will be asking: Which theory is the *more* successful: Milgrom's or the standard model? That would be a hard question to answer and I will not try to answer it. As philosophers of science have often pointed out, it is difficult in principle to judge the relative merits of two competing theories or research programs. Often it is the case that one research program is much more developed than the other, and so it is not clear which theory from each program should be singled out for the comparison. Also, a new research program is often initiated in response to anomalies that occur in one physical regime, and at least in its early stages, such a research program is likely to make successful predictions only in that regime. Both considerations apply to a comparison of the standard cosmological model with Milgromian theory.

But while a critique of the standard cosmological model is no part of the purpose of this book, there is a natural connection to be made with that model. One way to define a 'novel prediction' is by comparison with a rival theory. Alan Musgrave (1974, p. 15) expressed this as follows: "In assessing the evidential support of a new theory we should compare it, not with 'background knowledge' in general, but with the old theory which it challenges." The 'old theory' in this case is, of course, the standard cosmological model. And so it makes sense to judge the novelty of Milgrom's predictions by asking to what extent each predicted fact is improbable from the standpoint of the standard cosmological model. By proceeding in this way, I hope to give the reader the information she needs to reach an independent conclusion about the evidential support for Milgrom's hypothesis, prediction by prediction, and about the degree to which Milgrom's predictions constitute anomalies for the 'rival research program.'

I sometimes had the feeling while writing this book that I was belaboring the obvious. *Of course*, my inner voice would insist, *the fact that Milgrom's theory has made so many successful predictions implies that it is a theory worth considering. Why does it take a whole book to establish something so obvious?* But outside the (still small) community of Milgromian researchers, the point is, apparently, far from obvious. Here is a quote from standard-model cosmologist Joseph Silk (2004, p. 69): "The modified Newtonian dynamics (MOND) theory seeks to modify the law of gravity in order to dispense with the need for dark matter. There is little in the way of compelling theory or data to support such a position." That is *all* that Silk has to say about MOND in his review of 'dark matter theory.' I count two incorrect assertions in his two short sentences (and I have no idea what Silk means by 'a theory supporting a theory'). In the recent text *The Philosophy of Cosmology* (Chamcham et al., Cambridge University Press, 2017), the words 'Milgrom' and 'MOND' do not appear at all. I could give many other examples (and I do so, elsewhere in this book). I do not claim to understand the reasons for this attitude of studied indifference on the part of standard-model cosmologists. Perhaps, in the final analysis, their attitude will come to be seen as justified. But until that happens, I hope my book will motivate young scientists to think about the possibility of cosmology outside the standard model.

The analysis in this book is based on the scientific literature published before the end of 2017. The published record is an essential resource, of course, but it did not take me long to realize that it was not enough: I needed also to talk to people who had seen, up close, 'how the sausage is made.' This was particularly true, I came to appreciate, with regard to the standard-model literature (the Milgromian literature is much more transparent). Conversations with Pavel Kroupa, Stacy McGaugh, Marcel Pawlowski and Monique Spite were essential in this regard – I simply could not have written this book without their help in showing me where the bodies are buried, methodologically speaking. All of these people had more important things to do and I am grateful for their time. Federico Lelli kindly provided me with data (some unpublished) that were used in compiling Table 8.2 and in plotting Figure 5.2. I thank Cambridge University Press for their permission to quote at length from Lakatos's essays on the methodology of scientific research programs, as they appear in the 1978 volume edited by John Worrall and Gregory Currie. And I am deeply grateful to my editor at Cambridge, Vince Higgs, who saw (even before I did) the need for a book like this and who was supportive in every way possible.

1

The Epistemology of Science

The concordance model [of cosmology] is now well established, and there seems little room left for any dramatic revision of this paradigm.

(Olive et al. *(2014))*

The evidence for the dark matter of the hot big bang cosmology is about as good as it gets in natural science.

(Peebles (2015))

The trouble about people – uncritical people – who hold a theory is that they are inclined to take everything as supporting or 'verifying' it, and nothing as refuting it.

(Popper (1983))

There is a tendency, among both scientists and non-scientists, to assume that our current scientific theories are correct in some fundamental sense: that they embody deep and established truths about the physical universe. No one denies that the theories might benefit from further refinement, particularly in regimes where they have not been well tested, and everyone would acknowledge that there are things in the universe that remain to be discovered and understood. But it is widely assumed that the theories of physics, chemistry and biology that are set out in the current textbooks are unlikely to change in any fundamental way. After all, the argument goes, these theories are the basis for the spectacular material progress of the modern world: for the design of airplanes and computers, the production of serums and antibiotics, the manufacture of plastics and synthetic fibers, the successful prediction of spacecraft trajectories and the weather. It is almost impossible to imagine (the argument goes) that these theories could be so successful unless they were essentially correct.[1]

But the history of science suggests otherwise. Almost all of the theories that were at one time viewed as correct have been abandoned. And what is even more

---

[1] "we are strikingly good at making science-based interventions in nature.... this success in intervention is incomprehensible unless we suppose that the claims we are putting to work in our practical activities are correct (or, at least, approximately correct)" (Kitcher, 1995, p. 659); "it is reasonable to believe that the successful theories in mature science – the unified theories that explain the phenomena without ad hoc assumptions ... are, if you like, approximately true" (Worrall, 2007, p. 153–154).

striking is the manner in which theories change. There are certainly periods, within any scientific discipline, when the dominant theory undergoes only gradual modifications, without much change to the underlying assumptions. But such periods tend to last only so long; they are separated by revolutions during which the old assumptions are thrown out and a radically new set are brought in. As every student of physics knows, there were a number of such episodes in the early part of the twentieth century: classical mechanics and electromagnetism were replaced by quantum electrodynamics, Newton's theory of gravity and motion was replaced by Einstein's. The new theories were not simply improvements over the theories they replaced. The changes were so radical that even basic concepts like mass and time altered their meanings in fundamental ways.

That is not to deny that there are aspects of scientific progress that are genuinely cumulative. The universe is vast, and the longer we observe it, the more we learn about its composition and structure. Additions to knowledge of this sort are what the popular science writers usually have in mind when they talk about 'scientific discoveries.' But what lends science its particular prestige is not the accumulation of knowledge about what exists: it is the (apparent) ability of science to make correct predictions about things that no one had previously observed. Scientific theories contain *universal hypotheses*: statements or laws (often presented in mathematical language) that are claimed to be valid at all places and for all times, and that can be used to generate predictions even in situations that have never been encountered before. For instance: 'The gravitational force between two point objects varies as the inverse square of their separation'; 'the entropy of an isolated system never decreases'; 'the wavelength of a particle varies inversely with its momentum.'

Where do such hypotheses come from? It is tempting to believe that they are arrived at through induction: that they are generalizations from what is observed. But a few minutes' thought shows that that can not possibly be correct. Discrete instances do not imply universal laws; a finite set of observations is always consistent with an infinite number of different theories. Not only is induction insufficient to the task: it is fair to say that induction does not exist. The fallacy of induction has been discovered and rediscovered many times, going back at least to the fourth century BCE and the Greek philosopher Pyrrho of Elis.[2] Modern discussions of the 'problem of induction' usually adopt the formulation of the eighteenth century philosopher David Hume: "we have no reason to draw any inference concerning any object beyond those of which we have had experience" (Hume, 1739–40/1978, Book I, Part III, section xii). As an illustration, Hume invoked the impossibility of predicting the future: "For all inferences from experience suppose, as their foundation, that the future will resemble the past" (Hume, 1748/1975, section 4.2, 37–38).

---

[2] Pyrrho of Elis (*c.* 360–275 BCE) left no writings; the sole surviving texts from the Pyrrhonian movement are those of Sextus Empiricus (*c.* 160–210 BCE). Richard H. Popkin (2003) traces the history of Pyrrhonian skepticism from its revival in fifteenth-century Europe until the early eighteenth century.

Just because the Sun rose yesterday, and on all previous days for which records exist, there is no logical basis to assume that it will rise tomorrow (and of course it may not).

Hume's 'problem of tomorrow', to adopt the phrase of Karl Popper (1983, Part I, 4.III) – the lack of any basis in logic for assuming the regularity of nature – is one aspect of the problem of induction. But what is equally relevant to the epistemology of science is a different aspect: the logical impossibility of generalizing from a limited set of observations to an unrestrictedly general law, and (what is almost the same thing) the impossibility of *verifying* a universal law (whatever its provenance) given known instances of its success.

The essential point here is that even an incorrect theory can generate correct predictions. Take a simple example: today is Saturday, and someone proposes the hypothesis "Today is Sunday." That hypothesis is false, but from it necessarily follow any number of true statements, including "Today is not Monday," "The English word for this day of the week begins with the letter S," "It is illegal to sell packaged liquor after 9:00 pm today in Milwaukee" etc. Anyone so inclined could confirm the correctness of an unlimited number of such predictions ("It is not noon on Monday," "It is not 12:01 on Monday" etc.). This example may seem too simple or contrived to be relevant to the justification of scientific theories. But then, consider the fact that for two hundred years Newton's theory of gravity and motion was found again and again to yield accurate predictions, even to the extent of correctly predicting the existence and location of a new planet (Neptune). And yet we now know (or at least believe) that Newton was wrong: not wrong in a minor or trivial way, but deeply, fundamentally, conceptually wrong. Einstein's theory correctly predicts the same facts as Newton's, but interprets them as instances of a quite different set of hypotheses. And it would be foolish to assume that Einstein's theory, as well-corroborated as it is,[3] will not itself be replaced one day by another theory, perhaps a theory that differs as much from Einstein's as Einstein's differs from Newton's.

These arguments are convincing enough, but they do not bring us any closer to explaining the success of science. And if the inductive method – which since the time of Francis Bacon was widely (though mistakenly) seen as the principle that separates science from non-science[4] – does not exist, then what basis do we have for calling some theories 'scientific' and others just speculation?

---

[3] Here and throughout this book, 'corroborate' has the meaning adopted by Karl Popper after about 1958, roughly, 'provide evidential support for' (Popper, 1983, section 29). 'Corroboration' differs from 'confirmation'; the latter implies demonstration or proof of correctness. Following Popper and Hume, it is reasonable to believe that a *prediction* of a theory can be confirmed, but theories themselves can only be corroborated, never confirmed. E.g. Magee (1997, p. 188): "it is possible sometimes to be sure of a direct observation, but not of the explanatory framework that explains it."

[4] E.g. Lakatos (1974, p. 161): "at least among philosophers of science, Baconian method [i.e. inductivist logic of discovery] is now only taken seriously by the most provincial and illiterate."

Karl Popper, in his *The Logic of Scientific Discovery* (1959), claimed to have found the answer to both questions.[5] Popper granted the correctness of Hume's analysis: induction, he said, does not exist, and therefore it can be invoked neither as a basis for the growth of knowledge, nor as a criterion of demarcation between science and pseudoscience. But, he said, induction is not needed. Popper began by emphasizing the logical asymmetry between proof and disproof. While no number of observations can ever prove the validity of a universal law, a single observation that *conflicts* with the law is sufficient to *dis*prove it. The hypothesis 'All swans are white' can not be true if even a single black swan exists.

Of course, this argument – what logicians call *modus tollens* – was well known to Hume. But Popper went a big step further. All knowledge, said Popper (still in agreement with Hume), is uncertain and must always remain so. But if a hypothesis is testable, there exists at least the possibility that it can be shown to be incorrect and replaced with another, better one: "For it may happen that our test statements may refute some – but not all – of the competing theories; and since we are searching for a true theory, we shall prefer those whose falsity has not been established" (Popper, 1972, p. 8). Popper emphasized that the most useful tests are those carried out with the *intent* of falsifying a theory; for instance, experiments that test a prediction that conflicts with the experimenter's prior expectations. As long as a new theory holds up to such tests, Popper said, we are justified in considering the theory viable. Whereas if a prediction is shown to be false, the theory has been disproved, and it can be replaced. In this manner, via "conjectures and refutations," knowledge can grow.

Popper's view of epistemology is called 'critical rationalism.'[6] Critical rationalism is opposed to – for instance – inductivism, and to logical positivism, the belief that the only meaningful statements are those that are *verifiable* through observation. Critical rationalists deny that theories are verifiable. They assert that theories should be judged on the basis of how well they stand up to attempts to refute them.

But where do the hypotheses that we are testing come from? Popper was adamant on this point: it simply does not matter. Theories can come from anywhere.[7] What

---

[5] The 1959 publication date of *The Logic of Scientific Discovery*, the English translation of *Logik der Forschung*, is misleading. The German text was published in 1934. The original manuscript, in two volumes, was titled *Die Beiden Grundprobleme der Erkenntnistheorie* and was completed in early 1932; it was scheduled for publication in 1933 but the publisher (Springer) objected to its length. A new manuscript, which consisted of extracts from the two unpublished volumes, was also rejected by the publisher. Popper (1974, p. 67) gives credit to his uncle, Walter Schiff, who "ruthlessly cut about half the text" resulting in the 1934 publication of *Logik der Forschung*. Popper (1972, p. 1, n. 1) has said that he discovered the solution to the problem of induction around 1927.

[6] Here Popper uses 'rationalism' to mean the opposite of 'irrationalism' (and not the opposite of 'empiricism'); he defines it as "an attitude of readiness to listen to critical arguments and to learn from experience" (Popper, 1945, p. 225). Paul Feyerabend (1975, p. 172), in a discussion of Popperian epistemology, writes: "rational discussion consists in the attempt to criticize, and not in the attempt to prove or to make probable."

[7] E.g. Popper (1959, p. 32): "there is no such thing as a logical method of having new ideas, or a logical reconstruction of this process." Peter Urbach (1978, p. 102) notes that many philosophers and scientists have endorsed Popper's view of the irrationality of scientific theorizing, including Albert Einstein, Carl Hempel, William Whewell and Hans Reichenbach.

does matter, crucially, is that a theory be testable. And this argument led Popper to his famous 'criterion of demarcation': *falsifiability* is the quality that separates science from non-science. If no experiment can be imagined that will disprove a theory, then all observations are consistent with it: it might as well be true as false and there is no basis for calling it 'scientific.'[8] And equally, *any* hypothesis that makes testable predictions (and which also satisfies certain other basic conditions, such as consistency) can legitimately claim to be scientific, irrespective of (for instance) how wide or narrow its domain of applicability.

Popper was quite aware that falsifying a theory is not always a straightforward proposition. An experiment rarely tests one hypothesis in isolation. The prediction that a quantity will have a certain measured value almost always involves a set of assumptions about the measuring apparatus and the experimental design, and if the measured value differs from the prediction, one can never be completely certain where the fault lies. In addition, the interpretation of an experiment often requires assumptions about the validity of various other scientific hypotheses in addition to the hypothesis being tested; it may take a series of cleverly designed experiments to ferret out which of the hypotheses has been falsified by a conflicting measurement.[9] But Popper insisted that – in spite of practical problems like these – it is the responsibility of the scientist to adopt a methodology that maintains falsifiability: "*criteria of refutation* have to be laid down beforehand: it must be agreed which observable situations, if actually observed, mean that the theory is refuted" (Popper, 1963, p. 38, n. 3).

§§

Philosophers are divided over whether Popper's demarcation criterion – which requires that scientific theories be testable, or refutable, or falsifiable – is really the best way to distinguish science from non-science (Laudan, 1983; Grünbaum, 1989; Hull, 2010). But even philosophers who object to falsifiability as a criterion of demarcation are likely to acknowledge the usefulness of a falsificationist *approach* to the testing of scientific hypotheses. The essential point (which Popper often made) is that scientists who are looking for evidence to support a theory can always

---

[8] Of course one can ask whether this is anything more than a *definition* of science. David Miller (2014b) notes that Popper's goal was not to certify certain hypotheses as 'scientific' and others as 'unscientific.' Rather, it was to determine whether an empirical investigation is worth undertaking. Miller quotes from Popper (1983, p. 174): "my 'problem of demarcation' . . . was not a problem of classifying or distinguishing some subject matters called 'science' and 'metaphysics'. It was, rather, an urgent practical problem: under what conditions is a *critical appeal to experience* possible–one that could bear some fruit?"

[9] The idea that theories are related to experimental results via a web of auxiliary hypotheses is probably obvious to most practicing scientists. Philosophers of science, on the other hand, never seem to tire of reiterating this point, often in the context of a critique of Popper's demarcation criterion (e.g. Suppes, 1967; Schaffner, 1969; Hempel, 1973; Grünbaum, 1976). For Popper's view of these critiques see "Replies to my critics: difficulties of the demarcation proposal" in Book 2 of Schilpp (1974). Anthony O'Hear (1980, chapter VI) presents a balanced discussion and concludes sensibly: "A genuinely scientific method of investigation, then, is one which proposes testable theories and which takes the tests seriously."

find it. Whereas a scientist who is skeptical will pay more attention to the anomalies, and will be less inclined to miss any indications that a new and better theory is needed.

The case can be made that a 'verificationist' approach to theory testing is dangerous in a much more insidious way. Francis Bacon, writing in 1621, saw this clearly:

> The human understanding when it has once adopted an opinion (either as being the received opinion or as being agreeable to itself) draws all things else to support and agree with it. And though there be a greater number and weight of instances to be found on the other side, yet these it either neglects and despises, or else by some distinction sets aside and rejects; in order that by this great and pernicious predetermination the authority of its former conclusions may remain inviolate (Bacon, 1621/1863, XLVI).

What Bacon described in the seventeenth century is nowadays referred to, by psychologists and sociologists, as 'confirmation bias.' Here is a more modern definition:

> The human organism tends to seek, embellish, and emphasize experiences that support rather than challenge already held beliefs. As an information processor, it filters experiences in a confirmatory manner – highlighting those that are consistent with its conceptual biases and ignoring or discrediting those that ... [are] ... not (Mahoney and DeMonbreun, 1977, p. 229).

That scientists might "neglect and despise" evidence that contradicts their cherished beliefs will be easy for anyone to understand. What non-scientists might find more difficult to accept is the many ways in which a scientist – despite the best of intentions, and without any conscious desire to 'fudge the data' – can actively (even if subconsciously) distort his experimental results to bring them in line with expectations. But most scientists (at least, the more self-aware ones) who have engaged in theory-testing are aware of this pitfall.

§§

Suppose that a theory has been refuted, and another theory is proposed to take its place. If the new theory is to be reckoned an improvement over the old one, it should do certain things: for starters, it should correctly explain the experimental results that brought down the previous theory. But this is a weak condition, because it is often possible to modify a theory by some ad hoc device that targets nothing but the anomaly. A famous example is the use of epicycles in the Ptolemaic model of the solar system: any discrepancy between the predicted and observed motions of a planet could be accommodated by adding more epicycles to the planet's trajectory. Such changes are objectionable – not simply because they are ad hoc – but because they result in a theory that tells us nothing more about the world than we already knew. And the aim of science should always be to increase our knowledge.

Popper expressed this in the following way: A condition for progress is that "the new theory should be *independently testable*. That is to say, apart from explaining all the *explicanda* which the new theory was designed to explain, it must have new and testable consequences (preferably consequences of a *new kind*); it must lead to the prediction of phenomena which have not so far been observed" (Popper, 1963, p. 241). Popper defined the 'empirical content' of a theory as its class of potential falsifiers: statements that, if confirmed, would refute it: "the more a statement forbids, the more it says about the world of experience" (Popper, 1959, p. 119). Popper understood that the total number of testable consequences of a theory is undetermined, and probably infinite. But it is the increase in predictive power that matters, and it is often possible to judge unambiguously whether a given modification to a theory represents an increase in its empirical content.

This line of thinking led Popper to a conclusion which at first sight seems (like so many of Popper's conclusions) obviously wrong. The best theories, Popper said, are the *least probable* ones: the ones that, on their face, have the smallest likelihood of being correct. The idea is simple: A good theory says a lot about the world; and the greater a theory's predictive power, the greater the number of ways in which its predictions can turn out to be false. The only sort of statement that is guaranteed to be correct is a tautology, but a tautology has no predictive power; it tells us nothing about the world. Whereas a theory that purports to predict the position and velocity of every particle in the universe, at every time, would be a theory with enormous content and power, but it would also be trivial to falsify and would almost certainly turn out to be incorrect.

The idea that the best theories are the ones that are least likely to be correct is counter-intuitive, but perhaps only until we remember another condition that an acceptable theory must satisfy: it must not (yet) have been falsified. Bold theories that withstand all attempts to falsify them: that is the goal. Popper emphasized again and again the importance of *bold* theorizing. Scientists, he said, "invariably prefer a highly testable theory whose content goes far beyond all observed evidence to an *ad hoc* hypothesis, designed to explain just this evidence, and little beyond it, even though the latter must always be more probable than the former" (Popper, 1983, p. 256). Of course, most bold theories will turn out to be wrong – but that is precisely Popper's point; and because bold theories are the most easily falsified, their incorrectness will quickly become apparent. Furthermore, by testing the predictions of bold theories, scientists will be motivated to look in places they might otherwise not have thought to look, leading to novel insights that could motivate a more successful theory.

§§

One reason that Popper's criterion of falsifiability is hard for many people to accept is that it seems to relegate experiments to a purely negative role: that of disproving theories. But almost everyone, whether inductivist or not, wants to believe that it is possible for experiments to *support* theories, by showing that a theoretical prediction was correct. And in fact it is easy to find examples from the history of science where the experimental or observational confirmation of a prediction led scientists to strongly endorse a new theory; and in at least some of those cases, scientists made (we would now say, with the benefit of hindsight) the 'correct' inference: they endorsed the 'right' theory on the basis of its experimental success.

But if induction is a fallacy, then it is very hard, from a strictly logical point of view, to connect a theory's predictive success to its correctness. Even incorrect theories can make correct predictions, and there will always be an infinite number of theories (most of them yet undreamed of) that can correctly explain any finite set of observations. Only one theory, at most, from that infinite set can be correct, and so there is simply no basis, logically speaking, for claiming that one's pet theory is *the* theory that is supported by the data.

In fact the situation is far worse even than this. For not only can many theories explain the same experimental results. One can also show that *any* observation of *anything* that does not contradict a theory is equally confirming of it, regardless of whether the observation targets a prediction of the theory.

This surprising result is usually[10] attributed to the logician and philosopher Carl Hempel and it is sometimes called 'Hempel's paradox' or the 'paradox of confirmation'[11] – although in fact there is no paradox, in the sense of logical inconsistency; the result is simply extremely counter-intuitive. The proof is simple and goes as follows:

Consider a universal hypothesis such as 'All ravens are black.' This can be written symbolically as the conditional statement

H: If A then B,

where A = raven and B = black. By *modus tollens*, hypothesis H is precisely equivalent to hypothesis H', where

H' : If not B then not A,

i.e. 'All non-black things are non-ravens.'

---

[10] The origin of the theorem is not clear. A common reference is to Hempel (1937) but the theorem does not appear there; Hempel first presented the theorem in print some years later (Hempel, 1945). In the meantime, the 'paradox' had been pointed out by Janina Hosiasson-Lindenbaum (1940), who would seem to deserve at least partial credit. In her paper (p. 136), Hosiasson-Lindenbaum attributes the result to Hempel without giving a reference ("C. G. Hempel has stated the following paradox"). According to Hempel (1965, p. 20, n. 25), "Dr. Hosiasson's attention had been called to the paradoxes by my article "Le problème de la vérité" [i.e. (Hempel, 1937)] ... and by discussions with me." Henry Kyburg (1970, p. 166) sums up this confusing set of circumstances as follows: "The oddities that are referred to as the "paradoxes of confirmation" were first noted by Janina Hosiasson-Lindenbaum in 1940; they were christened by Carl Hempel in 1945."

[11] Another name one sometimes sees is 'the paradox of the ravens.' This seems to be an instance of the rule that favorable results in logic are associated with swans, unfavorable results with ravens.

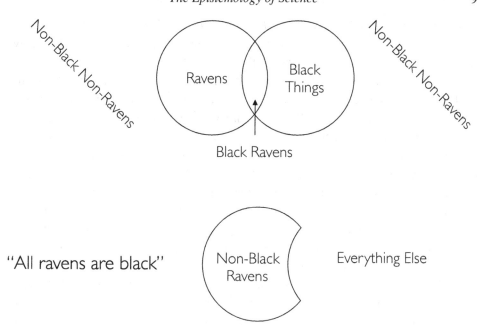

Figure 1.1 A graphical representation of Hempel's theorem (Hosiasson-Lindenbaum, 1940; Hempel, 1945). The top panel divides the universe of things into four groups. Black ravens occupy the intersection of 'Ravens' and 'Black Things.' The bottom panel illustrates the way in which observation of a thing supports the hypothesis 'All ravens are black.' Observation of a non-black raven falsifies the hypothesis; observation of anything else (black ravens, black crows, red robins, yellow mushrooms etc.) is equally confirming of the hypothesis. In other words: from a purely logical standpoint, a theory is supported by any observation that does not refute it. Hempel's theorem implies that – if scientists wish to claim that a theory is 'supported' by confirmed predictions – they need to give a reason why confirmed predictions are more evidentiarily valuable than any other sort of observation. Karl Popper argued that confirmed, *novel* predictions – prediction of facts that were first discovered in the process of attempting to falsify the theory – are the *only* kind that can lend support to a theory, and that the degree of support increases in proportion with the prior improbability of the predicted fact.

Now assume, as an inductivist would, that observation of a black raven supports hypothesis H. (Exactly what is meant here by 'support' is unimportant, as will become clear in a moment.) It must therefore also be true that observation of a non-black non-raven supports hypothesis H'. But H' is precisely equivalent to H. QED: observation of a non-black non-raven – for instance, a red robin, or a yellow mushroom – supports the hypothesis that all ravens are black.

This result is represented graphically in Figure 1.1. It takes only a few more lines of analysis to show that black non-ravens are no different than non-black non-ravens in terms of their ability to provide evidential support of H. In other words: *every hypothesis is supported by anything that does not contradict it; the only sort of observation that fails to support a hypothesis is one that disproves it.* Hempel's proof provides striking support for Popper's argument that the only, evidentiarily relevant sorts of observational facts are those that refute a hypothesis.

One is tempted to object to Hempel's theorem on the ground that 'all ravens are black' is a statement only about ravens, hence only the color of *ravens* can be relevant to assessing its truth. But as Figure 1.1 suggests, a universal conditional such as 'all ravens are black' in fact says something about every possible thing in the universe: namely that it is either not a raven, or it is black. We should therefore not be too surprised if every thing in the universe is equally capable of confirming it.

If this result doesn't strike you as devastating to the idea of theory corroboration, you may not have fully grasped its implications. Here is a concrete illustration. A colleague comes to you and says, proudly, "I just spent a year analyzing data from the *Planck* satellite observatory, and the data confirm a prediction that I made about the cosmic microwave background. What a tremendous success for my theory!" You would be perfectly justified, from a *logical* point of view, in responding, "That's very nice; but looking out my window right now, I can see a red robin; and since your theory does not rule out the existence of red robins, my observation is just as genuine a confirmation of the theory as yours. So I don't understand why you expect me to be impressed."

The logic of Hempel's proof is unassailable, but very few people – scientists or philosophers – are willing to accept the conclusion that verified *predictions* are no more corroborating of a theory than any other sort of data. And indeed there is an impressively large body of philosophical literature that is directed toward finding a way round the seeming paradox. One approach (endorsed by Hosiasson-Lindenbaum) starts from the premise that a confirmed prediction only increases the *probability* that the hypothesis on which the prediction was based is correct. Another approach (which Hempel endorsed) begins by supposing that the evidence implies the correctness of only a *weakened version* of the hypothesis.

These end runs around Hempel's theorem are essentially inductivist, or (what is a slightly better term in this context) verificationist. Popper, characteristically, did not even try to evade the implications of Hempel's proof: "Thus an observed white swan will, for the verificationist, support the theory that all swans are white; and if he is consistent (like Hempel), then he will say that an observed black cormorant also supports the theory that all swans are white" (Popper, 1983, p. 235). But Popper argued that there was one special set of circumstances under which observation of a white swan could be seen as supporting the all-swans-are-white hypothesis, while observation of a non-white non-swan would not. Suppose, Popper said, that one's background knowledge – one's earlier theory about swans, together with the existing corpus of data relating to cygnine coloration – had less to say about the color of swans than the new theory. That previous theory might have said that swans can be either white or brown; or that swans come in all colors with equal probability; or perhaps there was simply no basis in the existing theory or data to believe anything definite about the color of swans. Armed now with a bold new theory that claims 'all swans are white,' the scientist will try to falsify it. *Based on his existing knowledge*, he has every reason to expect that he will succeed in falsifying the new theory, since nothing in his prior experience would have led him to expect that the color white is so favored by swans. If the scientist then *fails* to falsify the new theory – if he finds nothing but white swan after white swan – he is justified, Popper argued, in being impressed, since his new theory has *correctly predicted something that previously would have been considered unlikely.*

Based on this argument, Popper (1959, Chapter 10; 1983, Chapter IV) was led to reformulate the question 'Does an observation *E* support a hypothesis *H*?' as 'Does *E* support *H* in the presence of background knowledge *B*?' The answer to the latter question, he said, is 'yes,' as long as two conditions are satisfied:

(i) *E* follows from the conjunction of *H* and *B*;
(ii) *E* is improbable based on *B* alone.

Popper defined 'background knowledge' as knowledge that existed "previous to the theory which was tested and corroborated."[12] Popper noted that "it is not so much the number of corroborating instances which determines the degree of corroboration as the *severity of the various tests* to which the hypothesis in question can be, and has been, subjected" (Popper, 1959, p. 267).[13] The more improbable a prediction – the greater its novelty based on the background knowledge – the more strongly the new hypothesis is corroborated when the prediction is verified. The *most* severe test would be one that has the potential to contradict, i.e. falsify, the previous theory. Popper called such a test a "crucial experiment" or a "crucial test." But even predictions that are not inconsistent with an old theory can still corroborate a new theory, as long as, and to the extent that, they are novel predictions.

How best to define qualifiers such as 'improbable' or 'novel' is, of course, not obvious. Popper required that an experimental test consist of a "genuine attempt" to refute the theory, but he acknowledged that there was no objective way of judging whether an attempt was genuine.[14] In the years since Popper's proposal, a number of more concrete definitions of novelty have been proposed; these are discussed in detail in Chapter 2. But there is one sort of test that many philosophers agree does *not* satisfy the novelty condition. That is when the theory being tested contains parameters, and the parameters are determined from the same observations

---

[12] If the reader is reminded here of Bayes's theorem, she is in good company. There is a substantial literature that, building on Popper's insights, attempts to find a Bayesian 'solution' to Hempel's 'paradox'; see e.g. Earman (1996), Howson and Urbach (2006), Crupi et al. (2010). Such attempts have been partially successful at best (e.g. Miller, 2014a). Popper himself was skeptical of Bayesian approaches to theory confirmation: in his words, Bayes's theorem "is not generally applicable to hypotheses which form an infinite set – as do our natural laws" (in Schilpp, 1974, Book 2, p. 1185–1186, n. 68).

[13] Miller (2014a, p. 106): "Sitting around complacently with a well-meant resolve to accept any refutations that happen to arise is a caricature of genuine falsificationism."

[14] Popper (1983, p. 236): "sincerity is not the kind of thing that lends itself to logical analysis."

that constitute the test.[15] An example is the standard, or 'concordance,' model of cosmology which contains roughly a half-dozen parameters that are adjusted in order to give the best correspondence between the theoretical predictions and a particular set (also numbering roughly a half-dozen) of observational results. There is, of course, nothing illegitimate in determining a theory's parameters from data, but it is problematic (a critical rationalist would likely say) to claim that such a procedure constitutes *corroboration* of the theory. Data that are used to set the parameters of a theory do not *corroborate* a theory; they only *complete* the theory; and of course those data can be considered part of the 'background knowledge' that was used in theory construction. Corroboration would consist (for instance) of using the theory – with its parameters now fixed – to generate *new* predictions and demonstrating that they are correct.

§§

Scientific theories contain universal laws, but they can also contain more prosaic elements; for instance, statements that something *exists*. In some cases, these statements refer to entities the existence of which is not (or at least, is no longer) debated; for instance, the list of fundamental particles (and their measured properties) that are part of the standard model of particle physics. But existential statements can also refer to entities that are conjectural. In such cases, existential statements take the form of hypotheses. Examples are statements in the standard model of cosmology about dark matter and dark energy.

There is an interesting logical symmetry between universal statements (such as scientific laws) and existential statements. As discussed above, universal statements are falsifiable (at least, if they are well formulated) but they are not verifiable. Existential statements, on the other hand, *are* verifiable: one can verify the statement 'positrons exist' by finding a positron. And existential statements are *not* falsifiable: one could look forever for a positron without finding one, but never be certain that a positron didn't exist in some place that hadn't yet been searched.

The symmetry goes deeper. The negation of a universal statement is an existential statement, and the negation of an existential statement is a universal statement. The negation of 'All swans are white' (a universal statement) is 'It is not true that all swans are white' or equivalently 'There exists at least one non-white swan.' That is an existential statement and it is capable of being verified. Or consider the existential statement 'There is an element with atomic weight 52.' The negation of this statement is 'There is no element with atomic weight 52' – a universal hypothesis that can be falsified, by finding such an element, but never verified.

15 E.g. Worrall (1978b, p. 48): "one can't use the same fact twice: once in the construction of a theory and then again in its support"; Zahar (1973, p. 102–3): "Very often the parameters can be adjusted so as to yield a theory T^* which 'explains' the given facts ... In such a case we should certainly say that the facts provide little or no evidential support for the theory, since *the theory was specifically designed to deal with the facts*."

These considerations of logic led Popper to another of his counter-intuitive insights: that the universal hypotheses that appear in any scientific theory can be recast as negations of existential hypotheses: as statements that say that something does *not* exist, or *never* happens. For instance, the law of conservation of energy could be recast as 'There is no such thing as a perpetual motion machine.' Thus, said Popper, "natural laws ... do not assert that something exists or is the case; they deny it. They insist on the non-existence of certain things or states of affairs, proscribing or prohibiting, as it were, these things or states of affairs: they rule them out. And it is precisely because they do this that they are *falsifiable*"(Popper, 1959, p. 69).

But: If scientific theories can contain both universal and existential hypotheses, and if existential hypotheses can not be falsified, does that mean that Popper's demarcation criterion ('scientific theories are falsifiable') is invalid? Popper answered as follows: In isolation, existential statements – statements such as 'There are positrons' – say almost nothing; they have no predictive or explanatory power; they are 'metaphysical.'[16] To the extent that an existential hypothesis is *useful*, it is always part of a theoretical *system* that says something more about the postulated thing than simply that it exists: about its properties, its behavior, where it can be found etc. And these additional statements render the existential hypothesis (or rather: the theoretical system containing the hypothesis) falsifiable.[17] Popper gave a concrete example: the existence of hafnium (atomic number 72) was postulated before its discovery. But "all attempts to find it were in vain until Bohr succeeded in predicting several of its properties by deducing them from his theory. But Bohr's theory and those of its conclusions which were relevant to this element ... are far from being isolated purely existential statements. They are strictly universal statements" (Popper, 1959, p. 69–70).

The existence of hafnium has been verified by laboratory experiments. That is not the case with regard to dark matter or dark energy. This raises the important question: What sort of thing constitutes corroboration of an existential hypothesis, before its correctness has been intersubjectively established? What does it mean to say (as many standard-model cosmologists *do* say) "I believe, based on the evidence, that dark matter exists," given that no one has ever detected a dark matter particle in the laboratory?

A critical rationalist like Popper might answer as follows. To be scientific, a statement about dark matter must necessarily appear as a component of a system of testable, universal hypotheses. In the absence of confirmation, that is, detection of a dark matter particle, corroboration of the dark matter hypothesis would consist

---

[16] Alfred Ayer (1946, p. 41) gives a definition of 'metaphysical statement' that seems consistent with Popper's use of the term: "We may accordingly define a metaphysical sentence as a sentence which purports to express a general proposition but does, in fact, express neither a tautology nor an empirical hypothesis."

[17] Stanislaw Lem (1970, p. 10): "In science, truth is not a quality of singular scientific statements; it depends on the whole system."

of a demonstration from experiment or observation of the correctness of some prediction that is derived from that set of hypotheses.[18]

So much is straightforward. But there is a possible source of ambiguity here. There is always more than one way to construct the set of hypotheses that make up a given scientific theory. To the extent that this is true with regard to the statements that relate to hypothetical entities such as dark matter, the sense in which the existence of such entities is corroborated by experiment or observation may depend on exactly how the theory is formulated.

Popper (1963, p. 239) makes an interesting point about situations like this:

This, incidentally, speaks in favour of operating, in physics, with highly analysed theoretical systems – that is, with systems which, even though they may fuse all the hypotheses into one, allow us to separate various groups of hypotheses, each of which may become an object of refutation by counter examples.

To the final sentence might be added: 'or of corroboration by confirmed predictions.' It is probably fair to say that most descriptions of the standard cosmological model in textbooks and review articles (and, perhaps, in the minds of cosmologists) have 'fused into one' the various hypotheses about dark matter. This fact can make it difficult to evaluate statements that assert evidential support for its existence.

Following Popper's recommendation, we can try to (re)state some of the hypotheses relating to dark matter in the standard cosmological model, in a way that maximizes their independent testability. Consider first the following hypothesis:

DM-1: In any galaxy or galactic system for which the observed motions are inconsistent with the predictions of Newton, the discrepancy is due to gravitational forces from dark matter in and around the galaxy(ies).[19]

Since the early 1980s, and continuing up until the present day, standard-model cosmologists have routinely assumed the correctness of this hypothesis, regardless of anything else they may believe or postulate about dark matter.

A second hypothesis is:

DM-2: Beginning at some early time, the universe contained a (nearly) uniform dark component, which subsequently evolved, as a collisionless fluid, in response to gravity.

This second hypothesis is adopted by standard-model cosmologists who consider the evolution of the galaxy distribution on large spatial scales, or compute the power

---

[18] J. O. Wisdom (1968) makes a similar point about the *falsification* of existential statements that appear in the context of a broader theory; he proposes the expression "refutation by theory" to describe this situation.

[19] 'Predictions of Newton' here means predictions based on the Newtonian gravitational field computed from the observed distribution of matter. I am here assuming (as most standard-model cosmologists currently do) that no *known* form of matter can comprise the dark matter; in other words, that the discrepancies attributed to dark matter are not explainable in terms of some type of normal matter that happens to be 'dark.'

spectrum of temperature fluctuations in the cosmic microwave background. As in the case of DM-1, the correctness of DM-2 is typically assumed by these scientists, regardless of what else they might postulate or believe about dark matter. Logically, the assumption that DM-2 is correct implies nothing about the correctness of DM-1, or *vice versa*, although standard-model cosmologists routinely 'fuse' the two hypotheses into one.

A third assumption is often (though not always) made by standard-model cosmologists:

DM-3: The dark matter component that appears in DM-1 (or in DM-2, or in both) consists of elementary particles.

Now consider how these hypotheses might be refuted or corroborated. One thing to notice is that testable statements derivable from DM-1 constitute a completely distinct set from testable statements derivable from DM-2. DM-1 can only be used to make predictions about actual, observed galaxies,[20] while DM-2 says nothing at all about individual galaxies. The two hypotheses are empirically, as well as logically, independent – which, of course, was one of the reasons for stating them in this way. It follows that observational data that corroborate, or falsify, DM-1 can contain nothing of any evidentiary relevance to DM-2, and *vice versa*. It is perfectly possible for a particular set of data to corroborate DM-2 (for instance), while another set of data falsifies, or fails to corroborate, DM-1. And given that the existence of dark matter (not to mention its nature) is speculative, such a situation can not be ruled out *a priori*.

Not only is this *possible*: it is a pretty fair description of the current state of affairs. Hypothesis DM-2 has been corroborated (though not, of course, confirmed) by observations of large-scale structure and of the cosmic microwave background.[21] But rotation-curve data can not corroborate hypothesis DM-1, since, as Mordehai Milgrom (1989a, p. 216) has noted, "The DMH [dark matter hypothesis, i.e. DM-1] simply states that dark matter is present in whatever quantities and space distribution is needed to explain away whichever mass discrepancy arises." Another way to express this is via the argument mentioned above: data (that is, a galaxy rotation curve) that are used to determine the parameters of a hypothesis (in this case, the parameters that specify the distribution of the galaxy's dark matter) do not provide evidential support for the hypothesis. Stated yet another way: one can not invoke a hypothetical entity to explain anomalous data, then turn around and claim that those same data constitute evidence for the existence of the entity.[22]

A potentially testable prediction *does* follow from the conjunction of DM-1 and DM-3. If the dark matter consists of elementary particles, then the mass density

---

[20] That is: about the *dark matter* in actual, observed galaxies
[21] See Chapter 6.
[22] I belabor the point, because textbooks and review articles on cosmology do routinely argue in just this way.

of those particles near the Sun is known (modulo degeneracies in reproducing the known Milky Way rotation curve using different assumed spatial distributions of dark matter) as is their approximate velocity distribution. Laboratory experiments on the Earth could therefore corroborate the joint hypothesis by detecting the particles. But any such experiments suffer from a lack of knowledge about the cross section for interaction of the putative particles with the normal matter in the detectors, rendering the prediction essentially unfalsifiable. And of course, all attempts to detect the putative particles have so far failed.[23]

In this respect, the dark matter hypothesis is in a state similar to that of the atomistic hypothesis at the end of the nineteenth century. Popper notes that the hypothesis that atoms exist was, for a long time, too vague to be refuted. "Failure to detect the corpuscles, or any evidence of them, could always be explained by pointing out that they were too small to be detected. Only with a theory that led to an estimate of the size of the molecules was this line of escape more or less blocked, so that refutation became in principle possible" (Popper, 1983, p. 191). Popper's statement is perfectly applicable to the (particle) dark matter hypothesis if one replaces 'size of the molecules' by 'cross section of interaction of the dark matter particles with normal matter.'

As we will see in Chapters 6 and 7, Milgromian theory postulates rather *different* explanations for the anomalous observations on galactic and cosmological scales, observations which in the standard model are explained by a single postulated entity, 'dark matter.'

§§

Popper's scheme of conjectures and refutations implies that theories will evolve along a sequence, with each refuted version replaced in turn by another version, hopefully having more empirical content – that is, capable of generating more testable predictions – than the version it replaced. But to the two requirements of falsifiability and greater content, Popper added a third desideratum: a new theory "should pass some new, and severe, tests" (Popper, 1963, p. 242). That is: some of its novel predictions should be experimentally verified.

The idea here is again very simple. Scientists should not be in the business, Popper said, of "merely producing theories so that they can be superseded" (Popper, 1963, p. 245). Consider, said Popper, a sequence of theories, each of which explains the observations or experiments that brought down its predecessor, and each of which makes some new predictions. Now suppose that the novel predictions are

---

[23] Liu et al. (2017, p. 215): "there has been no solid evidence of a real event yet . . . one cannot ignore the importance of those null searches which have been setting tighter constraints to many theoretical models and which may eventually direct us on a completely different path towards understanding this mysterious component of our Universe." Note that the failure to detect dark matter particles has not, apparently, shaken Liu et al.'s conviction that dark matter exists.

always immediately refuted. There would be no reason to believe, said Popper, that such a sequence of theories represents an approach to the truth. A true theory, after all, makes nothing *but* successful predictions, and it is reasonable to require some reassurance that we are moving in the direction of that theory. "If we are content to look at our theories as mere stepping stones", said Popper, "then most of them will not even be good stepping stones":

Thus we ought not to aim at theories which are mere instruments for the exploration of facts, but we ought to try to find genuine explanatory theories: we should make genuine guesses about the structure of the world ... if we should cease to progress in the sense of our third requirement – if we should only succeed in refuting our theories but not in obtaining some verifications of predictions of a new kind – we might well decide that our scientific problems have become too difficult for us because the structure (if any) of the world is beyond our powers of comprehension (Popper, 1963, p. 245).

As an example of a corroborated novel prediction, Popper cited the bending of starlight by the gravitational force from the Sun, which led many scientists to accept the correctness of Einstein's theory of general relativity.

In fact, Popper's conclusion about the privileged role of confirmed novel predictions is one that scientists have independently arrived at, again and again, though not necessarily via the same chain of reasoning as Popper's. Here are three examples, all pre-dating Popper; many more could be given. The astronomer John Herschel wrote in 1842:[24]

The surest and best characteristic of a well-founded and extensive induction ... is when verification of it springs up, as it were, spontaneously, into notice, from quarters where they might be least expected ... Evidence of this kind is irresistible and compels assent with a weight which scarcely any other possesses.

William Whewell wrote in 1847 that if a theory

of itself and without adjustment for the purpose, gives us the rule and reason of a class of facts not contemplated in its construction, we have a criterion of its reality, which has never yet been produced in favour of a falsehood (Whewell, 1847, Vol. 2, p. 67–68).

And the physicist Norbert Robert Campbell wrote in 1921:

A true theory will not only explain adequately the laws that it was introduced to explain; it will also predict and explain in advance laws which were unknown before. All the chief theories in science (or at least in physics) have satisfied this test (Campbell, 1921, p. 87).

As we will see, there *is* a research tradition in cosmology – the one originated by Mordehai Milgrom in 1983 – that has repeatedly been successful in just this privileged way, predicting again and again (to use Whewell's words) "a class of

---

[24] Herschel (1842), Sect. 180. Quoted in *Theories of Scientific Method: The Renaissance Through the Nineteenth Century*, E. Madden (ed.) University of Washington Press, 1960, p. 177. Note Herschel's assumption that scientific theories are arrived at via induction.

facts not contemplated in its construction." By contrast, the standard cosmological model has rarely succeeded in making successful novel predictions; instead it has repeatedly been forced to 'play catch-up,' finding post hoc explanations for unexpected discoveries rather than predicting them in advance. (In many cases, those discoveries *were* predicted in advance by Milgromian researchers; they were 'unexpected' only from the standpoint of standard-model researchers.) Given scientists' supposed predilection for theories that (in Campbell's words) "predict and explain in advance laws which were unknown before," one might reasonably ask why the standard cosmological model is currently so dominant, while Milgrom's theory is so marginalized. I will return to this question in Chapter 9.

§§

It is clear from his writings that Popper had a sophisticated appreciation of the different ways that theories can evolve: sometimes via the addition of auxiliary hypotheses, as in the case of Ptolemy's equants or 'dark matter,' and sometimes via radical or revolutionary changes, as when Newton's theory of gravity and motion was replaced, wholesale, by Einstein's. But in his arguments about the criteria for scientific progress (or as he often called it, the "growth of knowledge"), Popper did not distinguish strongly between these different modes of theory change, and it seems likely that he intended his three criteria of progress to apply to all of them.

The approach to theory appraisal that will be followed in this book is due to Popper's colleague Imre Lakatos. Lakatos accepted many of Popper's arguments about evidential support and about the necessary conditions for scientific progress. For instance, Lakatos emphasized, as did Popper, that the success of a theory is measured not by the total number of successful predictions, but only by successful *new* predictions: "the only relevant evidence is the evidence anticipated by a theory" (Lakatos, 1970, p. 38). But Lakatos argued, based on the historical record (and in agreement with Thomas Kuhn), that scientists will go to extreme lengths to avoid modifying the fundamental assumptions that underlie their theories. Scientists, he said, almost always respond to anomalies – that is, falsifications – by adding auxiliary hypotheses, and leaving the "hard core" of the theory unchanged:

Scientists have thick skins. They do not abandon a theory merely because facts contradict it. They normally either invent some rescue hypothesis to explain what they then call a mere anomaly or, if they cannot explain the anomaly, they ignore it, and direct their attention to other problems (Lakatos, 1973, p. 4).

The result is a series of *connected* theories, linked together by an essentially fixed set of fundamental assumptions (the hard core), which Lakatos called a "research program." Lakatos argued that the proper unit of appraisal for scientific progress is the research program, rather than the isolated theory. By failing to distinguish clearly between theories and research programs, Lakatos argued, Popper had been unable to explain the *continuity* of science: the fact that theories often retain a recognizable character over time in spite of changes.

Lakatos died in 1974 at the age of 51. Had his short life continued just a few years longer, Lakatos might have applied his method of appraisal to the standard cosmological model – or as he might have called it: the standard cosmological research program.[25] No one, it seems, has yet taken the time to do that, nor will the attempt be made in this book. But it is clear that the development of the standard cosmological model since about 1970 adheres quite nicely to Lakatos's basic template. There is a fixed core, which includes Einstein's theory of gravity and the standard model of particle physics.[26] When the predictions of the theory have been refuted,[27] the response has almost always been to add an auxiliary hypothesis to the theory, a hypothesis explicitly designed to maintain the integrity of the hard core in the face of the anomalous data. The postulates relating to 'dark matter' and 'dark energy' came about in this manner. Of course, nothing in the foregoing sentences should be read as implying that the evolution of the standard cosmological model has been *progressive* in the sense understood by Popper or Lakatos.

Lakatos emphasized that two or more competing research programs are often pursued at the same time in a given field, typically by different sets of scientists, before one research program finally succeeds in supplanting the other(s). For instance, with regard to theories of matter prior to the early twentieth century, there were continuity theories, atomistic theories, and theories that tried to combine the two. Much the same is true today in the field of cosmology. There is a research program, begun by the physicist Mordehai Milgrom in 1983, that has evolved side-by-side with the standard cosmological research program. Milgrom's research program is the topic of most of the remainder of this book.

---

[25] Nor did Popper, who outlived Lakatos, have much to say about theories of cosmology; in Helge Kragh's (2012, p. 332) words, "one looks in vain in [Popper's] main works for discussions of the science of the universe."

[26] Throughout this book, 'standard-model cosmologist' refers to an adherent of the standard, or concordance, or ΛCDM cosmological model. Such cosmologists typically assume the correctness of the standard model of particle physics as well.

[27] Pavel Kroupa (2012) gives a timeline showing the major failures of the standard cosmological model.

2

The Methodology of Scientific Research Programs

I state my case, even though I know it is only part of the truth, and I would state it just the same if I knew it were false, because certain errors are stations on the road to truth.

(Musil (1961))

It is the *evolution* of a theory over long periods of time and not its shape at a particular moment that counts in our methodological appraisals.

(Feyerabend (1975))

All theories . . . are born refuted and die refuted.

(Lakatos (1973))

Of the many ways that a scientist can respond to an anomaly – an experiment or observation that contradicts a theory – Karl Popper identified a particular class of responses that were strictly forbidden. Suppose, he said, that a scientist sets out to test the hypothesis 'All swans are white,' and discovers a black swan. On the one hand, he can decide that his theory has been refuted and construct a new theory that allows for the existence of black swans, and that postulates something new and testable about them. On the other hand, he might simply say: "This is not a black *swan*; it is a black *waterfowl*; therefore my original hypothesis still stands." Or he could choose to ignore the existence of the black swan; or he could claim that the swan is white, but only *appears* black through a trick of the light. Or perhaps it was the scientist's assistant who made the observation, and the scientist could declare (arbitrarily) that the assistant made a mistake, or was lying.

Popper (1959, p. 82) coined the term "conventionalist stratagem" to describe this type of response: a response intended to evade the consequences of a falsifying observation and keep a theory intact. The term 'conventionalist' is nowadays fairly obscure; Popper was referring here to the idea, usually attributed to Poincaré, that any theory can be made consistent with any data by a sufficiently clever redefinition of terms. (Popper later adopted the expression "immunizing stratagems" to describe such evasive tactics.) From a strictly logical standpoint, there is no compelling reason to eschew conventionalist stratagems; but Popper argued that in order to maintain falsifiability, conventionalist tactics needed to be avoided, and that "The only way to avoid conventionalism is by taking a *decision*: the decision not to apply its methods. We decide that, in the case of a threat to our system, we will not save it by any kind of *conventionalist stratagem*" (Popper, 1959, p. 82–84).

What Popper was proposing here was a *methodology of science*. It is not sufficient, he said, for a scientific theory to be falsifiable: "*Only with reference to the methods applied* to a theoretical system is it at all possible to decide whether we are dealing with a conventionalist or an empirical theory" (Popper, 1959, p. 82). In other words: scientists who deliberately evade the consequences of a falsifying observation are *not doing science*, and the theories they produce are *not scientific*, regardless of whether those theories satisfy the (logical) condition of falsifiability.

Popper was perfectly aware that scientists sometimes fail to behave in the proper manner. He recalled a "very famous physicist"

who, although I implored him, declined to look at a most simple and interesting experiment ... It would not have taken him five minutes, and he did not plead lack of time. The experiment did not fit into his theories, and he pleaded that it might cost him some sleepless nights (in Schilpp, 1974, Book 2, p. 1185, n. 58).

(Popper added: "And I know more than one philosopher who refuses to look at a simple proof or disproof.") But Popper regarded instances like these as regrettable exceptions: "We, and those who share our attitude, will hope to make new discoveries ... Thus we shall take the greatest interest in the falsifying experiment. We shall hail it as a success, for it has opened up new vistas into a world of new experiences" (Popper, 1959, p. 90).

§

It would be difficult to overstate how different Popper's view of science was from Thomas Kuhn's. Kuhn, in his *The Structure of Scientific Revolutions* (1962) and some later works,[1] argued not only that scientists sometimes avert their gaze from the falsifying result. Kuhn claimed in addition that the sort of science that most scientists do, most of the time – what Kuhn called "normal science" – would be essentially *impossible* if scientists responded to anomalies in the way that Popper required:

No part of the aim of normal science is to call forth new sorts of phenomena; indeed those that will not fit the box are often not seen at all. Nor do scientists normally aim to invent new theories, and they are often intolerant of those invented by others (Kuhn, 1962, p. 24).

---

[1] There is general agreement that Kuhn's work tended over time toward a more rational, i.e. Popperian, view of science. For instance, in a review of the second edition of *The Structure of Scientific Revolutions*, Alan Musgrave (1971, p.296) notes that "In his recent writings, then, Kuhn disowns most of the challenging ideas ascribed to him by his critics ... the new, more real Kuhn who emerges ... [is] but a pale reflection of the old, revolutionary Kuhn."

And elsewhere Kuhn declared that normal science "often suppresses fundamental novelties because they are necessarily subversive of its basic commitments" (Kuhn, 1962, p. 5).

Kuhn reached these dismal conclusions (he said) based on an examination of the historical record, but his argument was essentially a logical one. Science (said Kuhn), like all rational discourse, requires a common language and set of assumptions. Kuhn called this shared framework a "paradigm,"[2] and he argued (in much the same way as cultural relativists and postmodernists[3]) that meaningful communication is only possible between scientists who share, uncritically, the same paradigm: "it is precisely the abandonment of critical discourse that marks the transition to a science" (Kuhn, 1970, p. 6). Kuhn acknowledged that scientists frequently engage in 'testing,' but he said that these were almost never tests of a *theory*; rather they were tests of the scientist's skill at (re)interpreting the theory so as to reconcile it with the data. He called this activity "puzzle solving," and said that a failure to solve a puzzle reflected on the scientist, not the theory: "in the final analysis it is the individual scientist rather than current theory which is tested" (Kuhn, 1970, p. 5).

Most scientists will admit that Kuhn's description of science-as-practiced, although unflattering, rings true in many respects (as did Popper, although he suggested that Kuhn's claims for the dominance of 'normal science' applied much more to science as practiced after the First World War than before; in Schilpp, 1974, Book 2, p. 1146). And there is no question that Kuhn's observations forced philosophers of science to give greater attention to the day-to-day activities of real scientists. But Kuhn never stated clearly how his "new image of science" could differentiate the practice of science from the other human activities that also take place within a shared paradigm. As Popper remarked,

> Kuhn and I agree that astrology is not a science, and Kuhn explains why from his point of view it is not a science. This explanation seems to me entirely unconvincing: *from his point of view* astrology should be accepted as a science. For it has all the properties which Kuhn uses to characterize science: there is a community of practitioners who share a routine, and who are engaged in puzzle solving ... we may find, in a couple of years' time, the great foundations supporting astrological research. From Kuhn's sociological point of view, astrology would then be socially recognised as a science (in Schilpp, 1974, Book 2, p. 1146).

Or as Paul Feyerabend (1970, p. 200) put it: "Every statement which Kuhn makes about normal science remains true when we replace 'normal science' by 'organized crime'."

---

[2] At least, that is one way to interpret Kuhn's term. Margaret Masterman (1970, p. 59) finds that "On my counting, [Kuhn] uses 'paradigm' in not less than twenty-one different senses."

[3] The editors of *A Postmodern Reader* include, between selections of Jean Baudrillard and Cornel West, Kuhn's "The Resolution of Revolutions" and preface Kuhn's article with the approving words: "Kuhn argues that what is at stake at such moments of change is learning to "see science and the world differently." ... Seeing within a different paradigm means being in a place where we can make conceivable that which is not already presentable within our prevailing paradigm's "rules of the game"" (Natoli and Hutcheon, 1993, p. 307).

Kuhn's 'new image of science' is indistinct in other ways. It is very well to point out that scientists can be indifferent to experimental anomalies. But what constitutes *support* for a theory, and on what basis do scientists make this judgment? Recall that Popper gave a carefully reasoned answer to the first question: theories are supported by confirmed, novel predictions. One can disagree with Popper's conclusion, or argue based on the historical record that scientists use a different criterion; but in light of Hempel's theorem (anything that does not contradict a hypothesis is, logically, equally confirming of it), scientists (and philosophers of science) need *some* criterion for separating the wheat from the chaff, evidentiarily speaking. Kuhn has little to say on this essential question.[4] When listing the features of a theory that are relevant in corroborating it, Kuhn mentions "accuracy of prediction, particularly of quantitative prediction . . . and the number of different problems solved"(Kuhn, 1962, postscript, p. 206). But Ptolemy's epicycles and equants were capable of making predictions with arbitrary accuracy. And with regard to solved problems, Kuhn does not distinguish between problems which a theory was specifically designed to solve, and those which only appear after a theory's construction; nor did it appear to matter, to him, whether a solution arises organically from the theory, or consists of (as Popper would say) a conventionalist stratagem.

In the words of Paul Feyerabend (1970, p. 198):

Whenever I read Kuhn, I am troubled by the following question: are we here presented with *methodological prescriptions* which tell the scientist how to proceed; or are we given a *description*, void of any evaluative element, of those activities which are generally called 'scientific'? Kuhn's writings, it seems to me, do not lead to a straightforward answer.

Or as John Kadvany (2001, p. 151) put it: "Kuhn never clearly identified just what should be *done* with the new image of science."

§§

It fell to Imre Lakatos to develop a description of science that incorporated Popper's logical and epistemic insights; maintained a demarcation between science and pseudoscience; and accommodated (or neutralized) the apparent threats to scientific rationality posed by Kuhn's interpretation of the historical record.

Lakatos was born Imre Lipsitz, a name which he changed during the Nazi occupation of Hungary to the less Jewish-sounding Imre Molnar, and again after the start of the Russian occupation to the more working-class Imre Lakatos ("Locksmith" in Hungarian). After the war, Lakatos was politically active and was made a secretary in the Ministry of Education. On returning to Hungary from a visit to Moscow in 1949, he was arrested (exactly why is not clear) and spent more than three years

---

[4] Larry Laudan (1984, p.73) makes a similar point: "Kuhn has failed over the past twenty years [i.e. since 1964] to elaborate any coherent account of consensus formation, that is, of the manner in which scientists could ever agree to support one world view rather than another."

in prison. After his release in 1954, Lakatos began studying mathematics with Alfréd Renyi, and translated György Pólya's *How to Solve It* into Hungarian. He also began to turn away from Marxism, and when the Hungarian uprising of 1956 was put down by Soviet troops, Lakatos fled Hungary for Vienna. From 1960 until his early death in 1974 he taught alongside Karl Popper at the London School of Economics.

Before moving to London, Lakatos completed a doctoral thesis in King's College, Cambridge, on the philosophy of mathematics, entitled *Essays in the Logic of Mathematical Discovery*. Four papers based on this work were published during 1963–1964; but Lakatos considered the work unfinished, and it was not until after his death that John Worrall and Elie Zahar published a volume, *Proofs and Refutations* (1976), that included selections from the thesis, the four published papers, and commentary speculating how Lakatos might have further developed his arguments had he lived.

In *Proofs*, Lakatos argued that mathematical theorems are not derived simply by deductive reasoning starting from some set of fixed postulates, as the textbooks usually imply. When proving a theorem, he said, mathematicians will start from a general idea, or hunch, as to what it is they are trying to prove, then 'stretch' the definitions of fundamental terms, or modify the statement of the theorem, as needed to resolve difficulties and allow the proof to go forward. Lakatos borrowed the term 'heuristic' from Pólya to describe this process of conceptual growth through conjectures and refutations. As an example, Lakatos considered Euler's formula for simple polyhedra, $V - E + F = 2$ (V = number of vertices, E = number of edges, F = number of faces). Lakatos pointed out that some geometrical objects fail to satisfy Euler's theorem; for instance, a solid cube with a cubic space inside it has $V - E + F = 4$. Given a counter-example like this, the mathematician can abandon the theorem; restate the theorem in such a way as to account for the confounding object; or change the definition of some term or terms, e.g. 'polyhedron' or 'vertex.' Lakatos showed that by considering objects for which $V - E + F$ does *not* equal two, one is led to a rule for the 'Euler characteristic' $V - E + F$ that works for a much larger class of solids than the simple polyhedra.

Pólya had argued in *How to Solve It* that there were rational methods, or "heuristics," available to the mathematician for generating theorem-candidates. For instance, theorems in plane geometry often have counterparts in higher dimensions. Lakatos used 'heuristic' to mean, roughly, the process of critical argument that leads to a change in mathematical concepts or language, or a shift to a new conceptual framework. He did not claim that there was a *unique* mode of mathematical discovery; instead he sought to extract the heuristic used, case by case, by examination of particular historical episodes. In this respect, Lakatos was departing from the Popperian view that discovery and justification are two quite distinct things, and that only the latter is subject to rational analysis. Lakatos also differed from Popper by adopting a more nuanced view of the role of falsification; as in the Euler proof described above, he argued that counter-examples can sometimes be useful, by indicating the way forward.

After joining Popper in London, Lakatos turned his attention to the philosophy of science. Like Popper, Lakatos was uncomfortable with Kuhn's view of science as an enterprise detached from criticism. In his words:

The clash between Popper and Kuhn is not about a mere technical point in epistemology. It concerns our central intellectual values, and has implications not only for theoretical physics but also for the underdeveloped social sciences and even for moral and political philosophy (Lakatos, 1970, p. 9).

In two long essays – "Falsification and the methodology of scientific research programmes" (1970) and "History of science and its rational reconstruction" (1971)[5] – Lakatos showed how to solve Popper's problem of demarcation in a way that respects the historical record. In outline, his procedure was as follows:

Scientists and philosophers have a pretty good idea (Lakatos said) about which episodes of intellectual history correspond to 'good' science (Newton, Einstein) and which do not (Marx, Freud). Suppose that the goal is to test a general hypothesis about what constitutes the methodology associated with good, or rational, science. For instance, one might postulate that scientists produce their theories by generalizing from experimental data; that is, via induction. In the same way that scientific *theories* can be tested and potentially falsified, so can hypotheses about the methodology of science. The trick is to inspect the historical record and look for episodes during which a scientific theory evolved in the postulated manner (the 'rational' episodes) and the episodes during which it did not (the 'irrational' episodes). Lakatos called such an exercise "rational reconstruction," and, like all historiography, a certain amount of interpretation is involved. If it turns out that the rational episodes dominate the non-rational ones, then the postulated methodology is reasonable: it has survived the attempt to falsify it, and one can proceed to apply the test to another period of successful science. Whereas if the postulated methodology turns out to conflict with the historical record – if most episodes of what we consider 'good' science cannot be described as 'inductivist' without doing violence to the historical record – then induction is probably not a good description of how science actually works.

Lakatos tested three hypotheses about what constitutes rational scientific practice: inductivism, conventionalism (in the original sense of that term due to Poincaré), and Popper's falsificationism. (He did not test any hypothesis of Kuhn because Kuhn never provided one.) As historical episodes, Lakatos chose the Copernican and Newtonian revolutions – clearly episodes of 'successful'

---

[5] Reprinted as Chapters 1 and 2 of *The Methodology of Scientific Research Programmes*, Philosophical Papers Volume 1 (Cambridge University Press, 1978), edited by John Worrall and Gregory Currie. In citing Lakatos from these two essays, I will adopt the pagination of that edited volume. I am grateful to Cambridge University Press for their permission to quote liberally from this volume.

science. Lakatos argued that all three hypotheses failed the historiographical test. For instance, under falsificationism, one would expect scientists to immediately abandon a theory once it has been falsified; continuing to work on a falsified theory would be irrational. But Lakatos found (as had Kuhn) that theories are engulfed in an "ocean of anomalies" from the start, and that scientists nevertheless continue to work on them; in fact a scientist who abandoned falsified theories would not be able to do science at all. Thus, falsificationism was 'falsified' by the historical record.

Having shown that none of the extant hypotheses about what constitutes rational science could be made to fit the historical record, Lakatos's next step was to formulate a new hypothesis and test it. His proposal – the "methodology of scientific research programmes" – was still (like everything in Popper) essentially critical-rationalist: the rational scientist is assumed to postulate theories, generate predictions, and test them via experiment or observation. And following Popper, Lakatos defined evidential support purely in terms of novel predictions: "the only relevant evidence is the evidence anticipated by a theory" (Lakatos, 1970, p. 38). But Lakatos broke with Popper by arguing (as he had in *Proofs*) that failures of prediction are not fatal: what matters more is how scientists *develop* a theory in response to a falsifying instance. (Popper, it seems, never forgave Lakatos for this apostasy.) The sequence of theories that results from this developmental process Lakatos called a "research program," and he argued that the proper unit of appraisal is the entire program, not any single theory taken from it. Lakatos also carried over from his earlier work in philosophy of mathematics the idea of a heuristic that guides the scientist's work and suggests (among other things) how theories should be modified in response to anomalies.

In his two long essays, Lakatos tested his proposed methodology against some well-known episodes from the history of science, and argued that he was able to correctly distinguish 'good' science from 'bad' science (or, as he termed it, "progressive" versus "degenerating" science). A number of more thorough historiographical appraisals of this sort were carried out by his students and published after his death, many of them in the volume *Method and Appraisal in the Physical Sciences* (C. Howson, ed., 1976). More appraisals have been published since then: not only in the physical sciences, but in economics, demographics, biology etc. This large body of work can be interpreted as corroboration (for the most part) of Lakatos's proposed methodology. But it is important to recognize the dual role played by the historical record in the original formulation of the *Methodology*. Not only did Lakatos appeal to history to test his proposed methodology. It is also clear that Lakatos was guided in *formulating* that methodology by his knowledge of the history of science, and of the ways in which the extant demarcation proposals failed when confronted with the historical record. Quite a bit of 'conjecture and refutation' had already taken place in Lakatos's thinking before he committed his *Methodology* to paper. In the words of John Kadvany:

The methodology of scientific research programmes is not baldly "proposed" or simply stated by Lakatos. The methodology is developed dialectically as the synthesis and culmination of a sequence of successively more sophisticated and powerful methodologies. Lakatos's philosophy of science, his "methodology of scientific research programmes," is an elaboration of Popper and a merciless attack by way of historicizing Popper's ideas (Kadvany, 2001, p. 157).

The reasoning that led Lakatos to his *Methodology* is described in more detail in the next two subsections of this chapter.

Lakatos did not intend for his *Methodology* to be the final word. Just as scientific theories evolve in response to counter-examples, so should theories of scientific methodology evolve, in response to historiographical studies that are found to be in conflict with them. At the time of his death Lakatos was still working on such an analysis, which he gave the provisional title "The Changing Logic of Scientific Discovery."[6] Nevertheless, one major change in the *Methodology* was initiated while Lakatos was still alive: Elie Zahar proposed a change in the definition of 'novel' prediction which Lakatos subsequently endorsed. An even broader definition of this important concept has been developed since then, as discussed later in this chapter. And as we will see in Chapter 7, there is widespread agreement that Lakatos's conception of what he called the "hard core" of a research program should be modified. But for the most part, the description of the *Methodology* that follows in this chapter is the same as originally put forth by Lakatos.

§§

Lakatos began by critiquing the idea of falsification. From a logical standpoint, he said, a finite number of experimental *facts* can not conclusively establish the falsity of a theoretical *proposition*, any more than they can establish its correctness. This consideration calls into question what Lakatos labelled "dogmatic falsificationism":

For the truth-value of the 'observational' propositions cannot be indubitably decided: *no factual proposition can ever be proved from an experiment.* Propositions can only be derived from other propositions, they cannot be derived from facts . . . If factual propositions are unprovable then they are fallible. If they are fallible then clashes between theories and factual propositions are not 'falsifications' but merely inconsistencies. Our imagination may play a greater role in the formulation of 'theories' than in the formulation of 'factual propositions', but they are both fallible. Thus we *cannot prove theories and we cannot disprove them either* (Lakatos, 1970, p. 15–16).

---

[6] Mentioned in a letter of 10 January 1974 from Lakatos to Paul Feyerabend; see Motterlini (1999, p. 355).

Even if experimental facts are granted the power to establish the truth value of a (falsifying) *proposition*, Lakatos (following Duhem and Quine) argued that any finite number of observations is still inadequate to falsify any *hypothesis* that is sufficiently complex to deserve the name 'scientific':

A proposition might be said to be scientific only if it aims at expressing a causal connection ... 'all swans are white', if true, would be a mere curiosity unless it asserted that swanness *causes* whiteness. But then a black swan would not refute this proposition, since it may only indicate *other causes* operating simultaneously. Thus 'all swans are white' is either an oddity and easily disprovable or a scientific proposition with a *ceteris paribus* clause and therefore undisprovable (Lakatos, 1970, p. 18–19).[7]

Lakatos noted that objections similar to these were raised already by Popper, who proposed to deal with them via a set of methodological conventions, an approach which Lakatos labeled "methodological falsification." In the case of basic statements (singular statements about observable facts), which are needed in order to decide whether a theory is falsifiable, Popper proclaimed that "basic statements are accepted as the result of a decision or agreement; and to that extent they are conventions" (Popper, 1959, p. 106). Lakatos recast this as "the list of 'accepted' falsifiers is provided by the verdict of the experimental scientists" (Lakatos, 1970, p. 24). In the case of *ceteris paribus* clauses, Lakatos pointed out that Popper likewise required a (set of) decision(s) to separate the theory being tested from the unproblematic 'background knowledge.' Lakatos argued that the inevitable arbitrariness of these conventional decisions means that "we may well end up by eliminating a true, and accepting a false, theory" (Lakatos, 1970, p. 24):

Methodological falsificationism represents a considerable advance beyond both dogmatic falsificationism and conservative conventionalism. It recommends risky decisions. But the risks are daring to the point of recklessness and one wonders whether there is no way of lessening them (Lakatos, 1970, p. 28).

Lakatos's solution was to invoke the historical record. Both dogmatic and methodological falsificationism, he argued, are poor descriptions of how scientists in the past actually went about deciding whether to adopt or eliminate theories. As an illustrative example, Lakatos turned again and again to Newton's theory of gravity and motion:

When it was first produced, it was submerged in an ocean of 'anomalies' (or, if you wish, 'counterexamples'), and opposed by the observational theories supporting these anomalies. But Newtonians turned, with brilliant tenacity and ingenuity, one counterinstance after another into corroborating instances, primarily by overthrowing the original observational theories in the light of which this 'contrary evidence' was established. In the process they themselves produced new counter-examples which they again resolved (Lakatos, 1970, p. 48).

---

[7] *Ceteris paribus* (or *caeteris paribus*) is a Latin phrase meaning 'other things equal.' In this context, it refers to a statement of conditions that are assumed to obtain when applying an otherwise universal law or hypothesis.

Newton's theory was eventually replaced by Einstein's, but not, Lakatos argues, because Newton's theory was falsified; rather because Einstein's theory

explained everything that Newton's theory had successfully explained, and it explained also *to some extent* some known anomalies and, in addition, forbade events like transmission of light along straight lines near large masses about which Newton's theory had said nothing but which had been permitted by other well-corroborated scientific theories of the day; moreover, *at least some* of the unexpected excess Einsteinian content was in fact *corroborated* (for instance, by the eclipse experiments) (Lakatos, 1970, p. 39).

Lakatos saw reflected in these examples two ideas already in Popper: (*i*) a theory can be 'saved' by invoking auxiliary hypotheses, as long as they are not 'conventionalist,' i.e. as long as they constitute an increase in empirical content (degree of falsifiability); and (*ii*) one theory may be preferred over another even if the first has not been conclusively falsified. The first idea is illustrated by the persistence of Newton's theory in spite of anomalies; the second by the adoption of Einstein's theory, for reasons other than the falsification of Newton's. What makes for scientific progress, Lakatos argued, is the *manner in which a theory is developed* over time.[8] The proper entity to consider when appraising scientific progress is not a theory in isolation, but what Lakatos called a "research program": a *series* of theories in which

each subsequent theory results from adding auxiliary clauses to (or from semantical reinterpretations of) the previous theory in order to accommodate some anomaly, each theory having at least as much content as the unrefuted content of its predecessor (Lakatos, 1970, p. 33).

Theories change, but a research program maintains its identity over time. Popper, Lakatos felt, had sometimes acknowledged a distinction between theories and series of related theories, but his focus on the former had kept him from producing a convincing explanation for the continuity of science over time.

But what is it that remains *fixed* in a research program, in spite of the changes? Lakatos (1974, p. 146) reminds us of Popper's insistence that a scientist be prepared to state under what conditions he would abandon his most basic assumptions; to the extent that a Freudian or a Marxist is unwilling to do this, said Popper, their theories are unscientific. But, Lakatos argued, much the same could be said of Newtonian scientists, who (at least until shortly before the replacement of Newton's theory by Einstein's) were equally unwilling to abandon the central tenets of their theory: Newton's laws of gravity and motion. Lakatos (1970, p. 48) described Newtonian gravitational theory as "possibly the most successful research programme ever"; clearly, Lakatos said, the maintenance by Newtonian scientists of a fixed, irrefutable, "hard core" to their research program was not sufficient to render it unscientific. Here Lakatos made a substantial break with Popper, for whom

---

[8] Note that Lakatos would consider Newton's and Einstein's theories as belonging to different research programs.

conventions determined the acceptance of singular statements only, not of universal ones (Popper, 1959, section 30). Lakatos argued that the decision (i.e. convention) to define certain *universal* statements as unfalsifiable is a generic feature of scientific research programs.[9]

Lakatos noted that Niels Bohr, already in a paper from 1913, explicitly set out the postulates constituting the hard core of his research program: in Bohr's own words (Bohr, 1913, p. 874–875),

1. That energy radiation [within the atom] is not emitted (or absorbed) in the continuous way assumed in the ordinary electrodynamics, but only during the passing of the systems between different "stationary" states.
2. That the dynamical equilibrium of the systems in the stationary states is governed by the ordinary laws of mechanics, while these laws do not hold for the passing of the systems between the different states.
3. That the radiation emitted during the transition of a system between two stationary states is homogeneous, and that the relation between the frequency v and the total amount of energy emitted E is given by $E = hv$, where h is Planck's constant.
4. That the different stationary states of a simple system consisting of an electron rotating round a positive nucleus are determined by the condition that the ratio between the total energy, emitted during the formation of the configuration, and the frequency of revolution of the electron is an entire multiple of $\frac{h}{2}$. . . .
5. That the "permanent" state of any atomic system, i.e. the state in which the energy emitted is maximum, is determined by the condition that the angular momentum of every electron round the centre of its orbit is equal to $\frac{h}{2\pi}$.

In other research programs, said Lakatos (though without giving examples), the hard core "develops slowly, by a long, preliminary process of trial and error" (Lakatos, 1970, p. 48, n. 4).[10] But substantial change in the hard core would mean abandoning the research program altogether.

Lakatos made much of the fact that Bohr's hard core postulates were *inconsistent*; in effect, Bohr "grafted" his research program onto Maxwell's theory of electromagnetism. Lakatos saw inconsistencies also in the hard cores of Prout's theory of atomic weights and of Copernican astronomy, and even (in its early days) in Newtonian theory, before the acceptance of 'action-at-a-distance.' Of course, to the extent that a theory corresponds to reality, it must be consistent. Lakatos argued, based on examples from the historical record, that "As the young grafted

---

[9] Although Lakatos appears never to use the term, *fideism* is the epistemic principle that he is here ascribing to scientists. Richard H. Popkin notes that fideism can be usefully defined in a number of ways, but that "there is . . . a common core, namely that knowledge . . . is unattainable without accepting something on faith" (Popkin, 2003, p. xxii). Or in Pierre Jurieu's more succinct seventeenth century formulation, "Je le crois, dis-je de cette manière: parce que je le veux croire" (Jurieu, 1687, p. 248–249). That scientists behave in the way that Lakatos describes can hardly be gainsaid, but one is left wondering on what basis a scientific community decides to assign certain assumptions to their hard core and not others. This question is revisited in Chapter 7.
[10] An instance from the standard cosmological model (or rather, its associated research program) would be the dark matter hypothesis, which was not part of the model prior to about 1980, but which since then has acquired the status of an unchallengeable assumption (Merritt, 2017).

programme strengthens, the peaceful co-existence comes to an end, the symbiosis becomes competitive and the champions of the new programme try to replace the old programme altogether" (Lakatos, 1970, p. 56–57).

§

To the extent that the hard core of a research program is taken as invariant, experimental or observational anomalies (i.e. falsifications) must be dealt with via auxiliary hypotheses. "It is this protective belt of auxiliary hypotheses which has to bear the brunt of tests and get adjusted and re-adjusted, or even completely replaced, to defend the thus-hardened core" (Lakatos, 1970, p. 48). Here again, the idea has a basis in Popper, who noted that, logically, any falsification could be dealt with by an addition to the theory. As discussed in Chapter 1, Popper required that, to be acceptable, changes must satisfy two extra conditions (in addition to preserving falsifiability): they must be content-increasing, and at least some of the theory's new predictions should eventually be confirmed. Lakatos adopted essentially the same requirements in defining what he called "progressive problemshifts":[11]

Let us take a series of theories, $T_1, T_2, T_3, \ldots$ [in a given research program] where each subsequent theory results from adding auxiliary clauses to (or from semantical reinterpretations of) the previous theory in order to accommodate some anomaly, each theory having at least as much content as the unrefuted content of its predecessor. Let us say that such a series of theories is *theoretically progressive (or 'constitutes a theoretically progressive problemshift')* if each new theory has some excess empirical content over its predecessor, that is, if it predicts some novel, hitherto unexpected fact. Let us say that a theoretically progressive series of theories is also *empirically progressive (or 'constitutes an empirically progressive problemshift')* if some of this excess empirical content is also corroborated, that is, if each new theory leads us to the actual discovery of some new fact. Finally, let us call a problemshift *progressive* if it is both theoretically and empirically progressive, and *degenerating* if it is not. We *'accept'* problemshifts as 'scientific' only if they are at least theoretically progressive; if they are not, we *'reject'* them as 'pseudoscientific' (Lakatos, 1970, p. 33–34).[12]

That theories can be, and will be, falsified, in the sense understood by Popper, is taken for granted by Lakatos. But "We regard a theory in the series 'falsified' when it is superseded by a theory with higher corroborated content" (Lakatos, 1970, p. 34). Lakatos emphasized – as did Popper – that the success of a theory is measured not by the *total* number of successful predictions, but only by its success

---

[11] "The appropriateness of the term 'problemshift' for a series of theories rather than of problems may be questioned. I chose it partly because I have not found a more appropriate alternative – 'theoryshift' sounds dreadful – partly because theories are always problematical, they never solve all the problems they have set out to solve" (Lakatos, 1970, p. 34, n. 2).

[12] Note that Popper's definition of empirical content (the class of potential falsifiers; see Chapter 1) differs from that of Lakatos; see e.g. Popper (1963, p. 385) for a critical comparison of the two definitions. Lakatos's usage (which many authors adopt) is retained in what follows.

at making *new* predictions: "the only relevant evidence is the evidence anticipated by a theory" (Lakatos, 1970, p. 38).

A research program is defined in terms of its hard core, but Lakatos identified one more element that is characteristic of research programs, the "positive heuristic." He began by noting (as he had in *Proofs*) that scientists do not respond to anomalies in a random fashion; rather,

The order is usually decided in the theoretician's cabinet, independently of the *known* anomalies. Few theoretical scientists engaged in a research programme pay undue attention to 'refutations'. They have a long-term research policy which anticipates these refutations. This research policy, or order of research, is set out – in more or less detail – in the *positive heuristic* of the research programme . . . the positive heuristic consists of a partially articulated set of suggestions or hints on how to change, develop the 'refutable variants' of the research-programme, how to modify, sophisticate, the 'refutable' protective belt (Lakatos, 1970, p. 49–50).

Again: *"It is primarily the positive heuristic of his programme, not the anomalies, which dictate the choice of [the theorist's] problems."* (Lakatos, 1971, p. 111). For instance, in the case of Bohr's research program,

His first model was to be based on a fixed proton-nucleus with an electron in a circular orbit; in his second model he wanted to calculate an elliptical orbit in a fixed plane; then he intended to remove the clearly artificial restrictions of the fixed nucleus and fixed plane; after this he thought of taking the possible spin of the electron into account, and then he hoped to extend his programme to the structure of complicated atoms and molecules and to the effect of electromagnetic fields on them, etc., etc. All this was planned right at the start: the idea that atoms are analogous to planetary systems adumbrated a long, difficult but optimistic programme and clearly indicated the policy of research (Lakatos, 1970, p. 60–61).

But, Lakatos noted,

Not all developments in [Bohr's] programme were foreseen and planned when the positive heuristic was first sketched. When some curious gaps appeared in Sommerfeld's sophisticated models (some predicted lines never did appear), Pauli proposed a deep auxiliary hypothesis (his 'exclusion principle') which accounted not only for the known gaps but reshaped the shell theory of the periodic system of elements and anticipated facts then unknown (Lakatos, 1970, p. 67).

Lakatos argued that the existence of a positive heuristic is more than just an historically recurring *feature* of research programs. He proposed also that it is an essential factor in judging the *progressivity* of a research program. Research programs should evolve autonomously, he argued, and not simply in response to anomalies. "Heuristic progress" requires that "the successive modifications of the protective belt must be in the spirit of the heuristic" (Lakatos and Zahar, 1976, p. 179). It is probably fair to say that Lakatos did not justify this requirement quite as convincingly as his other conditions for progressivity. Read in context, Lakatos seems to be saying that heuristic progress is a concomitant of heuristic *power* – the ability of a research program to anticipate novel facts and "to explain their refutations in the course of their growth" (Lakatos, 1970, p. 52).[13]

In any case, Lakatos explicitly included the notion of heuristic progress as an additional element when defining the distinction between progressive and degenerating problemshifts:

A research programme is said to be *progressing* as long as its theoretical growth anticipates its empirical growth, that is, as long as it keeps predicting novel facts with some success ('*progressive problemshift*'); it is *stagnating* if its theoretical growth lags behind its empirical growth, that is, as long as it gives only *post hoc* explanations either of chance discoveries or of facts anticipated by, and discovered in, a rival programme ('*degenerating problemshift*') (Lakatos, 1971, p. 112).

The positive heuristic thus plays two roles. It presents a sort of road map for a program's future theoretical development; and it also (in part) sets the standards by which the program's success is to be evaluated. As an example of an auxiliary hypothesis that violates the condition of 'heuristic progress,' Lakatos cited the introduction of equants into the Ptolemaic research program (Lakatos and Zahar, 1976, p. 181–182).

Perhaps the closest that Popper came to Lakatos's idea of a positive heuristic was when he suggested that a "new theory should proceed from some simple, new, and powerful, unifying idea about some connection or relation (such as gravitational attraction) between hitherto unconnected things (such as planets and apples) or facts (such as inertial and gravitational mass) or new 'theoretical entities' (such as field and particles)" (Popper, 1963, p. 241).

Lakatos also defined what he called the "negative heuristic" of a research program as the methodological requirement that the assumptions underlying the hard core are not to be exposed to falsification. To avoid confusion, in the remainder of this book, I will use the term 'heuristic' in place of Lakatos's 'positive heuristic,' and I will not use the term 'negative heuristic.'

Lakatos's three criteria for progressivity (increased content, corroborated excess content, growth according to the heuristic) imply that there are distinct ways in which the development of a theory can *fail* to satisfy the full set of conditions, i.e. can be judged ad hoc or degenerating:

I distinguish three types of *ad hoc* auxiliary hypotheses: those which have no excess empirical content over their predecessor ('*ad hoc$_1$*'), those which do have such excess content but none of it is corroborated ('*ad hoc$_2$*') and finally those which are not ad hoc in these two senses but do not form an integral part of the positive heuristic ('*ad hoc$_3$*') (Lakatos, 1971, p. 112, n. 2).

---

[13] Glass and Johnson (1989, p. 71) suggest the following, more Kuhnian, justification for Lakatos's requirement of heuristic progress: "Since these modifications [of the protective belt] are guided by the positive heuristic, then they will involve phenomena that can be explained within the scope of the programme's theories. This, in turn, means that the empirical confirmation of the predictions flowing from these modifications will have precise meaning in the sense that it can be readily related to the explanation provided by the programme's theories."

These three categories are mutually exclusive but not comprehensive; for instance, an auxiliary hypothesis could be *ad hoc*$_1$ and also fail to satisfy the condition of heuristic progress.[14]

§

In summary: according to Lakatos, changes to a theory in a research program should satisfy the following conditions if the research program is to be judged progressive: (1) A new theory should account for the successes of the previous theory, but it should also predict some new facts: it should have excess empirical content. (2) At least some of the novel predictions should be validated experimentally. (3) Theory changes should conform to the heuristic of the research program; they should not be driven purely by unanticipated discoveries. Conditions (1) and (2) are sometimes expressed, jointly, as 'incorporation with corroborated excess content.' Condition (3) was called by Lakatos "heuristic progress."

The importance of falsifiability – the first of Popper's conditions for progressivity, and the basis of Popper's demarcation criterion – is acknowledged by Lakatos, in the sense that acceptable changes are expected to make predictions that are susceptible to empirical testing; but given that theories are always embedded in an "ocean of anomalies," what mattered most to Lakatos was how the anomalies are dealt with as a theory evolves. By shifting the focus from theories to research programs, Lakatos claimed that he had found a demarcation criterion that "no longer rules out essential gambits of actual science" (Lakatos, 1974, p. 148).

§§

The prediction of novel facts, and their experimental confirmation, were proposed as requirements for theory progress by both Popper and Lakatos. Recall from Chapter 1 that Popper required new theories to "pass some new, and severe, tests" (Popper, 1963, p. 242) and defined a test as "severe" if it confronts a prediction that is "highly improbable in the light of our previous knowledge (previous to the theory which was tested and corroborated)" (Popper, 1963, p. 220). According to Popper, a fact that was known prior to the construction of a theory can not support it; a theory

---

[14] Just such a case can be found in the standard cosmological model: the 'dark energy' postulate is *ad hoc*$_1$ (it does not increase the theory's empirical content since the assumed properties of dark energy are unspecified; they can be, and in fact are, varied at will in response to new data) and, in addition, the hypothesis was made purely in response to unanticipated observational facts, hence it violates the condition of heuristic progress. 'Dark matter' is a slightly more nuanced case. Consider the hypothesis DM-1 of Chapter 1. If one includes as part of the dark matter hypothesis a particular elementary particle, or well-defined class of particles, which are said to constitute the dark matter, then the hypothesis can be made refutable, but the lack of experimental corroboration (so far) means that the hypothesis is still *ad hoc*$_2$. Whereas if the particle properties such as cross section for interaction with normal matter are left unspecified, the hypothesis is not refutable and hence is *ad hoc*$_1$. In either case, the dark matter hypothesis is also ad hoc in the sense of not satisfying the heuristic of the program: the hypothesis was added purely in response to unanticipated observational facts.

derives support only from facts that are discovered in the process of attempting to falsify it. Lakatos endorsed essentially the same view when he characterized a novel fact as "hitherto unexpected" (Lakatos, 1970, p. 33) and "inconsistent with previous expectations" (Lakatos and Zahar, 1976, p. 184).

Deciding which facts are 'novel' is essential for assessing the progressivity of research programs under Lakatos's *Methodology*. But since the mid-1970s, a number of objections have been raised to Popper's and Lakatos's criteria for what constitutes novel facts, or empirical support more generally. In what follows, I briefly summarize these critiques, and present a (by no means complete) list of four criteria that have been proposed for novelty, or for the conditions under which a confirmed prediction constitutes evidential support for a theory. In subsequent sections I will apply all of these criteria (or as many as are relevant) when assessing progressivity of the Milgromian research program. Of course, the reader is free to assign different degrees of epistemic significance to the different criteria.

Following Popper, I propose a first criterion for novelty as P, where

P: A predicted fact is novel if it was unknown prior to the formulation of the theory.[15]

Lakatos himself raised an objection to criteria, like P, that have a temporal or historical basis: they exclude many facts that are widely seen, by scientists, as having provided evidential support. For instance, the Balmer series of hydrogen was known before Bohr published his model of the hydrogen atom; knowledge of Mercury's anomalous apsidal precession pre-dated the theory of relativity; Kepler's laws were known to Newton, etc. Examples like these led Lakatos to suggest that some previously known facts can still be considered 'novel,' as long as a new theory explains them in a "novel way" (Lakatos, 1970, p. 70).

Elie Zahar (1973, p. 101–104) objected to this revision as being too liberal, since, he argued, a new theory will often explain old facts in a new way. Simple accommodation of a fact by a theory does not constitute support; theories can always be "cleverly engineered to yield the known facts." What matters, Zahar argued, is whether a fact "belong[s] to the problem-situation which governed the construction of the hypothesis" – i.e. whether the theory was "specifically designed to deal with" the fact. In judging the novelty of a fact, argued Zahar, "one has to take into account the way in which a theory is built and the problems it was designed to solve." Following Zahar, a second definition of novelty is

Z: A novel fact is one that a theory was not specifically designed to accommodate.

---

[15] If one takes seriously Popper's suggestion that novel facts must be discovered *during an attempted falsification*, the date associated with the *prediction* would be relevant to judgments of novelty. Predictions are sometimes only made (or at least, published) long after the theory that entails them. In the following chapters I will highlight instances where corroboration of a fact occured after a theory was published, but before a prediction was made. The reader is welcome to take Popper's argument into account when assessing the degree of evidential support in these cases.

As an example of how a fact can be used in theory construction, Zahar cited the determination of a theory's parameters from data: "In such a case we should certainly say that the facts provide little or no evidential support for the theory" (Zahar, 1973, p. 102–103). Lakatos subsequently endorsed Zahar's revised criterion (Lakatos and Zahar, 1976, p. 184–185). A fact that is novel according to *P* will also be novel according to *Z*; the converse is not necessarily true.[16]

A number of authors have criticized Zahar's criterion on the grounds that it can be ambivalent or difficult to apply. Clark Glymour (1980, p. 99) noted that it is not always clear what constitutes the 'use' of a fact in constructing a theory. Michael Gardner (1982) pointed out that a fact may belong to the problem-situation addressed by the theory but not be used in the construction of the theory, or vice versa. Alan Musgrave (1974) and Martin Carrier (1988) complained that the historical record is often ambivalent with regard to which data and experiments played supporting roles in a theory's development.

Musgrave (1974, 1978) thought to see a way around these difficulties. He noted that both Popper and Lakatos defined the severity of a test in terms of prior expectations, and that those expectations can derive both from known facts, and from a pre-existing theory. Musgrave argued that novelty should be defined with reference to the predictions of the "best existing competing theory": "in assessing the evidential support of a new theory we should compare it, not with 'background knowledge' in general, but with the old theory which it challenges" (Musgrave, 1974, p. 15). Following Musgrave,

M: A corroborated fact provides evidential support for a theory if the best rival theory fails to explain the fact, or if the fact is prohibited or improbable from the standpoint of that theory.

Musgrave's criterion has a potentially fatal flaw: it does not forbid ad hoc explanations (Worrall, 1978a, p. 331). Suppose that two theories are each incapable of explaining a known fact; according to Musgrave's criterion, whichever theory is *first* to explain the fact (via an auxiliary hypothesis, say) is the theory that gains support from the fact. Popper and Lakatos were both aware of this possibility: Popper labelled such adjustments "conventionalist" unless they increase a theory's empirical content; Lakatos required in addition that the adjustments comply with the positive heuristic of the research program. John Worrall (1978b, p. 51, 59–60)

---

[16] In summarizing their comparative appraisal of different theories of scientific change (including that of Lakatos), Laudan *et al.* (1988) select essentially this criterion as the one that is most widely adopted by scientists when judging evidential support: "there is strong agreement that ... theories are expected to solve problems they were not invented to solve" (Laudan et al., 1988, p. 40). See also the discussion near the end of Chapter 1 where it is noted that scientists prior to Popper have often endorsed the same criterion. Of course, Lakatos is making a slightly stronger claim: not only that theories *should* make successful novel predictions, but that such are the *only* sorts of prediction that are capable of providing evidential support.

emphasized the Lakatosian point of view, arguing that the difficulty with ad hoc adjustments vanishes if one requires that theory modifications be made according to the program's heuristic, rather than in response to unexpected facts.

Martin Carrier (1988, 2002) chose a different tack to deal with the problem of ad hoc adjustments. Recalling a suggestion of Lakatos that "A given fact is explained scientifically only if a new fact is also explained with it" (Lakatos, 1970, p. 34) Carrier proposed

C: "A hypothesis explains a fact in a non-ad-hoc manner, if it simultaneously explains at least one additional independent fact that either constitutes an anomaly for the rival theory or that falls beyond its realm of application" (Carrier, 1988, p. 206).

Carrier's criterion can find evidential support even in a fact that a theory was specifically designed to explain, as long as the theory is modified in such a way as to make new, and corroborated, predictions.

In what follows, I will consider Musgrave's criterion (M) to be applicable to a fact only in cases where it is demonstrable that the theory was not expressly modified to explain the fact; in other words, cases where theory development was driven by the program's heuristic (Worrall, 1978a). I will consider Carrier's criterion to be applicable only in the complement of cases, i.e. only in cases where it is demonstrable that the theory *was* expressly designed to explain the fact. As we will see, both definitions find application in the Milgromian research program.

§

Recall that in assessing the progressivity of a research program, Lakatos (following Popper) distinguished between 'theoretical' and 'empirical' progressivity:

Let us say that such a series of theories is *theoretically progressive* (or '*constitutes a theoretically progressive problemshift*') if each new theory has some excess empirical content over its predecessor, that is, if it predicts some novel, hitherto unexpected fact. Let us say that a theoretically progressive series of theories is also *empirically progressive* (or '*constitutes an empirically progressive problemshift*') if some of this excess empirical content is also corroborated (Lakatos, 1970, p. 33–34).

Much of the discussion of novel predictions in the philosophy of science literature, as cited in the previous section, is concerned with what Lakatos called 'empirical progressivity': that is, with predictions that are both novel, *and* that have been confirmed. That is, of course, the situation of most interest to scientists as well. But one sometimes encounters predictions that are arguably novel (in the sense of reflecting excess content, or being unexpected) but which – while not yet refuted – have not been confirmed or even corroborated.

In assessing whether such predictions contribute to the *theoretical* progressivity of the Milgromian research program, some of the novelty criteria (P, Z, M, C) can still usefully be applied, but with the understanding that the 'fact' under consideration is only predicted, and not yet confirmed. Thus, criterion P could become:

A prediction is novel if the previous theory variant in the research program would not have made the same prediction.

In the case of Zahar's criterion: it is hard to imagine that a theory would ever be designed to explain a fact before that fact is known! I will assume that criterion Z is simply not applicable in this case. Criterion C, likewise, seems inapplicable to the case in which there is no fact to explain.

Finally, Musgrave's criterion is worth retaining, if '[corroborated] fact' is replaced by 'prediction.' Modified in this way, the criterion judges whether the prediction is one that the 'rival theory' fails to make.

§

Applying criteria M and C requires that we nail down what is meant by the terms "rival theory" and "(independent) fact." For the latter, Carrier (1988, p. 217) proposed "an empirical regularity which occurs repeatedly if certain circumstances are realized." More concretely, he suggested "a law-like relation between two variables that are observationally or experimentally detectable," together with a *ceteris paribus* clause of the form "either there do not exist any further variables which might influence the experimental outcome or such variables are kept constant." Thus, the ideal gas law, $pV = RT$, embodies three independent facts, obtained by fixing one of the three variables (p, V, T) and adjusting the *ceteris paribus* clause accordingly. Most (though not quite all) of the novel predictions to be discussed in the following chapters fit Carrier's definition.

What constitutes a "rival theory"? Musgrave and Carrier note that there are two possible choices: (*i*) the immediate precursor within the *same* research program, or (*ii*) the current best theory in a *competing* research program. Lakatos appeared to endorse both possibilities when he stated that a theory is acceptable "only if it has corroborated excess content over its predecessor (or rival)" (Lakatos, 1970, p. 31–32). Popper's definition of "previous knowledge" ("previous to the theory which was tested and corroborated') would also seem to apply equally well to the current, 'best' theory from a rival research program.

The aim of the present work is not *comparative* research program evaluation. However, there would seem to be no reason, when judging the novelty of a fact from the perspective of the Milgromian research program, not to interpret 'rival theory' as a theory from a rival research program. That is the convention I will adopt in what follows when applying the criteria M and C. And from the standpoint of Milgromian theory, there is no difficulty in identifying what constitutes the rival theory: it is the standard, or concordance, or ΛCDM cosmological model.[17]

§§

Of course, the standard cosmological model as it exists today is only the latest in a *series* of theories that comprise what might be called the 'standard cosmological research program.' Standard-model cosmologists have shown themselves adept at modifying their theory in response to anomalies large and small; dark matter, dark energy, and inflation are some of the auxiliary hypotheses that have been added to the theory since its inception in the 1960s. And cosmologists have sometimes done so without apparent regard for Popper's strictures against conventionalist strategems (Kroupa et al., 2012; McGaugh, 2015; Merritt, 2017). This situation creates the potential for ambiguity when it comes to applying criteria of novelty like Musgrave's or Carrier's. Suppose that a theorist working in the Milgromian research program predicts a fact, and the fact is subsequently confirmed by observational astronomers (as, indeed, has happened many times). At some point after the prediction is made, a standard-model theorist may set about trying to explain it, and she may even succeed – perhaps before the fact's observational confirmation, perhaps after; perhaps by way of unproblematic modifications to the theory, perhaps via a conventionalist stratagem. In situations like these, how should one judge whether the fact is 'improbable' or 'prohibited' from the standpoint of the 'rival theory'?

The situation is complicated by the approach that standard-model theorists take to dealing with anomalies that crop up on the spatial scales corresponding to single galaxies – the same regime in which Milgromian dynamics makes many of its most testable (and surprising) predictions. Simply put, standard-model cosmologists have historically shown little interest in using data from galaxies to *test* their theory; they direct their efforts instead to finding ways of augmenting or adjusting the theory to *accommodate* each newly observed fact – virtually all of which constitute anomalies for the theory when they first come to light. There is nothing very surprising about this; both Lakatos and Kuhn recognized that scientists often proceed in this way. But if challenged, a standard-model cosmologist will protest that his theory adjustments are not ad hoc; he prefers to view them as idealized representations of physical processes that must actually occur but that are too 'small-scale' to model directly, and therefore need to be put into the simulation codes 'by hand.' In an understated passage, Milgrom notes the methodological

---

[17] Specifications of the current, standard cosmological model in textbooks are remarkably similar; Merritt (2017) gives a reasonably comprehensive list of current textbooks at the graduate level.

difficulties that are created by this practice: in attempting to reproduce the known properties of galaxies, he says, standard-model theorists

treat very complicated, haphazard, and unknowable events and processes taking place during the formation and evolution histories of these galaxies. The crucial baryonic [i.e. non-dark-matter] processes, in particular, are impossible to tackle by actual, true-to-nature, simulation. So they are represented in the simulations by various effective prescriptions, which have many controls and parameters, and which leave much freedom to adjust the outcome ... The exact strategies involved are practically impossible to pinpoint by an outsider, and they probably differ among simulations. But, one will not be amiss to suppose that over the years, the many available handles have been turned so as to get galaxies as close as possible to observed ones (Milgrom, 2016b, p. 3).

It is not hard to find examples in the galaxy-formation literature of auxiliary hypotheses that are manifestly ad hoc.[18] For instance, Vogelsberger et al. (2014; the "Illustris Project"), in seeking to link normal matter to dark matter in their simulated galaxies, include a recipe that ties the energy of stellar winds (a phenomenon that originates in the atmospheres of individual stars) directly to the local velocity dispersion of the dark matter particles. There is no plausible, physical connection between the two quantities nor do the authors claim to provide one. As Kroupa (2015b) wryly observes: "This would imply, essentially, that the table in my dining room would know it exists in the MW [Milky Way] DM [dark matter] halo rather than in the DM halo of the Large Magellanic Cloud."

Setting parameters in this way may be reasonable if the goal is to answer questions such as "How might galaxies form under the standard cosmological model?" But it goes without saying that such a practice renders the simulations useless in terms of judging whether the underlying theory is correct.

Even in studies where the 'sub-grid physics' is represented in a less ad hoc manner, standard-model simulation codes always contain a large number (tens or hundreds) of adjustable parameters. Success in these simulation studies is judged – not in terms of corroborated, novel predictions – but rather in terms of the degree to which a choice of parameters *can be found* that generates simulated galaxies which reproduce, in a *statistical sense*, the *known properties* of real galaxies. It sometimes happens (as detailed in the following chapters) that years or even decades of code refinement are required before a known fact is successfully 'explained' in this way. As Kuhn noted, these are tests of the theorist, not the theory.

Yet another point – perhaps the most important – needs to be made. Theories in the Milgromian research program make predictions that can be tested using data from *individual galaxies*, predictions that are independent of the (unknown) origin or history of that galaxy. Calculations of galaxy formation under standard-model assumptions can not do this, *even in principle*. The best the latter can hope to

---

[18] Additional examples are presented in the chapters to follow, especially Chapters 4 and 7. The reader is referred to the entries 'ad hoc elements of the standard cosmological model' and 'feedback, ad hoc prescriptions for' in the Index.

achieve are statistical statements about, for instance, the typical mass and concentration of a 'dark matter halo' that surrounds a galaxy of a given mass and angular momentum. In response to a question such as "What is the predicted dependence of rotation velocity on radius in the galaxy NGC 598?," Milgrom's theory makes a unique and quantitative (and hence testable) prediction, while the standard model is silent. In instances like these, it would be reasonable to conclude that the Milgromian prediction is unambiguously 'novel' with respect to expectations that derive from the competing research program, since the latter makes *no* prediction.

There is one possible caveat to that statement. Milgromian predictions often take the form of 'exact' or 'functional' relations: an observable quantity is predicted to be related to another via a deterministic formula. An example, discussed in detail in the next chapter, is the so-called 'baryonic Tully–Fisher relation' that relates the total mass of a galaxy to its asymptotic rotation velocity. Suppose that such a prediction is subsequently corroborated or confirmed by observations (as a number have been). 'Corroboration' here means a demonstration that the relevant *observables* exhibit zero or nearly zero variance about the predicted relation, aside from scatter due to observational errors. Now, the fact that the relation between these observables is functional might never have been detected in the absence of the Milgromian prediction, since a standard-model cosmologist would have had no reason to plot the appropriate variables one against the other, or to carry out a sufficiently careful analysis of errors. But once the Milgromian prediction has been made, and corroborated, it becomes part of the 'background knowledge,' and a standard-model theorist may set about trying to explain it – typically by adding some prescription for feedback to her model in order to reduce the variance. Success in this enterprise will have been achieved when the theorist has managed to drive the predicted variance to some very small value, since that is what is observed. And having managed to reduce the scatter in her predicted relation to such a small value, the theorist will then be in a position to make (effectively) unique predictions about *individual* galaxies. While it is not clear that standard-model cosmologists have ever succeeded in such an effort, it is easy to document that an enormous amount of time and energy has been, and continues to be, devoted to attempts of this sort.

Given this complicated situation, how should one decide whether a fact predicted by the Milgromian research program is 'prohibited' or 'improbable' as viewed from the standpoint of the standard cosmological research program? I will argue as follows:

Suppose that standard-model cosmologists – having been made aware of a (corroborated) Milgromian prediction, and having decided that their current theory does not explain it – set about trying to explain it, and that they fail. In this case I will judge the fact to be anomalous (and hence improbable) from the standpoint of the 'rival theory.' (A crude measure of the *degree* to which the fact is improbable might be the time interval over which standard-model theorists have sought, unsuccessfully, to explain it.)

Suppose on the other hand that – at the time the Milgromian prediction is first made – the then-current version of the standard model predicts the same fact. It may already have done so; or it may be found to predict the fact, immediately once the question is put to it for the first time. In such cases, I will argue that the fact does not constitute an anomaly for the rival theory, and that corroboration of the fact does not constitute evidential support for the Milgromian program according to the criteria of Musgrave or Carrier. (The novelty criteria of Popper or Zahar may still be satisfied.)

Finally, suppose that – some time after the Milgromian prediction and its obser-vational confirmation – attempts to explain the fact under the standard model have finally been successful. In cases like this, I maintain, a judgment call is required. If the theory was modified purely in response to the predicted fact (and espe-cially, if those modifications were 'fine-tuned,' manifestly ad hoc or otherwise unreasonable); if the explanation is achieved at the cost of weakening the explana-tion of some other known fact; if the modification(s) make the theory inconsistent (or increase its inconsistency); if the claim that the fact is successfully explained is unconvincing, controversial or incoherent: to the degree that these conditions are satisfied, it may still be reasonable to conclude that the fact constitutes an 'anomaly' for the rival theory (claims by standard-model theorists notwithstanding) and therefore that its prediction and corroboration provide evidential support for the Milgromian program.

3

The Milgromian Research Program

Theories which effect the overthrow of a comprehensive and well-entrenched point of view ... are initially restricted to a fairly narrow domain of facts, to a series of paradigmatic phenomena which lend them support, and they are only slowly extended to other areas.

(Feyerabend (1975))

The Milgromian research program is a response to the riddle of the 'missing mass,' an anomaly that arose in the late 1970s in observations of disk galaxies.[1] The speed, V, at which stars or gas clouds orbit at distance R about the galaxy center[2] is predictable using Newton's laws of gravity and motion given the observed distribution of mass in the galaxy. Near the centers of most large galaxies (roughly speaking, galaxies of the mass of the Milky Way or greater), the Newtonian prediction, V_{Newton}, is found to be reasonably correct, i.e. $V_{\text{obs}}(R) \approx V_{\text{Newton}}(R)$. But deviations always begin to appear far out in the disk: the observed orbital speed is found to be systematically larger than the speed predicted by Newton, i.e. $V_{\text{obs}} > V_{\text{Newton}}$. At distances sufficiently great that essentially all of the gravity-producing mass is enclosed, Newton's laws predict that the rotation speed should depend on radius in the same way as for circular motion about a point mass, i.e. $V(R) \approx (G\,M_{\text{gal}}/R)^{1/2}$, where G is the gravitational constant and M_{gal} is the total mass of the galaxy. Observations reveal instead that rotation curves (plots of V vs. R) are 'asymptotically flat,' that is, the orbital speed in any galaxy tends to a constant value (different in different galaxies), $V_{\text{obs}}(R) \rightarrow V_\infty$, at large radii. In some observed galaxies, the flat part of the rotation curve can be traced out to distances greater than 100 kpc from the center; by comparison, the Sun is located less than 10 kpc from the center of the Milky Way.[3] In no galaxy is there

---

[1] Intimations of a 'mass discrepancy' based on the observed kinematics of groups and clusters of galaxies were noted much earlier, e.g. Zwicky (1937), Kahn and Woltjer (1959).

[2] Cylindrical coordinates are natural when describing galaxy disks: R represents distance from the disk (z) axis and ϕ is the azimuthal angle. Here and in what follows, the symbol r will be used when describing distance from the coordinate origin, as in spherical-polar coordinates.

[3] kpc = kiloparsec $\approx 3.09 \times 10^{21}$ cm.

evidence that the rotation curve is tending toward the expected, Newtonian form at large distances.

Standard-model cosmologists deal with this anomaly in precisely the manner that Imre Lakatos might have predicted. The hard core of their research program – in particular, Newton's theory of gravity and motion – is left intact, and the anomaly is addressed via an auxiliary hypothesis: galaxies are assumed to be embedded in 'dark matter haloes,' approximately spherical systems composed of some subtance that does not interact significantly with radiation but which does generate (and respond to) gravitational forces. The greater-than-predicted orbital speed of the observed matter is accommodated by assuming that the dark matter around any galaxy is distributed in whatever way is needed to yield the observed rotation curve. Under the dark matter hypothesis, ordinary matter (stars, gas, dust etc.) would comprise only a small fraction of the total mass of a typical galaxy, perhaps ten percent or less, and most of the dark matter would be situated far from the galaxy center.

In Chapter 1, this auxiliary hypothesis of the standard cosmological model was called DM-1 and was written as

DM-1: In any galaxy or galactic system for which the observed motions are inconsistent with the predictions of Newton, the discrepancy is due to gravitational forces from dark matter in and around the galaxy(ies).

Starting around 1980, and continuing since then, standard-model cosmologists have universally *assumed* the correctness of DM-1, regardless of what else they may believe, or assume, about the nature of the 'dark matter.'

In his first paper from 1983, "A modification of the Newtonian dynamics as a possible alternative to the hidden mass hypothesis," Milgrom argued that the "hidden mass hypothesis" (that is, DM-1) is uncomfortably ad hoc:

The hidden mass hypothesis (HMH) explains the dynamics in galaxies and systems of galaxies by assuming that much of the mass in these systems is in, as yet, unobserved form ... This hypothesis has not yet encountered any fatal objection. However, in order to explain the observations in the framework of this idea one finds it necessary to make a large number of ad hoc assumptions concerning the nature of the hidden mass and its distribution in space (Milgrom, 1983a, p. 365).

In that paper, and in two other papers from the same year (Milgrom, 1983b,c), Milgrom initiated a new research program. Milgrom postulated that one element of the hard core of the standard cosmological model – the relation between the Newtonian gravitational field produced by a given mass distribution, g_N, and the acceleration, a, experienced by test mass moving in that field (a relation that Milgrom called "standard dynamics") – was incorrect. Milgrom did not specify a particular mathematical replacement for Newton's law of motion or gravity. Instead, he wrote down three postulates that specify how Newtonian dynamics is to be modified in regions where the gravitational acceleration is sufficiently small.[4] Here is how Milgrom stated his three foundational postulates in the second of his three papers from 1983:

(*a*) Standard dynamics breaks down in the limit of small accelerations; (*b*) In the limit of small accelerations, the acceleration of a test particle, in a gravitating system, is given by $(a/a_0)\boldsymbol{a} \approx \boldsymbol{g}_N$, where $\boldsymbol{g}_N$ is the conventional gravitational acceleration and a_0 is a constant with the dimensions of acceleration; (*c*) The transition from the Newtonian regime to the low acceleration asymptotic regime is determined by the acceleration constant a_0 (in the sense that the transition occurs within a range of accelerations of order a_0 around a_0) (Milgrom, 1983b, p. 371).

At first sight, postulate (*a*) appears to be fully entailed by postulate (*b*). But elsewhere Milgrom formulates postulate (*a*) in a way that makes explicit its independent content: i.e. that Newtonian dynamics is *retained* in the limit of *large* accelerations. For instance,

A correspondence principle is required that guarantees restoration of Newtonian physics when we formally take $a_0 \to 0$ in all the equations of motion (Milgrom, 2009a, p. 1630).

In his first paper from 1983, Milgrom stated the first of the three postulates in yet another way:

The Newtonian dynamics of a gravitating system (and perhaps of an arbitrary one) break down in the limit of small accelerations (Milgrom, 1983a, p. 367).

The parenthetical clause refers to the possibility that the second postulate might apply to motion that occurs in response to non-gravitational (e.g. hydrodynamical) forces, in addition to purely gravitational forces.

The first two postulates imply the asymptotic flatness of galaxy rotation curves. Sufficiently far from the center of a galaxy, the Newtonian gravitational acceleration has magnitude $\left|\boldsymbol{g}_N\right| \approx GM_{\mathrm{gal}}/r^2$, with M_{gal} the total mass of the galaxy and r the distance measured from the galaxy center. Milgrom's second postulate, $(a/a_0)\boldsymbol{a} \approx \boldsymbol{g}_N$, then implies

$$a \approx \left(a_0 GM_{\mathrm{gal}}\right)^{1/2} r^{-1}. \tag{3.1}$$

---

[4] Newton's second law of motion equates acceleration (rate of change of velocity) with the force acting on the accelerating body divided by its mass. Perhaps for this reason, astrophysicists are in the habit of using the term 'acceleration' to mean both 'rate of change of velocity' and 'gravitational force per unit of mass' and trusting the reader to infer which of the two meanings is being invoked. Under Milgromian dynamics, the relation between (gravitational) force and acceleration is different and one must be careful to specify which meaning of the word 'acceleration' is intended. In the Newtonian regime, and (approximately) in the transition regime between Newtonian and Milgromian dynamics, the two meanings of 'acceleration' are equivalent even in Milgromian dynamics. When discussing the weak-field ($a \ll a_0$) regime, I will consistently use 'acceleration' to mean 'rate of change of velocity' and 'gravitational acceleration' to refer to $\boldsymbol{g}_N$. (Milgrom sometimes uses 'acceleration' when referring to the latter quantity.) Even in the weak-field regime, certain statements in terms of inequalities (e.g. 'motion changes when the acceleration is much less than a_0') can remain correct under either definition of 'acceleration.'

Equating this expression with the centripetal acceleration of a test mass moving in a circular orbit of radius $r = R$, or V^2/R, yields

$$\frac{V^2}{R} = \frac{\left(a_0 G M_{\text{gal}}\right)^{1/2}}{R} \quad \text{i.e.} \quad V = \left(a_0 G M_{\text{gal}}\right)^{1/4}, \tag{3.2}$$

so that V is independent of R.

In his three papers from 1983, Milgrom elaborated on the content of his three postulates:

- The quantity a_0 is assumed to be a fundamental constant of Nature, the same everywhere and in every physical system. The possibility is left open that a_0 may change over cosmological time (as does, for instance, the 'Hubble constant' in the standard cosmological model).
- In the second postulate, the gravitational acceleration g_N is assumed to be generated by a distribution of mass that is static and spatially symmetric (disklike, spherical). For configurations of lower symmetry, a more general form for the relation between a and g_N is anticipated.
- By "conventional gravitational acceleration," Milgrom means the quantity $-\nabla\phi_N$, i.e. minus the spatial gradient of the (Newtonian) gravitational potential ϕ_N; the latter is to be computed in the manner dictated by Newton, by summing the potentials of the bits of mass that comprise the galaxy or galactic system.

Milgrom acknowledged in his first paper from 1983 (p. 366) that his postulates were constituted in such a way that "the rotation curve of a finite galaxy becomes asymptotically flat" in the absence of any hypothesized dark component. Asymptotic flatness of galaxy rotation curves is therefore not a 'novel fact' according to Zahar's criterion, since Milgrom's postulates were explicitly designed to explain this fact. On the other hand, it is clear that the prediction of asymptotic rotation-curve flatness requires only the first two postulates – not the third. When discussing the novelty of predicted facts, and their evidential support, it will sometimes be useful in what follows to consider postulates (*a*) and (*b*) together, and separately from, postulate (*c*).

The empirical content of postulate (*c*) becomes clear when the following argument is made (Milgrom, 1983a): Postulate (*c*) states that a_0 is the only constant with dimensions of acceleration in the theory. It follows on dimensional grounds that a must be related to g_N via a relation $a = f(g_N/a_0)g_N$ with $f(x)$ some as yet unspecified function of $x \equiv g_N/a_0$. Regardless of the functional form of f, a prediction of the three postulates is therefore that the acceleration (as measured, for instance, from rotation curves) will be a universal function of the Newtonian prediction, hence of the mass distribution. There will be no dependence on the 'initial conditions' (the details of galaxy formation and evolution), the type or composition of the galaxy, the galaxy's environment (unless other galaxies are close enough to disturb it gravitationally) and so forth. This prediction might seem unsurprising, until one remembers that – according to the standard cosmological model – the gravitational acceleration is determined essentially by the *dark matter*, not the observed matter, and there is no *a priori* reason to expect the dark matter to be distributed in a way that strictly respects the distribution of the observed matter.

§§

In his papers from 1983 and later, Milgrom often makes a distinction between his three postulates, on the one hand, and a "theory" (or "basic," "ultimate," "complete," "full-fledged," "fundamental" theory) from which the postulates could be derived as limiting cases, on the other hand. For instance:

However, these assumptions [the three postulates] are not sufficient for describing the dynamics of an arbitrary N-body system, for which we shall need a theory (Milgrom, 1983a, p. 367).

As explained in Paper I [i.e. Milgrom (1983a)], the formulation given by equation (1) [postulate (*b*) above] cannot be considered a theory, but only a successful phenomenological scheme for which an underlying theory is needed (Bekenstein and Milgrom, 1984, p. 8).

Moreover, there are strong signs that MOND, as we now perceive it, is an effective, approximate theory. If so, clearly some principles, even if obeyed by the more fundamental theory, might be spontaneously broken by the effective MOND theory that we may have to satisfy ourselves with temporarily (Milgrom, 2015, p. 111).

Now, from a critical-rationalist standpoint, the three postulates of Milgrom's hard core *already* constitute a scientific theory, since they are universal statements with empirical content: testable propositions can be derived from them. What Milgrom understands by the term '[complete] theory' is, apparently, a theory that has been under development for a long time and that consequently has the potential to explain a broad variety of observed facts.[5]

Many readers will no doubt already have noticed the similarities between Milgrom's three postulates from 1983, and the five postulates with which Niels Bohr initiated his research program in 'old' quantum mechanics in his epochal paper from 1913. As the historian Max Jammer (1966, p. 87) has noted, Bohr's "intention was not to give a satisfactory answer to a definite question but to search for the right question to ask." That is: Bohr did not set out to explain spectral laws, so much as to find a way to relate those laws to the known properties of the elements, as a first step toward a more complete (and consistent) theory. Much the same spirit is evident in Milgrom's early papers.

---

[5] Standard-model cosmologists often raise the objection that Milgromian dynamics is 'not a complete theory.' Most likely, Einstein's theory of gravity is their basis of comparison. But the standard cosmological model supplements Einstein's theory with a raft of auxiliary hypotheses, e.g. 'dark energy,' 'inflation' etc. In any case, the relativistic versions of Milgrom's theory presented in Chapter 6 are no less 'complete' than Einstein's.

There are other parallels between Milgrom's papers from 1983 and Bohr's paper from 1913:

– In crafting his postulates, Bohr was motivated by the known hydrogen spectral series, as it had been captured, already in 1885, by Johann Balmer's famous formula. In a similar way, Milgrom designed his postulates to reproduce the known, asymptotic flatness of galaxy rotation curves.
– Bohr's postulates as given in his 1913 paper apply only to a small subset of atomic systems: in his words, to "the different stationary states of a simple system consisting of an electron rotating round a positive nucleus," i.e. an unperturbed hydrogen atom (or an ionized atom of higher atomic number). In the same way, as we have seen, Milgrom intended that his postulates from 1983 should apply only to highly symmetric, isolated and stationary self-gravitating systems, not to galaxies or galactic systems in general.
– Bohr "grafted" (Lakatos's word) his postulates, inconsistently, onto Maxwell–Lorentz electromagnetic theory.[6] In the same way, Milgrom assumed that all of the components of the standard cosmological model, *c.* 1980, that were not directly impacted by his three postulates should remain in effect.

Of course, in 1913, there was no competing theory that claimed to explain the anomalous spectrum of hydrogen. In 1983, by contrast, standard-model cosmologists had recently adopted an auxiliary hypothesis ('dark matter'; that is, DM-1) to explain galaxy rotation curves. Thus, from the very start, Milgrom's research program co-existed with another research program which claimed (with more or less justification) to be able to explain the same facts that Milgrom's theory explained. When Milgrom wrote of a 'complete theory,' it is likely that he was mentally comparing his research program, in its early state, with the standard cosmological model, which indeed was much more developed and able to be applied to a broader range of problems than his own.

§§

Many of Milgrom's statements about what should constitute a 'complete theory,' and about the steps by which such a theory might be achieved, can be interpreted from the standpoint of Lakatos's *Methodology* as statements about the heuristic of his research program.

There would seem to be no hard-and-fast rules for deciding what constitutes the heuristic of a research program. Scientists engaged in research may never articulate how theory development is expected to proceed; once encultured into a program, a scientist may not be consciously aware of its methodological guidelines

---

[6] Margenau (1950, p. 311): "Bohr's atom sat like a baroque tower upon the Gothic base of classical electrodynamics."

or strictures any more than a fish is conscious of water.[7] Before attempting to specify the heuristic of Milgrom's program, it seems appropriate to begin by reviewing the *sorts of things* that have been proposed by philosophers of science as constituting the heuristic of other scientific research programs.

Again quoting Lakatos (and recalling that his "postive heuristic" is identical to what is called here, simply, the 'heuristic'):

the positive heuristic consists of a partially articulated set of suggestions or hints on how to change, develop the 'refutable variants' of the research-programme, how to modify, sophisticate, the 'refutable' protective belt.... The positive heuristic sets out a programme which lists a chain of ever more complicated *models* simulating reality: the scientist's attention is riveted on building his models following instructions which are laid down in the positive part of his programme (Lakatos, 1970, p. 50).

Lakatos gives the following advice for separating the hard-core postulates from the directives of the heuristic:

Positive heuristic is thus in general more flexible than negative heuristic [i.e. hard core]. Moreover, it occasionally happens that when a research programme gets into a degenerating phase, a little revolution or a creative shift in its positive heuristic may push it forward again. It is better therefore to separate the 'hard core' from the more flexible metaphysical principles expressing the positive heuristic (Lakatos, 1970, p. 51).

Of course, over time, new techniques can emerge for elaborating a research progam, causing the heuristic to evolve. For instance, computer simulations are much more powerful and efficient now than they were in 1983.

John Worrall suggests:

The positive heuristic may include mathematics – for example, how theoretical assumptions should be formulated so that consequences may be drawn from them will be guided by the available mathematics; the heuristic may include hints on how to deal with refutations if they arise (e.g. 'Add a new epicycle!'); and it may include directions to exploit analogies with previously worked out theories (Worrall, 1978b, p. 69).

Lakatos's reconstruction of the positive heuristic of the Bohr research program was described in Chapter 2: according to Lakatos, it directed the researcher to develop a series of ever-more-realistic atomic models, based on the idea that atoms are analogous to planetary systems. Lakatos (1970, p. 50–51) described the heuristic of the Newtonian research program in a similar way, as a sequence of models that approximate ever more closely to observed systems of planets or stars. Lakatos noted:

Most, if not all, Newtonian 'puzzles', leading to a series of new variants superseding each other, were foreseeable at the time of Newton's first naive model and no doubt Newton and his colleagues did forsee them: Newton must have been fully aware of the blatant falsity of

---

[7] At least, until she violates them! See e.g. Kroupa (2012, p. 413, n. 12) *re* the consequences to a young cosmologist of departing from standard-model methodology.

his first variants ... Indeed, if the positive heuristic is clearly spelt out, the difficulties of the programme are mathematical rather than empirical (Lakatos, 1970, p. 51).

Peter Clark (1976, p. 45) argues that the heuristic of the atomic-kinetic research program directed the researcher to make assumptions about the properties of the particles constituting a gas, then to gradually weaken or eliminate those simplifying assumptions so as to increasingly approximate the properties of real gases. In the case of Einstein's research program, Elie Zahar (1973, p. 243) identified two elements of its positive heuristic (his italics): "(i) *a new law should be Lorentz-covariant and* (ii) *it should yield some classical law as a limiting case.*" Zahar contrasted Einstein's heuristic with Lorentz's, which he said "consisted in endowing the ether with such properties as would explain the behaviour both of the electromagnetic field and of as many other physical phenomena as possible" (Zahar, 1973, p. 242). And of course, one could mention here the standard cosmological research program, which, like Lorentz's, endows dark matter and dark energy "with such properties as would explain the behaviour" of the large-scale structure and the rotation curves of galaxies.

§

With these examples in mind, and being guided wherever possible by Milgrom's own words (particularly in his three papers from 1983) I propose the following as the heuristic of the Milgromian research program:

A series of dynamical theories is to be generated. All variants of the theory should imply the three hard-core postulates; the second postulate $((a/a_0)a \rightarrow g_N)$ is only expected to be accurately instantiated in the case of a static, symmetric, non-relativistic system. Theory development is guided by the following methodological directives:

Models should be constructed first as extensions of Newtonian dynamics, then as extensions of general relativity (i.e. invariant physical laws under differentiable coordinate transformations), and ultimately in a form that has no pre-ordained connection with Newton's or Einstein's theories.[8]

In addition to embodying the postulates of the hard core, theories should have as many of the following additional properties as possible:

(i) They should preserve the symmetries of space with respect to translations and rotations (the six isometries of a maximally symmetric three-dimensional space).

---

[8] E.g. Milgrom (2014, p. 2531): "Beyond the basic tenets, one wishes to construct an NR [non-relativistic] theory of dynamics based on them, and then extend the theory to a replacement of GR [general relativity]." Bekenstein (1992) proposed a more detailed list of guidelines for development of the relativisic theory; these are discussed in Chapter 6.

(ii) The equations of motion should be derivable from an action principle. This guarantees that the laws of conservation of energy and of linear and angular momentum are satisfied.

(iii) Milgrom's constant, a_0, should emerge from the theory; perhaps in a way that accounts for the approximate numerical coincidence between a_0 and the quantity cH_0, where H_0 is the Hubble constant.[9]

As in Newton's or Bohr's research programs, application of Milgrom's heuristic results in a series of theory variants, or models, each intended as a more precise and/or complete description of the real universe than the one before. In the initial phases of the program, testable propositions can be generated only for the simplest systems, e.g. static axisymmetric galaxies. Later variants permit predictions for ever-more complicated systems: pressure-supported (elliptical) galaxies; interacting galaxies; galaxy clusters; large-scale structure; and ultimately the universe as a whole.

§§

Popper defined the 'empirical content' of a theory in terms of the testable statements that could be derived from it. According to his definition, Milgrom's three postulates have a well-defined empirical content: because in addition to the asymptotic flatness of galaxy rotation curves, the postulates imply a number of other testable propositions. These predictions, their novelty, and their degree of observational corroboration are discussed in the next chapter. But while unambiguous predictions can be derived from the three postulates, there is nevertheless a sense in which the *interpretation* of the postulates is ambiguous, and this ambiguity necessarily affects how the theory is developed *beyond* the postulates – a situation that Milgrom clearly acknowledged already in his first paper from 1983.

Milgrom's second postulate states that a test particle of mass m experiences an acceleration a that is the solution of

$$m\boldsymbol{a} \times \frac{|\boldsymbol{a}|}{a_0} = m\boldsymbol{g}_{\mathrm{N}} \qquad (3.3)$$

in the low-acceleration regime. Since $m\boldsymbol{g}_{\mathrm{N}}$ is – by definition – the Newtonian gravitational force, it is natural to interpret Equation (3.3) as a modification of the Newtonian *inertia*, or of the 'inertial mass': that is, $m\boldsymbol{a}$ in Newton's second law is replaced by $m\boldsymbol{a} \times (a/a_0)$. This interpretation appears to be exactly what Milgrom had in mind. For instance, he wrote in the first of his 1983 papers:

---

[9] E.g. Milgrom (1983a, p. 370): "Hopefully, a theory can be found in which a_0 ... will turn out to depend on the distribution of mass in the universe (density) and its manner of expansion (very much in the vein of theories like that of Brans and Dicke)."

I allowed for the inertia term not to be proportional to the acceleration of the object but rather to be a more general function of it ... The force field F [i.e. g_N] is assumed to depend on its sources and to couple to the body, in the conventional way (Milgrom, 1983a, p. 366).

A theory that leaves Newton's law of gravity unchanged, but changes the definition of inertia, has come to be called a 'modified inertia theory.' Under this interpretation, it is reasonable (though of course not obligatory) to assume that Equation (3.3) obtains even if the force on the right hand side is *not* gravitational in origin.

But one could equally well choose to leave the definition of inertial mass unchanged. In that case, it is natural to write Equation (3.3) in the mathematically equivalent way as

$$ma = ma_0 \sqrt{\frac{|g_N|}{a_0}}\, e_N, \tag{3.4}$$

with e_N a unit vector in the direction of g_N. Equation (3.4) expresses the acceleration in terms of a modification of the *gravitational force* – it has the form of a 'modified gravity theory.'

To reiterate: predictions generated from Milgrom's three postulates are the same whether the second postulate is interpreted as the asymptotic form of a modified inertia theory or a modified gravity theory. But as we will see in Chapters 5 and 6, later versions of Milgrom's theory that derive the particle equations of motion from an action, and that respect the usual conservation laws as a consequence, fall naturally into just one or the other of the two categories. Not all predictions that can be generated from the two versions of the theory are the same.

Ideally, perhaps, the heuristic of the Milgromian research program would direct the theorist to develop both versions of the theory in tandem, making similar predictions that could be tested side by side. But in fact this has not occurred; instead, the lion's share of theory development has been directed toward modified gravity theories. The reason for this imbalance (or so it would appear) is that generating testable predictions from modified inertia theories is mathematically more complex than from modified gravity theories. For example, some predictions that take the form of explicit functional relations in the latter theories can appear as implicit solutions to integral equations in the former.[10]

Starting around 2009, Milgrom began presenting his postulates in a different way, following the realization that the equations of motion in the low-acceleration regime are symmetric under rescaling of time and length by the same factor. In simplified form, the argument runs as follows (Milgrom, 2001a, 2009a): Consider

---

[10] Urbach (1978, p. 108) suggests that "a research programme which makes its predictions in the form of equations which are insoluble ... is heuristically weaker, other things being equal, then [sic] one which incorporates an adequate mathematical theory."

a change of units such that length $l \rightarrow \lambda l$ and time $t \rightarrow \lambda t$, with no change in the units of mass, $m \rightarrow m$. Under these changes, acceleration scales as $a \rightarrow \lambda^{-1}a$ and Newton's constant as $G \rightarrow \lambda G$. The quantity Ga_0, which alone appears in the limiting theory (as shown above), is invariant. It follows that the asymptotic theory is invariant under this scaling: if a certain configuration is a solution of the equations, so is the rescaled configuration.

§

In the next four chapters, the progressivity of Milgrom's research program will be assessed, with each variant of the theory considered in turn. I will refer to the theory variants as T_0 (which I equate with the three foundational postulates from 1983), T_1, T_2 etc. By proceeding in this way, I do not mean to imply that the theories emerged, one after another, in strict temporal sequence. In fact they did not: for instance, a relativistic version of the theory was proposed as early as 1984 (Bekenstein and Milgrom, 1984). Thus I will be invoking Lakatos's notion of "rational reconstruction" of the historical record, but, I hope, only in a minimal way.

4

Theory Variant T_0: The Foundational Postulates

Theory variant T_0 is defined by Milgrom's three postulates from 1983. As we saw in the previous chapter, Milgrom has presented his postulates in a number of slightly different, but essentially equivalent, ways. In what follows I will adopt the following forms for these postulates:

I. Newton's second law relating acceleration to gravitational force is asymptotically correct when applied to motion for which the gravitational acceleration is sufficiently large, but breaks down when the acceleration is sufficiently small.

II. In the limit of small gravitational accelerations, the acceleration of a test particle, in a symmetric and stationary gravitating system, is given by $(a/a_0)\,\boldsymbol{a} \approx \boldsymbol{g}_{\mathrm{N}}$, where $\boldsymbol{g}_{\mathrm{N}}$ is the conventional gravitational acceleration and a_0 ('Milgrom's constant') is a constant with the dimensions of acceleration.

III. The transition from the Newtonian regime to the low-acceleration regime is determined by Milgrom's constant. The transition occurs within a range of accelerations of order a_0 around a_0.

(Recall the convention adopted throughout this book: 'acceleration' means 'rate of change of velocity' and 'gravitational acceleration' refers to $\boldsymbol{g}_{\mathrm{N}}$.)

The goal of this chapter is to judge the progressivity of the Milgromian research program at its initial stage. The requirement of heuristic progress ("successive modifications of the protective belt must be in the spirit of the heuristic") is not applicable, of course, to the foundational postulates themselves. But the theory even in this basic form makes a number of testable predictions, and on this basis its progressivity can be judged. Recall from Chapter 2 that a research program exhibits theoretical progressivity if it predicts some novel facts, and it exhibits empirical progressivity if, in addition, some of those novel predictions have been corroborated. As we will see, theory variant T_0 does indeed satisfy both requirements, and furthermore it does so based on multiple novel predictions.

The procedure to be followed, in this chapter and the next three, consists of the following steps:

(i) Derive testable (i.e. refutable) predictions from the theory variant under consideration.
(ii) Evaluate the novelty of the predictions, according to each of the four definitions of novelty given in Chapter 2. A given prediction may be judged novel according to some of the criteria and not novel with respect to others.
(iii) Document whether the novel predictions have been confirmed (or corroborated, or refuted, or none of these) by observation or experiment.

The theory variant is judged 'theoretically progressive' if at least one testable prediction exists that is novel according to at least one of the four criteria. It is judged 'empirically progressive' if at least one of its novel predictions has been corroborated. Of course, predictions that have not been corroborated, or that have been refuted, are also important to record. But recall that – according to Lakatos's *Methodology* – theories are *expected* to make some incorrect predictions ("All theories ... are born refuted"). It is always possible that a later version of the theory will deal successfully with anomalies that first appear in an earlier version.

§

Testable predictions derivable from T_0 include the following:

1. Asymptotic constancy of the circular orbital speed around an isolated disk galaxy;
2. A universal relation between asymptotic speed and total mass of a disk galaxy;
3. A unique relation between the acceleration a at any point in a disk galaxy and the Newtonian gravitational acceleration g_N due to the galaxy's mass.

Prediction number two has the form of a specified functional relation. In the case of prediction number three, the predicted relation is also universal and 'functional,' but the form of the function is not completely specified; as we will see, it must be inferred from rotation-curve data.

Existence of the so-called 'external field effect' (EFE) is also a prediction of T_0 (Milgrom, 1983a). I discuss the EFE at the end of this chapter, where I will argue that *testable* predictions deriving from the EFE only become feasible given a later variant of the theory.

§§

1. *Asymptotic constancy of the circular orbital speed around an isolated disk galaxy*: Milgrom's second postulate states that in the low-acceleration regime $(a/a_0)\,a \approx g_N$, where g_N is the gravitational acceleration as computed from

Newton's theory. This prediction is testable given measurements of the orbital speed of stars or gas in the outskirts of isolated disk galaxies. Sufficiently far from a galaxy's center, at distance r, one expects two conditions to hold: (*i*) the gravitational acceleration is small (compared with a_0), and (*ii*) the fraction of the galaxy's mass enclosed by a sphere of radius r approaches unity. Condition (*ii*), together with Newton's law of gravitation, implies

$$\boldsymbol{g}_{\mathrm{N}} \approx -\frac{GM_{\mathrm{gal}}}{r^2}\boldsymbol{e}_r, \tag{4.1}$$

where M_{gal} is the total mass of the galaxy and $\boldsymbol{e}_r$ is a unit vector in the direction of $\boldsymbol{r}$. The acceleration of a test mass in a circular orbit of radius R with speed V is $|\boldsymbol{a}| = V^2/R$. The second postulate then implies

$$\frac{a^2}{a_0} = \frac{1}{a_0}\left(\frac{V^2}{R}\right)^2 = \frac{GM_{\mathrm{gal}}}{R^2}, \quad \text{i.e.} \quad V^4 \equiv V_\infty^4 = GM_{\mathrm{gal}}\,a_0, \tag{4.2}$$

so that $V(R) = V_\infty$ is independent of distance (Milgrom, 1983a).

The prediction of asymptotic rotation-curve flatness has been strongly corroborated; indeed the fact that disk galaxy rotation curves are universally asymptotically flat is mentioned in virtually every current textbook on cosmology or galactic astrophysics. Is the fact a *novel* prediction of T_0? Not according to criterion P, since the asymptotic flatness of rotation curves had been well established prior to 1983 (as attested, for instance, by the 1979 review article by Faber and Gallagher). Nor is it novel under criterion Z since, as noted above, Milgrom has stated that his postulates were *designed* in order to give this result. And recall that I only consider Musgrave's criterion for novelty, M, to be applicable in cases where the theory was *not* expressly designed to explain the fact.

Finally, Carrier's criterion C states, "A hypothesis explains a fact in a non-ad-hoc manner, if it simultaneously explains at least one additional independent fact that either constitutes an anomaly for the rival theory or that falls beyond its realm of application." This criterion for novelty *is* clearly satisfied: because in addition to explaining (by design) the asymptotic flatness of rotation curves, Postulate II also predicts an additional fact: that the asymptotic *value* of the rotation velocity is equal to

$$V_\infty = \left(GM_{\mathrm{gal}}a_0\right)^{1/4}, \tag{4.3}$$

a function only of M_{gal}, the total mass of the galaxy. Such a relation was not known prior to 1983, nor was its existence predicted by standard-model cosmologists. Furthermore, as discussed in more detail below, standard-model cosmologists have never succeeded in explaining this fact, despite decades of effort. Hence Carrier's conditions for 'non-ad-hoc-ness' are fully satisfied here.

Table 4.1 *Corroborated excess content, theory variant T_0*

Theory variant	Prediction	Status	Novelty			
			P	Z	M	C
T_0	$V(R) \to V_\infty$	confirmed	no	no	–	yes
	$V_\infty = \left(a_0 G M_{\mathrm{gal}}\right)^{1/4}$	confirmed	yes	yes	yes	–
	$a = f\left(g_N/a_0\right) g_N$	confirmed	yes	yes	yes	–
	Renzo's rule	corroborated	yes	yes	yes	–

These conclusions are summarized in the first line of Table 4.1.

§§

2. *A universal relation between asymptotic speed and total mass of a disk galaxy*: that is, Equation (4.3). This prediction has been confirmed, as illustrated in Figure 4.1. The fact that asymptotic rotation speeds in disk galaxies depend as the one-fourth power on galaxy mass, with negligible (intrinsic) variance, is nowadays referred to as the 'baryonic Tully–Fisher relation,' or BTFR. That awkward name begs an explanation, which I will provide in a moment.[1]

The existence of a relation between asymptotic velocity and galaxy mass was not established until the late 1990s, so that criterion P ("A predicted fact is novel if it was unknown prior to the formulation of the theory") is clearly satisfied. And because the relation was not known in 1983, Milgrom could not have designed his postulates to explain it. Indeed, in his second paper from 1983, Milgrom explicitly states the relation as a *prediction* following from his postulates:

The relation between the asymptotic velocity (V_∞) and the mass of the galaxy (M) $\left(V_\infty^4 = M G a_0\right)$ is an absolute one (Milgrom, 1983b, p. 381).

(By 'absolute,' Milgrom (1983b, p. 377) means here "independent of galaxy type or any other property of the galaxy.") On this basis, it would be reasonable to conclude that Zahar's criterion for novelty – "A novel fact is one that a theory was not specifically designed to accommodate" – is satisfied as well.

But there is a complication here. Recall Zahar's argument that determination of the *parameters* of a theory from a data set renders that data set part of the 'background knowledge' that was used in the construction of the theory. Or to repeat John Worrall's concise formulation: "one can't use the same fact twice: once in the construction of a theory and then again in its support." The issue here, of course, is the parameter a_0, or 'Milgrom's constant,' which appears in Equation (4.3) and which must be determined from data.

---

[1] Milgrom sometimes refers to this relation as the 'mass–asymptotic speed relation,' or MASSR. That name is much more apt than BTFR but has not been widely adopted, at least not by standard-model cosmologists.

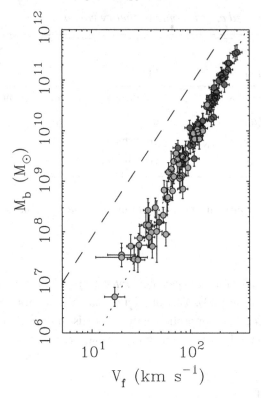

Figure 4.1 The baryonic Tully–Fisher relation (BTFR). Each point corresponds to a single disk galaxy. The vertical axis (M_b) is the summed mass in stars and gas (the subscript 'b' stands for 'baryonic,' cosmologists' shorthand for 'non-dark'). The horizontal axis (V_f) is the outer, flat rotation velocity as inferred from 21 cm radio telescopic observations: what is called here 'V_∞.' The dark/light circles are galaxies with stellar mass that is greater than/less than the gas mass; the latter are generally less precise in measured velocity, but also less susceptible to systematic errors due to the assumed stellar mass-to-light ratio. The dotted line is the relation predicted by theory variant T_0 assuming a value $a_0 = 1.2 \times 10^{-10}$ m s^{-2} for Milgrom's constant. The dashed line, which has a slope of 3 (not 4), is a straightforward prediction of the standard cosmological model using the 'concordance' ratio of normal to dark matter (see Chapter 6 for a description of how standard-model cosmologists determine this ratio). Figure reprinted with permission from Benoît Famaey and Stacy S. McGaugh, "Modified Newtonian dynamics (MOND): Observational phenomenology and relativistic extensions," *Living Reviews in Relativity*, 15, p. 20, 2012.

Since the value of a_0 is not predicted by the theory, one could take the reasonable point of view that the *prediction* here consists *only* of the functional relation between V_∞ and M_{gal}, i.e. $V_\infty \propto M_{gal}^{1/4}$. Recall Martin Carrier's definition of a 'fact': "a law-like relation between two variables that are observationally or experimentally detectable," together with a *ceteris-paribus* clause of the form "either there do not exist any further variables which might influence the experimental outcome or such variables are kept constant." In the case of Equation (4.3), the two related variables are asymptotic velocity and galaxy mass, and the *ceteris-paribus* clause can be stated as 'assuming a universal value of a_0.' On this view (which I adopt in what follows), the prediction of the fact $V_\infty \propto M_{\text{gal}}^{1/4}$ is novel based on Zahar's criterion.

But while T_0 makes no prediction about the *value* of a_0, it does predict that a_0 is the *only* dimensioned parameter in the theory; or equivalently, that every determination of a_0 from data must yield the *same value*. That prediction *is* testable; and the results of observational tests turn out to be so interesting, and important, that an entire chapter (Chapter 8) will be devoted to them. For the moment, I will note simply that the BTFR provides one of the cleanest and most direct routes to determination of Milgrom's constant – or more accurately, of the quantity Ga_0, which is how the constant appears in Equation (4.3).

Finally, Musgrave's criterion (*M*) – "A corroborated fact provides evidential support for a theory if the best rival theory fails to predict the fact, or if the fact is prohibited or improbable from the standpoint of that theory" – is also satisfied because, in addition to not being known prior to 1983, the predicted relation was (and remains) unexplained by standard-model cosmologists. Justifying this statement requires a review of the historical record, to which we now turn.

§

The name 'baryonic Tully–Fisher relation,' or BTFR, derives from the name of an earlier-established empirical correlation, the 'Tully–Fisher relation' (Tully and Fisher, 1977), between the (optical) luminosity, L, and the 'velocity width,' ΔV, of disk galaxies, which has the approximate form

$$L \sim (\Delta V)^n, \quad n \sim 3 - 4. \tag{4.4}$$

'Velocity width' is a term that originated in observations of disk galaxies at 21 cm wavelengths (the wavelength of emission of neutral hydrogen gas due to the hyperfine transition) using radio telescopes. When imaging an entire galaxy, the observable quantities are the total radio flux and its distribution with respect to frequency, or wavelength; the latter differs from the emitted wavelength of 21 cm due to (*i*) the galaxy's heliocentric radial velocity and (*ii*) motion of gas within the galaxy. The galaxy's bulk velocity, V_0, results in a change in the mean value of the observed wavelength according to Doppler's equation:

$$\lambda_{\text{obs}} = \lambda_{\text{emit}} (1 + V_0/c).$$

Motion of gas within the galaxy – both orbital motion and (possibly) random motions within gas clouds – results in a broadening of the observed wavelength distribution by some amount $\Delta\lambda$, which is related to the intrinsic spread of velocities by

$$\Delta\lambda \sim (\lambda_{\text{emit}}/c)\,\Delta V.$$

In the idealized case where the gas motions are attributable purely to circular orbits and the rotation curve is asymptotically flat, the velocity width (corrected for inclination) is approximately twice V_∞, hence $\Delta V/2$ is an approximate measure of the asymptotic rotation speed.

The observed relation between L and ΔV has a substantial scatter, over and above what can be attributed to observational uncertainties, and furthermore the relation does not appear to be describable as a single power law except over small intervals in L (Tully and Pierce, 2000). Considerable effort (Bernstein et al., 1994; Tully and Courtois, 2012; Sorce et al., 2013; Ponomareva et al., 2017) has gone into finding ways of (re-)defining the quantities L and ΔV in order to yield a correlation having a minimum scatter; one motivation is that the relation is useful as a distance indicator, and the smaller the scatter, the more accurate the distance estimates based on it. The tightest correlations are obtained when L is defined as L_{NIR}, the luminosity of the galaxy as measured in near-infrared filters (Aaronson et al., 1979). Since L_{NIR} is expected to correlate more strongly with stellar mass than L_{optical} (Peletier and Willner, 1991), the tightness of the observed correlation when defined in terms of L_{NIR} suggests that it is the mass in stars, rather than their luminosity, that is the controlling variable.

Milgrom's predicted relation (4.3) is superficially similar to the Tully–Fisher relation (4.4),[2] but it differs from it in a number of ways. It is unambiguously the galaxy's *mass*, not its luminosity, that appears in Milgrom's relation; furthermore that mass must include all contributions: stars, gas, dust etc. – what cosmologists refer to, collectively, as 'baryons.' This is important because in many galaxies – particularly low-luminosity, and low surface brightness, galaxies – most of the mass can be in the form of gas, and this mass makes essentially no contribution to L (Milgrom and Braun, 1988). Milgrom's predicted relation also has a definite slope (exactly 4, on a log-log plot of V_∞ vs. M_{gal}) and zero intrinsic scatter; the latter feature is, of course, not to be expected in any empirical correlation unless the observed quantities are related via a universal law. Finally, the normalization of the relation is fixed by (what are presumably) constants of nature, G and a_0, through the product Ga_0.

It was pointed out, in two studies from 1999, that the Tully–Fisher relation becomes tighter if luminosity is replaced by an estimate of the galaxy's total (non-dark) mass, including both stars and gas (Freeman, 1999; Walker, 1999). Neither Freeman nor Walker makes a connection in their papers with Milgromian dynamics,

---

[2] This is hardly surprising. Graduate students are told – with only slight hyperbole – that all empirical laws in astrophysics are either power laws ($y = C \times x^D$) or exponentials ($\log y = C + D \times x$). In fact the Tully–Fisher relation is neither, but so strong is the urge on the part of astrophysicists to adopt simple functional forms that the relation is almost always presented as in Equation (4.4), or as a 'power law with varying index.' Milgrom's predicted relation is, of course, precisely expressible as a power law.

nor is Milgrom's prediction acknowledged by either author. A study published the following year (McGaugh et al., 2000) presents results from a large sample of late-type (disk-dominated) galaxies for which both multi-band photometric data (stars) and 21 cm data (gas) were available; masses of the galaxies in the sample extend over five decades. These authors conclude that their data are consistent with a relation like Equation (4.3), with "modest" scatter. Interestingly, even though McGaugh et al. were apparently aware of Milgrom's prediction – they cite, obliquely, his first paper from 1983 – they stop just short of presenting their results as a *corroboration*, choosing instead to emphasize the difficulty of explaining the observed relation from the standpoint of the standard cosmological model:

The results presented here make sense in terms of a simple interpretation of the Tully–Fisher relation in which the mass of observed baryons is directly proportional to the total mass that in turn scales with the observed rotation velocity. This potentially includes the case in which the mass observed in baryons is the total mass (Milgrom 1983). Matching these observations is a substantial challenge for modern structure formation theories based on 'cold dark matter' (McGaugh et al., 2000, p. L102).

Numerous observational studies since 2000 have further corroborated Milgrom's prediction (Verheijen, 2001; McGaugh, 2005; Geha et al., 2006; Begum et al., 2008; Stark et al., 2009; McGaugh, 2011, 2012; Lelli et al., 2016c). Milgrom's prediction that the relation between M_{gal} and V_∞ is 'functional,' i.e. that it is characterized by zero intrinsic variance, would be extremely difficult to understand from the standpoint of the standard cosmological model, and all of the cited studies have sought to test the zero-variance prediction by minimizing systematic errors in the determination of M_{gal} and V_∞. There has in fact emerged a consensus that the data are consistent with a zero-intrinsic-scatter relation of the form of Equation (4.3) (as reviewed by Famaey and McGaugh (2012); Figure 4.1 is taken from that review article).

A zero intrinsic scatter in the BTFR would also be impossible unless Milgrom's constant a_0 were the same in every galaxy.[3] The observational studies cited above can be seen as corroboration of this prediction, and they also allow the value of a_0 to be estimated from the data. In their review, Famaey and McGaugh (2012) conclude that

$$a_0 \approx 1.2 \times 10^{-10} \, \text{m s}^{-2}. \tag{4.5}$$

Chapter 8 will present a more detailed discussion of how Milgrom's constant is determined from these and other data. As we will see, different data-based techniques yield a convergent value of a_0 that is close to the value of Equation (4.5).

Whether the BTFR constitutes an 'anomaly' for standard-model cosmologists is important to establish here, for two reasons. (*i*) As shown in the previous section,

---

[3] At least, among galaxies in the local universe. Rotation-curve studies typically do not include galaxies more distant than a few hundred megaparsecs.

the same postulates that predict asymptotic rotation-curve flatness also predict the BTFR. If the latter relation "constitutes an anomaly for the rival theory," then Carrier's criterion of non-ad-hoc-ness is satisfied for Milgrom's prediction of asymptotic rotation-curve flatness. (*ii*) The BTFR is itself a novel prediction, according to criterion M, only if it is improbable in the light of the rival theory.

§

Attempts by standard-model cosmologists to explain the BTFR did not begin to appear in print until some twenty years after the relation's prediction by Milgrom (van den Bosch et al., 2003; Mayer and Moore, 2004), and about four years after its observational corroboration by McGaugh et al.[4] Under the standard model, the quantity V_∞ is attributed almost entirely to the gravitational force from a galaxy's 'dark matter halo.' Simulations that follow the clustering of dark matter starting from infinitesimal density perturbations in the early universe routinely fail to produce haloes with asymptotically flat rotation curves;[5] a common proxy for V_∞ in these simulations is V_{max}, the peak rotation velocity in the simulated halo. Finding a proxy for M_{gal}, the galaxy's (non-dark) mass, is a tougher nut, especially in simulations (like the earliest ones) that did not include *any* non-dark matter. A standard workaround is to relate M_{gal} to the so-called virial mass, M_{vir}, defined as the *dark* mass contained within a sphere having a radius such that the mean interior density is a factor Δ_{vir} times the cosmological mean density $\rho_0 = \rho_{crit} = 3H_0^2/8\pi G$.[6] It turns out that a value $\Delta_{vir} \approx 100$ effectively distinguishes the gravitationally relaxed interior of a dark matter halo from the surrounding infalling material. Assuming a fixed value for Δ_{vir}, simulations then predict a mean relation

$$M_{vir} \propto V_{max}^\alpha, \quad \alpha \approx 3 - 4 \tag{4.6}$$

between V_{max} and M_{vir} (e.g. Bullock et al., 2001). The final step is to relate M_{vir} to the (non-dark) mass of the galaxy that (presumably) forms inside the dark halo. For instance one can write for this mass

$$M_{gal} = f_b f_d M_{vir} \tag{4.7}$$

---

[4] Discussions of the BTFR in the standard-model literature, both observational and theoretical, routinely fail to mention that the relation is a novel prediction of a rival research program. One presumes that this reticence is disingenuous since Milgrom's 1983 prediction of the relation is well known (e.g. Peebles, 2015, p. 12247, "MOND made a strikingly successful prediction"). For examples of this troubling phenomenon, see (in addition to the papers by van den Bosch et al. and by Mayer and Moore) Gurovich et al. (2010), Trujillo-Gomez et al. (2011), Catinella et al. (2012), Zaritsky et al. (2014), Bradford et al. (2015), Desmond and Wechsler (2015), Bradford et al. (2016), Sorce and Guo (2016), Sales et al. (2017), Cattaneo et al. (2017), and Karachentsev et al. (2017).

[5] See, for instance, Desmond (2017) who compares rotation curves of real and simulated galaxies. Desmond remarks (p. L38) that the larger fraction of non-flat (i.e. rising) rotation curves in the simulated galaxies "deserves attention in future studies."

[6] H_0 is Hubble's constant. Here I am assuming, as do most standard-model cosmologists, a cosmological model based on Einstein's equations with zero curvature.

where f_b is the 'cosmological baryon fraction' (the average ratio of normal to dark matter) and f_d is the average fraction of available baryons that end up, after the complex process of galaxy formation, contributing to M_{gal}. (The remaining baryons might take the form, for instance, of ionized gas surrounding the galaxy.) The quantity f_b is thought to be known;[7] reproducing the observed normalization of the BTFR then requires $f_d \ll 1$, i.e., only a small fraction of the available baryons can find their way into galaxy disks. For the standard-model cosmologist, then, explaining the BTFR consists of finding a plausible way to make f_d dependent on the properties of dark matter haloes so as to reproduce the known characteristics (slope, normalization, scatter) of the observed relation between M_{gal} and V_{max}.

The most common schemes invoke one or more auxiliary hypotheses that 'couple' the dark matter to the baryonic matter. One such hypothesis – manifestly ad hoc – was adopted in the 'Illustris' study by Vogelsberger et al. that was cited in Chapter 2. Another, less ad hoc, approach is to postulate that supernova explosions occur in regions of star formation and drive winds, and the degree to which matter is removed by the winds is made dependent on the depth of the dark matter potential well. Such a hypothesis has the effect of preferentially removing baryons from lower-mass haloes, thus steepening the $M - V$ relation toward the desired slope of four (see Figure 4.1). The greater difficulty comes in reproducing the observed, very small scatter of the BTFR (McGaugh, 2012). The formation and evolution of galaxies and dark matter haloes are, presumably, stochastic processes and a certain minimum variance is implied in observable quantities like M_{gal} and V_{max}. All attempts since 2004 (as reviewed by Del Popolo and Le Delliou, 2017; Bullock and Boylan-Kolchin, 2017) have failed to overcome one or more of these hurdles. The most recent such attempts (e.g. Sales et al., 2017) sometimes claim to reproduce the observed small scatter, but the predicted relation deviates from a pure power law at values of M_{gal} corresponding to dwarf galaxies.[8]

I conclude that the BTFR constitutes an anomaly for the standard cosmological research program. It follows that (*i*) Carrier's criterion of non-ad-hoc-ness is satisfied for Milgrom's prediction of asymptotic rotation-curve flatness, and (*ii*) the BTFR is itself a novel prediction according to Musgrave's criterion. These conclusions are summarized in the second line of Table 4.1.

§§

3. *A unique relation between the acceleration $\boldsymbol{a}$ at any point in a disk galaxy and the Newtonian gravitational acceleration $\boldsymbol{g}_N$ due to the galaxy's mass*: As discussed in Chapter 3, adding Milgrom's third postulate to the first two implies

---

[7] See Chapter 6.
[8] Note that the success or failure of simulations like these has no bearing on the validity or interpretation of Milgrom's prediction, since that prediction is independent of a galaxy's history.

a relation of the form $a = v\,(g_N/a_0)\,g_N$ between the acceleration measured locally in a disk galaxy, and the gravitational acceleration as predicted by Newton's law of gravity based on the observed distribution of matter. Milgrom's postulates do not specify the *form* of the function $v\,(g_N/a_0) \equiv v(y)$ except in the limits of large and small acceleration: in the former limit $v = 1$ (Postulate I), in the latter $v = y^{-1/2}$ (Postulate II). As noted by Milgrom (2016a), this prediction can be expressed in a number of functionally equivalent ways: as a relation between a and g_N, between a and a/g_N, between g_N and a/g_N etc.

Milgrom's prediction has been splendidly corroborated. Before assessing the novelty of the prediction, I will first consider separately the different ways in which the data have been used in the corroborating studies. As will become clear, the fact that only the *existence* of a function $v(g_N/a_0)$ is predicted – and not the functional form itself – gives an observer more freedom that she might otherwise have when choosing how to confront Milgrom's prediction with the data. Another way to state this is to note that prediction number three – unlike the first two predictions from T_0 – applies even to regions in a galaxy that are *not* in the asymptotic, low-acceleration regime, and the theory in this variant does not make definite predictions about the equations of motion in this regime.

3(a). *Galaxy rotation curves.* In his second paper from 1983 (p. 381), Milgrom states the prediction as follows: "Velocity curves calculated with the modified dynamics on the basis of the observed mass in galaxies should agree with the observed curves."

Rotation-curve data are traditionally presented as plots of circular orbital speed, V, versus distance, R, from the galaxy center. The acceleration, a, is derivable from these quantities as $a(R) = V^2(R)/R$. When testing Milgrom's prediction using rotation-curve data, a standard practice (first introduced by Milgrom himself: Milgrom (1983b)) is to introduce a 'transition function' $\mu(x)$ defined via $\mu(a/a_0)a = g_N$.[9] As in the case of the function $v(y)$, the three hard-core postulates say nothing about the form of $\mu(x)$ except for its asymptotic behavior:

$$\mu\,(x \gg 1) \approx 1, \quad \mu\,(x \ll 1) \approx x, \quad x \equiv \frac{a}{a_0}. \tag{4.8}$$

It is also trivial to show that

$$v(y) = \frac{1}{\mu(x)} \text{ where } y = x\mu(x). \tag{4.9}$$

The predicted relation between V and R is expressible in terms of μ as

$$x\mu(x) = \frac{g_N}{a_0}, \quad x = x(R) \equiv \frac{a(R)}{a_0} = \frac{V^2(R)}{a_0 R}. \tag{4.10}$$

---

[9] At first blush, the function $\mu(x)$ looks like an auxiliary hypothesis. In fact it plays no such role here; it is introduced for reasons of computational convenience and could be dispensed with, as discussed below.

The procedure is then as follows: (*i*) One postulates a functional form for $\mu(x)$ that is consistent with the limiting forms (4.8). (*ii*) The gravitational acceleration in the galactic disk, $g_N(R)$, implied by the observed distribution of mass (stars, gas) is computed from Newton's law. (*iii*) The functions $\mu(x)$ (assumed) and $g_N(R)$ (computed) are inserted into Equation (4.10), which is used to compute the predicted V at every observed R. (*iv*) The results are compared with the observed velocities and the goodness-of-fit is assessed. Under this procedure, corroboration of Milgrom's prediction would consist of demonstrating that the rotation-curve data (from every galaxy in the sample, say) is well fit by some (i.e. by any) single function $\mu(x)$ and by some single value of a_0.

This procedure succeeds uncannily well. The first studies (Kent, 1987; Lake, 1989; Begeman et al., 1991) adopted a transition function of the form

$$\mu_n(x) = \frac{x}{(1 + x^n)^{1/n}}, \quad x \equiv \frac{a}{a_0} \tag{4.11}$$

with $n = 2$; Kent (1987, p. 829) states that this choice for $\mu(x)$ "works somewhat better than others." Subsequent researchers have found that different functional forms can give essentially the same results (Brada and Milgrom, 1995; Milgrom, 1999; Zhao, 2008). Figure 4.2 shows the residuals of fits to the rotation curves of 78 nearby galaxies. In this case, the authors used a transition function taken from a family of functions first proposed by McGaugh (2008), motivated by a similar family of functions due to Milgrom and Sanders (2008). The McGaugh (2008) family has the form

$$\nu(y) = \left[1 - \exp\left(-y^{n/2}\right)\right]^{-1/n}. \tag{4.12}$$

Figure 4.2 was based on this transition function with $n = 1$.[10]

It is found – consistent with the Milgromian prediction – that rotation curves are successfully reproduced in galaxies of wildly different types: small and large; low and high surface brightness; gas-dominated and star-dominated etc. (Gentile et al., 2007; Sanders and Noordermeer, 2007).

What is especially striking is the success of Milgrom's prediction in the case of so-called 'dark matter dominated' galaxies: galaxies for which the observed V exceeds the value predicted by Newton's equations at all radii, even near the center (de Blok and McGaugh, 1997). In the standard cosmological model, rotation curves of such galaxies would be determined almost entirely by the dark matter distribution; the normal matter would be almost irrelevant. Whereas Milgrom predicts – and the observations confirm – that the rotation curve is predictable from the distribution

---

[10] Equation (4.12) has the virtue of giving results that may be more consistent with solar system data (McGaugh, 2008). The gradual transition between Newtonian and MOND regimes implied by Equation (4.11) implies detectable departures from Newtonian motion in the inner solar system ($a \approx 10^8 a_0$). The more rapid transition of Equation (4.12) implies that such departures would be exponentially smaller. On the other hand, the 'Pioneer anomaly' (Anderson et al., 1998) can be explained by assuming a transition function that varies more gradually, like that of Equation (4.11) (Milgrom, 2006; McCulloch, 2007).

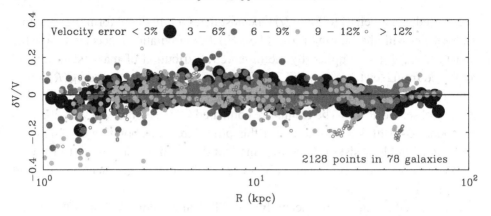

Figure 4.2 Residuals of Milgromian predictions for rotation-curve data of 78 nearby galaxies. The vertical axis is the difference between predicted and measured velocities, $\delta V = V_{\text{pred}} - V_{\text{obs}}$, divided by observed velocity. About two thousand individual resolved measurements are plotted; more accurate points are plotted using larger symbols. The bulk of the more accurate data are in good accord with the Milgromian prediction. The few deviant points are mostly at small radii, where noncircular motions make it difficult to infer circular velocities, and where finite observational resolution tends to 'smear' the data. Figure reprinted with permission from Benoît Famaey and Stacy S. McGaugh, "Modified Newtonian dynamics (MOND): observational phenomenology and relativistic extensions," *Living Reviews in Relativity*, 15, p. 61, 2012.

of observed matter alone (McGaugh and de Blok, 1998a,b). Discrepancies, which are rare, seem to be 'exceptions that prove the rule'; for instance, galaxies for which the distance estimate (needed to convert angular to actual distance) is particularly uncertain.[11]

3(b). *The mass discrepancy–acceleration relation (radial acceleration relation).* Recall that Milgrom's postulates imply the existence a functional relation of the form $a = \nu\,(g_N/a_0)\,g_N$ between the acceleration, a, measured locally in a disk galaxy, and the gravitational acceleration as predicted by Newton's law of gravity, g_N, based on the observed distribution of matter. Rotation-curve data do not confront Milgrom's prediction in a very convenient way, since the quantities that are predicted to be functionally related are not the measured V and R but rather a and g_N; hence the need to introduce a 'transition function.' It makes sense simply to

---

[11] One can ask whether each new galaxy provides a distinct opportunity for corroboration. Martin Carrier (1988, p. 218–219) argues 'no': "It is important to distinguish clearly between the empirical problem of the verification of a two-parameter-relation, that is the problem of establishing the facts, and the methodological problem of assessing the quality of an explanation of these facts. One can indeed question whether the relation Torricelli established between height of climb and specific weights is really sufficiently tested by an experiment involving only two substances. Further experiments with other liquids, however, do not constitute separate facts themselves, but only test whether one fact really proves correct. The same holds for the Puy-de-Dôme case [the measurement in 1648 by Florin Périer of the height of a column of mercury at three elevations]. The repetition of this experiment on other mountains or even in areas below sea-level serves only to establish one fact."

plot a and g_N against each other, since both quantities follow uniquely from the observations. Indeed doing so requires nothing more than the same data that would be used in carrying out the rotation-curve test.

The realization that the prediction can be cast in a testable form that does not require *a priori* specification of a transition function appears to have dawned only gradually on Milgromian researchers. Robert Sanders (1990) was apparently the first to approach the problem in this way. In plotting rotation-curve data from a sample of galaxies, Sanders chose as his vertical axis a quantity equivalent to a/g_N, calculated only at the last measured point of the rotation curve; he called this the "global mass discrepancy" or D. For a sample of disk galaxies, Sanders plotted D against R_{max} (the central distance of the last measured point) and also against $a = V^2(R_{max})/R_{max}$. He found that the latter correlation was substantially tighter than the former, although still not particularly tight. Sanders did not explain why he chose to plot only the outermost data points from his galaxies, but one supposes that he wished to include only data in the asymptotic, low-acceleration regime, the regime where the predicted rotation curve is independent of $\mu(x)$.

McGaugh (1999b) repeated the experiment using a somewhat larger and better-observed sample of galaxies. But unlike Sanders, McGaugh included on his plots rotation-curve data from *all* radii in his observed galaxies; he did not restrict the comparison to data points that lay only in the asymptotic, low-acceleration regime as Sanders had. In a subsequent, more detailed, study, McGaugh (2004) plotted the quantity a/g_N directly against a number of other quantities, including R, a and g_N. He noted (p. 655) that "when the mass discrepancy is plotted against acceleration, something remarkable happens. All the data from all the galaxies align." He found a similar "alignment" when g_N was chosen for the abscissa and noted that "The trend in the data in this figure is, in effect, the inverse of the MOND interpolation function $\mu(x)$." That is: he acknowledged that the transition function could be 'read off' from the same data used to construct the rotation curves; there was no need to assign a functional form to $\mu(x)$ when corroborating the Milgromian prediction.

In all studies like these, an algorithm is required to go from the observed surface brightness (due to stars), and the observed surface density of gas, to an estimate of the total matter distribution, and thence to g_N. The greatest freedom comes in relating stellar light to mass. McGaugh (2004) investigated how different assumptions about this ratio (called M/L) affect the form of the inferred correlation. In principle, M/L might (and probably does) have a different value at every location in every galaxy, since it depends on the detailed mix of stellar types, ages, compositions etc. McGaugh found that for any plausible M/L the correlation with a was tight.

The correlation found by McGaugh between a/g_N and a is now called the 'mass discrepancy–acceleration relation,' or MDAR. This name is explained in the following way: The quantity g_N is related to the Newtonian gravitational potential of a disk galaxy, $\Phi(x, y, z)$, as

$$g_N(R) = \left| \frac{\partial \Phi}{\partial r} \right|_{z=0} \approx \frac{GM(<R)}{R^2}$$

$$\equiv \frac{V_N^2(R)}{R}. \tag{4.13}$$

The derivative $\partial \Phi / \partial r$ is to be evaluated in the equatorial (mid-) plane of the galaxy; the second, approximate relation on the first line would be an equality in the case of a spherical galaxy.[12] The final line defines $V_N(R)$, which is the circular orbital velocity implied by the observed matter distribution under Newtonian dynamics. In addition to g_N as given by Equation (4.13), the other quantity that appears in Milgrom's predicted relation is $a = V^2(R)/R$. One can therefore write:

$$\frac{a(R)}{g_N(R)} = \frac{V^2(R)}{V_N^2(R)}, \tag{4.14}$$

which is the quantity plotted along the vertical axis by McGaugh (2004). In the standard cosmological model, the quantity on the right is identified with the 'mass discrepancy,' i.e. the ratio between the mass inferred from the rotation curve and the mass inferred from the observed stars and gas.

A methodological objection can be raised to presenting data in the form of the MDAR: both axes on such a plot depend on the measured V (through a in the case of the left hand side of (4.14)) and hence are not independent. A solution is simply to plot a vs. g_N. The correlation so obtained is called the 'radial acceleration relation,' or RAR. Such a plot appears to have first been published by Wu and Kroupa (2015). More recently, McGaugh et al. (2016) and Lelli et al. (2017) presented a re-analysis using 153 disk galaxies covering a very wide range in properties (size, luminosity, surface brightness). An important advance in these two studies was the use of near-infrared (3.6 micron) photometry for inferring the mass in stars; as noted above, the luminosity in the infrared is thought to be a better proxy than the optical luminosity for the stellar mass. Lelli et al. (2017, p. 18) conclude:

The observed scatter [in the RAR] is very small ($\lesssim$ 0.13 dex) and is largely driven by observational uncertainties: the radial acceleration relation has little (if any) intrinsic scatter. The tiny residuals show no correlation with either local or global galaxy properties.

Figure 4.3 shows the RAR as constructed from rotation-curve data by McGaugh et al. (2016).

Corroboration of Milgrom's prediction here means a demonstration that g_N and a are – as best as can be determined – functionally related (as opposed to simply being correlated). One can go a step further with data like these, and extract from them an estimate of the function $\nu(y)$ that (assuming the hard-core postulates are correct) relates a and g_N through $a = \nu(g_N/a_0)\, g_N$. Various functional forms have

---

[12] Under the 'dark matter hypothesis' of the standard cosmological model, the gravitational potential even of a disk galaxy is predicted to be nearly spherical due to the dominance of the 'dark matter halo.'

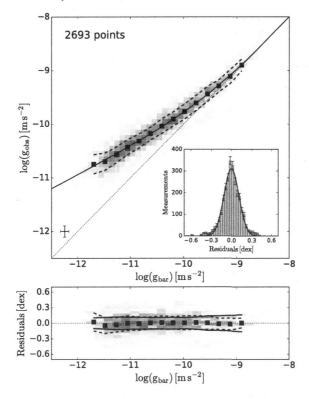

Figure 4.3 The radial acceleration relation (RAR), derived from rotation-curve data of 153 galaxies. The vertical axis plots the observed acceleration, $g_{obs} = V^2/r$, or what is called a in the text. The horizontal axis plots $g_{bar} = |\partial\Phi/\partial r|$, or what is called g_N in the text. The mean uncertainty on individual points is shown by the cross in the lower left corner. Large squares show the means of binned data. Dashed lines show the width of the ridge as measured by the rms in each bin. The solid curve is a fit of Equation (4.12) to the data. The inset shows the histogram of all residuals and a Gaussian of width $\sigma = 0.11$ dex; residuals are shown as a function of g_{bar} in the lower panel. The error bars on the binned data are smaller than the size of the points. The solid lines in the lower panel show the scatter expected from observational uncertainties and galaxy-to-galaxy variation in the stellar mass-to-light ratio. The data are consistent with a relation having negligible intrinsic scatter, as predicted by T_0. Figure reprinted with permission from Stacy S. McGaugh, Federico Lelli, and James M. Schombert, "Radial acceleration relation in rotationally supported galaxies," *Physical Review Letters*, 117, p. 201101, 2016. Copyright (2016) by the American Physical Society.

been found to work equally well in reproducing the observed RAR; Figure 4.3 shows that Equation (4.12), with $a_0 = 1.20 \times 10^{-10}$ m s^{-2}, provides an extremely good fit.

3(c). *'Renzo's rule'*. There is yet another way to test prediction number three from T_0, again using galaxy rotation-curve data, and again without the need to employ a 'transition function.' Differentiating Equation (4.10) with respect to radius gives

$$\frac{d}{dr}\left(\frac{V^2}{r}\right) = h(r)\frac{dg_N}{dr}, \quad h\left(x\left(r\right)\right) \equiv \left[\frac{d\left(x\mu\right)}{dx}\right]^{-1}. \tag{4.15}$$

The function $h(x)$ is, like the function $\mu(x)$, unknown. But its asymptotic forms follow from postulates I and II, namely:

$$h\left(x \gg 1\right) \approx 1, \quad h\left(x \ll 1\right) \approx \frac{1}{2x}, \quad x \equiv \frac{a}{a_0}, \tag{4.16}$$

and so it is reasonable to expect $h \gtrsim 1$. It follows that – in regions in a galactic disk where the Newtonian gravitational acceleration shows strong gradients – strong gradients should also be observed in the rotation curve. Of course just such a sensitivity is predicted by Newton's equations alone, but only in the absence of 'dark matter.' At locations in a galactic disk where $V \gg V_N$, the standard cosmological model postulates that most of the gravitational acceleration is due to the dark matter halo, and there is no obvious reason why $V(r)$ should be sensitive to the *local* details of the *baryonic* matter distribution.

And yet it is. The existence of such a correlation was first stated clearly by Renzo Sancisi in 2004 (italics added):

In the inner parts of spiral [i.e. disk] galaxies, of high or low surface brightness, there is a close correlation between rotation curve shape and light distribution. *For any feature in the luminosity profile there is a corresponding feature in the rotation curve and vice versa.* This implies that the gravitational potential is strongly correlated with the distribution of luminosity: either the luminous mass dominates or there is a close coupling between luminous and 'dark' matter (Sancisi, 2004, p. 233).

The italicized sentence is often called 'Renzo's rule.' Even so-called low surface brightness galaxies, which must be dark matter dominated everywhere according to the standard model, display the correspondence of bumps and wiggles described by Renzo's rule (e.g. Broeils, 1992).

Stated in the form of Equation (4.15), Renzo's rule is a 'fact' as defined by Carrier (a relation between V and g_N), and it is essentially the same 'fact' that is described by Equation (4.10). The separate evidentiary status of Renzo's rule in the minds of many Milgromian researchers is explained as follows: Under the dark matter hypothesis DM-1 of the standard cosmological model, correlations such as the BTFR and the MDAR/RAR can only be understood by supposing that dark and luminous matter obey an 'inverse' relation: higher densities of luminous matter appear to be associated with lower densities of dark matter and vice versa. Renzo's rule violates this expectation. Sancisi himself – who, it appears, was not aware of Milgrom's postulates in 2004 – wrote:

The unavoidable conclusion from the observed correspondence between the shapes of the rotation curves and those of the luminosity profiles is that the gravitational potential is strongly correlated with the distribution of luminous matter: either the luminous mass dominates or there is a close coupling between luminous and 'dark' matter (Sancisi, 2004, p. 239).

Sancisi's point is that either of these two possibilities is extremely unlikely given the assumptions of the standard cosmological model, and given the 'background knowledge' (e.g. the BTFR) that existed in 2004.

§

Having demonstrated that prediction number three – a unique relation between the acceleration, V^2/R, at any point in a disk galaxy and the Newtonian gravitational acceleration due to the galaxy's (normal) mass – is well corroborated, we turn now to the question of the prediction's novelty.

As we have seen, the earliest observational studies of relevance were rotation-curve measurements. Review articles published prior to 1983 (Faber and Gallagher, 1979; Bosma, 1983; Rubin, 1983) make no mention of a relation like the MDAR or the RAR. The focus was rather on the asymptotic flatness of observed rotation curves and the 'missing mass problem': that is, on the fact that rotation curves are *not* predictable (under Newtonian gravity) on the basis of the observed matter distribution. Sanders, in a historical review of this period, writes:

This is where the observational situation stood around 1982. In flat rotation curves, extending well beyond the visible disk, there was emerging evidence for a discrepancy between the visible and classical dynamical mass ... There were, in addition, scaling laws and photometric regularities of galactic systems: the Tully–Fisher law for spirals and Faber–Jackson for ellipticals; a characteristic surface brightness for spirals and for ellipticals, but there was no clear indication of a surface brightness dependence to the discrepancy in galaxies (Sanders, 2015, p. 128).

Sanders is perhaps guilty, in the final sentence, of overstating the situation slightly. One *can* find qualitative statements, in the pre-1983 literature, about the fact that the shapes of galaxy rotation curves seem to vary in a systematic way with galaxy type, the latter defined in terms of a galaxy's surface brightness distribution. For instance, Rubin et al. (1980, p. L150) wrote that "Rotation curves for [Hubble type] Sc galaxies show a systematic steepening of the central velocity gradient with increasing galaxy luminosity." But the pre-1983 literature does not seem to contain any claim that disk galaxy rotation curves are predictable based on the distribution of the observed matter.

Popper's condition, P, for novelty ("A predicted fact is novel if it was unknown prior to the formulation of the theory") is clearly satisfied here. Zahar's criterion, Z ("A novel fact is one that a theory was not specifically designed to accommodate") is also satisfied: Milgrom's theory could not have been designed to explain a fact if the fact was not yet known; there is no suggestion in Milgrom's three papers from 1983 that he was aware of such a relation in the data.

We are left with Musgrave's criterion, M, which asks whether the predicted fact is improbable or prohibited from the standpoint of the rival theory – that is: the standard cosmological model. And here it is useful to consider, separately, the three forms of the prediction as presented above.

§

Consider first rotation curves. Standard-model postulate DM-1 instructs the scientist to attribute any rotation-curve anomalies to the presence of dark matter; she is to assign to the dark matter whatever spatial distribution is required to reconcile the observed motions with Newtonian gravity. Unlike Milgrom's postulates, DM-1 makes no *predictions* about rotation curves; rather, it accepts a galaxy's observed rotation curve and its mass distribution as 'background knowledge' and uses that knowledge to make a statement about the galaxy's dark matter. No algorithm has ever been produced, under the standard cosmological model, that can predict the rotation curve of an observed galaxy.

As discussed in Chapter 2, it would be perfectly reasonable to rule, on the basis of these facts alone, that the Milgromian prediction of galaxy rotation curves is novel according to Musgrave's criterion, since the standard cosmological model is simply unable to make any such prediction.

I choose here to be more generous. While standard-model theorists are unable to predict the rotation curve of any observed galaxy, they have devoted (and continue to devote) enormous effort to *simulating* galaxy formation and evolution. Such simulations are rarely viewed, even by the simulators, as predictive; rather, the goal is almost always to adjust the simulation parameters in order to reproduce – in some statistical sense – the known properties of galaxies. Attempts by standard-model cosmologists to accommodate the baryonic Tully–Fisher relation, described earlier in this chapter, fall into this category.

But suppose that such a simulation has been carried out, and that it has resulted in a set of simulated disk galaxies that are judged to be reasonably similar in their properties to observed galaxies. If the spatial resolution of the simulation is adequate, the scientist could then analyze each simulated galaxy as if it were real, and ask (for instance) whether the rotation velocity in that 'galaxy' is related to the Newtonian acceleration due to the *normal* matter in the manner predicted by Milgrom at every radius. In other words: the scientist would be testing the Milgromian prediction in a suitably modified form, e.g. "Velocity curves calculated with the modified dynamics on the basis of the *non-dark* mass in *simulated* galaxies should agree with the rotation curves of the simulated galaxies."

In the last few years, the resolution of computer simulations has become good enough to permit such experiments. But interpreting the results as a test of the Milgromian prediction, even in its modified form, would be problematic, for a number of reasons. The simulated galaxies should be similar to observed galaxies, but 'similarity' is a notoriously vague concept; there is a multitude of ways to define similarity between two sets of objects, particularly objects that belong to such wildly disparate categories as simulated and real galaxies. The members of the two sets may be similar according to some criterion (the distribution of bulge-to-disk ratios, for instance) while being dissimilar according to another (e.g. the ratio of gas to stellar mass), and it is up to the researcher to decide which properties are 'fundamental.' Even more problematic: Suppose that 'similarity' is defined in terms of properties that appear, directly or indirectly, in Milgrom's predicted relations. For instance, standard-model theorists routinely *design* their galaxy formation codes to reproduce the baryonic Tully–Fisher relation, as discussed earlier in this chapter. To the extent that they are successful, the simulated galaxies will have rotation curves that match – by design – the Milgromian prediction in their outer regions.

Indeed, even simulated galaxies that 'look' very different from observed galaxies – for instance, by having unrealistic rotation curves – can accord with Milgromian predictions. For instance: Near the center of a simulated galaxy that is not 'dark matter dominated' – that is, in which $V^2/R > a_0$ – the standard-model scientist would calculate the same rotation curve that a Milgromian scientist would predict: that is: the rotation speed would be given by Newton's equations assuming that the gravitational force is due to the normal matter alone.

As it turns out, no standard-model cosmologist has yet (as of 2017) published the results of such a test. That is to say: no one has compared the rotation curves of their simulated galaxies with the Milgromian prediction, the latter based on the distribution of (non-dark) matter in the same simulated galaxies.[13]

But it is worth saying a little more. Even a quick survey of the standard-model literature reveals that an enormous effort *has* been devoted to 'getting rotation curves right' – that is: to adjusting the galaxy formation simulations so that the rotation curves of their simulated galaxies are not obviously inconsistent with what is observed. For instance: observed dwarf galaxies exhibit a wide variety of rotation curve shapes at fixed V_∞, while galaxy formation simulations (e.g. Oman et al., 2015) predict little variation, a consequence of the fact that dark matter haloes of a given mass should all have the same density profile. Shortly after this anomaly was identified, standard-model cosmologists identified an auxiliary hypothesis ('self-interacting dark matter') capable of resolving it (Creasey et al., 2017).

This anomaly is one of many that crop up when standard-model cosmologists attempt to simulate the formation of 'dark matter dominated' galaxies: that is, galaxies that have $V \gg V_{\text{Newtion}}$ at all radii. A Milgromian researcher would *expect* such galaxies to be problematic for standard-model theorists, since the latter would need to find a way for the dominant component by mass – the dark matter – to be accurately 'controlled' by the normal matter, and in just such a manner as to yield the rotation curve predicted, correctly, by Milgrom's theory. Furthermore, because

---

[13] Standard-model cosmologists *have* extracted rotation-curve data from their simulated galaxies and used those data in tests of Milgrom's postulates, as discussed in the next subsection. My point is that standard-model cosmologists appear never to have tested the prediction in the form that is under discussion here, that is, that *rotation curves* are predictable based on the distribution of normal matter alone.

the contribution of the normal matter to the gravitational potential is ignorable in 'dark matter dominated' galaxies, postulate DM-1 yields a particularly robust estimate for the dark matter distribution: there is relatively little 'wiggle room' due to uncertainties in the stellar mass-to-light ratio, the gas content, or the three-dimensional shapes of these galaxies.

One finds, generically, that the inferred density profile of the dark matter in such galaxies – the plot of ρ_{DM} versus radius – has a 'core,' the astronomer's term for a region of nearly constant density near the center. Whereas dark matter haloes that form in computer simulations of gravitational clustering always have 'cusps': density profiles that trend upward at small radii, typically as a power law, $\rho_{DM} \sim r^{-1}$, into the smallest radii that can be resolved. This failure of prediction was recognized already in the early 1990s (Flores and Primack, 1994; Moore, 1994) and is nowadays called the 'core-cusp problem' (de Blok, 2010).

The core-cusp problem is perhaps most striking in the case of dwarf galaxies, but even giant (disk) galaxies can be 'dark matter dominated' if their density of normal matter is sufficiently low, and the measured rotation curves of such galaxies are often better fit by assuming dark matter distributions with a core than with a cusp (de Blok et al., 2001; Gentile et al., 2004). Indeed the inferred 'core radius' in giant galaxies – the radius of the sphere inside of which the inferred dark matter density is nearly constant – can be of order 1 kpc, comparable to the scale length of the luminous disk (de Blok et al., 2001; Salucci, 2001).[14] It is fair to say that few, if any, galaxies have been found for which the application of postulate DM-1 *requires* the existence of a cusped density profile for the dark matter.

One proposed solution to the core-cusp problem was mentioned above: adding to the standard cosmological model the auxiliary hypothesis that the dark matter is 'self-interacting.' But the majority of the proposed solutions invoke gravitational interactions between the dark matter (assumed collisionless) and the normal matter (or 'baryons') that constitute the observed galaxy. It is easy to think of evolutionary pathways for the baryons that would decrease, *or* increase, the density of dark matter near the center of a galaxy. For instance, one widely adopted model (Blumenthal et al., 1986; Sellwood and McGaugh, 2005) approximates the formation of a galaxy as a slow contraction due to loss of energy from the cooling proto-galactic cloud. These 'adiabatic contraction' models predict that the dark matter density would increase together with that of the baryons as the gravitational potential well deepens, leading to even steeper density cusps than in the dark-matter-only simulations and worsening the core-cusp problem.

Of course the goal is to *decrease* the dark matter density. This too can be achieved in a variety of ways. For instance, if one postulates a 'clumpy' distribution of matter at some early time during the formation of a galaxy, the orbiting clumps can transfer

---

[14] That standard-model scientists should infer the presence of dark matter cores with these quantitative properties is correctly predicted by Milgrom's theory T_1, as discussed in Chapter 5.

some of their kinetic energy to the dark matter via gravitational scattering, causing the dark fluid to expand and lowering its density. A rotating stellar 'bar' can induce similar changes.[15]

The currently favored solution to the core-cusp problem is based on 'feedback': the deposition of thermal and kinetic energy into the gaseous, pre-galactic medium, either through supernova explosions in regions of star formation, or via energetic processes occurring near massive black holes. Feedback was mentioned earlier in this chapter; recall that it is invoked by standard-model cosmologists to remove gas from dark matter potential wells and to bring simulated galaxies more in agreement with the baryonic Tully–Fisher relation. But the same mechanism can alter the dark matter distribution, since removing baryons reduces the total attraction due to gravity and causes the dark fluid to expand (Mashchenko et al., 2006; Governato et al., 2010).

The physical processes that determine the efficiency of this sort of feedback would take place on spatial scales far smaller than the galaxy formation codes can resolve. A standard way to deal with this indeterminacy is to assume values for 'coupling factors' such as $f_{kin} \equiv E_{kin}/E_{SN}$, the fraction of the total energy produced in a supernova explosion (say) that is deposited as kinetic energy in the surrounding gas. To achieve the desired results, coupling factors in the simulations are sometimes pushed as high as 40% (Governato et al., 2010; Peñarrubia et al., 2012), much higher than is considered plausible by observers of actual supernovae, or by theorists who carry out high-resolution simulations of individual supernova explosions. Walch and Naab (2015), who find values for the coupling factors of only a few percent based on detailed modeling of single explosive events, note (p. 2757) that "Such very low coupling efficiencies cast doubts on many subresolution models for SN [supernova] feedback, which are, in general, validated a posteriori" – that is, which are justified on the ground that they produce the desired results.[16]

The ad hoc nature of the feedback prescriptions used in the galaxy simulation codes is widely acknowledged, even by the standard-model theorists who employ them. Here are three more examples. Schaye et al. (2015, p. 531) write that "Cosmological, hydrodynamical simulations have traditionally struggled to make stellar feedback as efficient as is required to match observed galaxy masses, sizes, outflow rates and other data." In discussing feedback from 'active galactic nuclei' (AGN), i.e. supermassive black holes, Naab and Ostriker (2017, p. 96) write that "almost all AGN feedback models – on cosmological scales – are of empirical

---

[15] Even the author has contributed, in a minor way, to this literature; see e.g. Merritt and Milosavljević (2002).

[16] That this is a correct intepretation of "validated a posteriori" was confirmed by a private communication from Stefanie Walch: "Galaxy formation simulations . . . tune parameters such that the simulations produce realistic-looking galaxy populations. In this sense the sub-grid models are 'validated' as 'realistic' models by plausibility arguments in comparison to observations. Historically these models result from trial and error experiments. The models themselves might easily be 'wrong' (in a strict physical sense) or assuming unrealistically high values for the coupling efficiencies – they still produce realistic galaxy properties and the authors claim success."

nature with accretion and energy conversion efficiencies adjusted, in a plausible fashion, to match observed scaling relations." (By "scaling relations," Naab and Ostriker mean relations like the BTFR.) And in their review of proposed solutions to the core-cusp problem, Bullock and Boylan-Kolchin (2017, p. 370) write that "while many independent groups are now obtaining similar results in cosmological simulations of dwarf galaxies ... this is not an ab initio ΛCDM prediction, and it depends on various adopted parameters in galaxy formation modeling."

I conclude that standard-model theorists have not yet found a convincing way to predict galaxy rotation curves based on their distribution of (non-dark) matter, either in the case of real or simulated galaxies. Musgrave's condition for novelty of the Milgromian prediction is therefore satisfied.

§

The second form of prediction number three from T_0 was a functional relation between $|g_N|$ and $|a| \equiv V^2/R$ – the 'mass discrepancy–accleration relation' (MDAR) or the 'radial acceleration relation' (RAR). Unlike in the case of rotation curves, standard-model theorists *have* tested this form of the prediction against their simulated galaxies. The first such study appeared in print in 2016 and it was followed almost immediately by several others (Santos-Santos et al., 2016; Ludlow et al., 2017; Keller and Wadsley, 2017; Tenneti et al., 2017). No doubt more will have appeared by the time this book is published.[17]

All of these studies consist of analysis of model galaxies formed using the latest generation of computer codes, codes designed to simulate the formation of disk galaxies in a standard-model context. As Milgrom (2016b, p. 3) notes, "one will not be amiss to suppose that over the years, the many available handles have been tuned so as to get galaxies as close as possible to observed ones." Milgrom is not suggesting here (nor is there any reason to believe) that standard-model theorists have adjusted their algorithms with the MDAR or the RAR specifically in mind. But (as documented earlier in this chapter, and in Chapter 2) it is clear that standard-model theorists *have* 'tuned' their computer codes in order to reproduce the baryonic Tully–Fisher relation, the observed, low central densities of dwarf galaxies, and other known, systematic properties of disk galaxies. To the extent that they have been successful in this enterprise, one expects the detailed kinematics of their model galaxies to fall, at least approximately, along relations like the MDAR and RAR, which encapsulate precisely those systematic properties. This situation

---

[17] One wonders what took standard-model cosmologists so long, and why they did not first investigate the Milgromian prediction in terms of rotation curves, given that they had been computing rotation curves for their simulated galaxies at least since the 1990s. My guess is that the remarkable observational studies by McGaugh and collaborators that appeared in 2016 and 2017 were responsible for the collective awakening. My own paper from 2017 may have played a minor role.

complicates the evaluation of Musgrave's condition for novelty – although as we will see, no standard-model study yet published has come very close to reproducing the observed MDAR or RAR.

The Santos-Santos et al. (2016) study was based on 22 model galaxies drawn from two, 'zoom-in' simulations of small cosmological volumes. The authors (p. 478) acknowledge that the feedback prescriptions in the simulation codes were adjusted to yield some known systematic properties of galaxies ("Further, in these types of simulations, feedback recipes are not well constrained, but are basically tuned . . . to match the constraints imposed.") Rotation-curve data from the simulated galaxies were plotted against $a \equiv V^2/R$ in two forms: (*i*) as V^2/V_N^2, with V_N the Newtonian velocity that would be predicted from all baryons, stars and gas; and (*ii*) as $V^2/V_\star^2$, with $V_\star$ the rotation velocity that would be predicted from the stars alone. Both plots exhibit a trend that is reminiscent of the MDAR but with considerable scatter and offset; furthermore the plot based on V_N exhibits greater scatter than the plot based on $V_\star$, contrary to what a Milgromian would predict. The authors note (p. 481) that "Better statistics are required to determine whether the deviations from the relation seen in the simulations are more or less prominent than seen in the observations."

Keller and Wadsley (2017) present a study based on a comparably small sample (18) of simulated galaxies. These authors plot a vs. g_N and compare their results with the RAR as determined from observations by McGaugh et al. (2016). Keller and Wadsley claim that their models are fully consistent with the observed relation ("We have shown here that the SPARC [i.e. McGaugh et al. (2016)] RAR can be produced by conventional galaxy formation in a ΛCDM universe"). This claim has met with scathing criticism (Milgrom, 2016b; Lelli et al., 2017). The model galaxies of Keller and Wadsley span a factor of only $\sim$15 in mass, which represents less than 0.05% of the mass range of the galaxies in the McGaugh et al. observed sample. Crucially, none of their simulated galaxies is 'dark matter dominated'; the only contributions to the $a < a_0$ part of the RAR come from outer rotation curves, precisely the region that standard-model cosmologists have targeted when adjusting their galaxy formation codes to reproduce the BTFR. The existence of a functional relation like the RAR is surprising precisely because regions of low g_N always exhibit the same $a = V^2/R$, regardless of whether those regions are near the center or far from it; the Keller and Wadsley sample of model galaxies does not confront this prediction.

Ludlow et al. (2017) present results based on a larger sample of model galaxies (the exact number is not specified) from two (the 'EAGLE' and 'APOSTOLE') cosmological simulations. Like Keller and Wadsley, Ludlow et al. claim success: "These observations [i.e. the RAR], consistent with simple modified Newtonian dynamics, can be accommodated within the standard cold dark matter paradigm". But an examination of their results reveals that the claim is unjustified:

1. As discussed earlier in this chapter, the Milgromian transition function $v(y)$ can be extracted directly from a plot of the RAR. Ludlow et al. find that the transition function implied by their simulated rotation-curve data requires a value of 2.6×10^{-10} m s^{-2} for Milgrom's constant. (In the version of the paper first submitted, the larger value 3.0×10^{-10} m s^{-2} was given.) For comparison, the value derived from real galaxies is about 1.2×10^{-10} m s^{-2}. There are various ways to estimate the uncertainty of the latter number (see Chapter 8), but probably the best way in the current context is to look at studies in which the same transition function is fit to the same sort of data; McGaugh et al. (2016) find a value 1.20 ± 0.02 (random) ± 0.24 (systematic) $\times 10^{-10}$ m s^{-2}. Thus, the Ludlow et al. value is enormously discrepant: by $\sim 70\sigma$ (using the random error) or $\sim 5.8\sigma$ (using the systematic error).[18]

2. Ludlow et al. claim that "All [simulation] runs are now consistent with the observed relation to within the observational scatter." But as discussed earlier in this chapter, the observed scatter in the RAR is due almost entirely to observational errors; as nearly as can be determined, the relation has zero *intrinsic* scatter, as Milgrom would predict. It is the latter result that Ludlow et al. should require from their simulated galaxies; or equivalently, they should 'observe' their model galaxies as if they were real and include the likely observational errors. In any case, the value that they find for the scatter (0.09 dex) is too large.

3. Ludlow et al. find a systematic offset between the RAR as defined by high-mass and low-mass galaxies. No such offset is observed in data from real galaxies.

Tenneti et al. (2017) analyze a larger sample (1594) of model galaxies from yet another cosmological simulation ('MassiveBlack-II'). Unlike the other authors just cited, Tenneti et al. do *not* claim success in recovering the observed RAR relation:

We find that radial accelerations contributed by baryonic matter only and by total matter are highly correlated, with only small scatter around their mean or median relation, despite the wide ranges of galaxy luminosity and surface brightness. We further find that the radial acceleration relation in this simulation differs from that of the SPARC [i.e. McGaugh et al. (2016)] sample, and can be described by a simple power law in the acceleration range we are probing (Tenneti et al., 2017, p. 1).

The fact that Tenneti et al. find a simple power law for the relation means that there is no indication in their simulated data of an acceleration scale. The slope of their inferred relation (on a log-log plot) differs from the slope of the observed RAR, both in the low- and high-acceleration regimes.

---

[18] Lelli et al. (2017) point out that this discrepancy in Milgrom's constant is consistent with the results of Oman et al. (2015), who concluded that the same simulated galaxies had a systematic over-abundance of dark matter compared with observed galaxies.

In summarizing the results from these studies, Lelli et al. (2017, p. 16) conclude: "In summary, several key properties of the radial acceleration relation are not reproduced by the current generation of cosmological simulations."

§

The third form of Milgromian prediction number three discussed above was 'Renzo's rule': "For any feature in the luminosity profile there is a corresponding feature in the rotation curve and vice versa."

One might think that – to the extent they are dark matter dominated – galaxies formed according to standard-model prescriptions could not possibly exhibit Renzo's rule. Yet Santos-Santos et al. (2016, p. 480) write, in regard to their model galaxies:

One can also see that features from the baryonic components are often reflected in the total-components curves. This is known as 'Renzo's Rule' (Sancisi 2004; McGaugh 2014), and has long been observed in real galaxies. In particular, these bumps and features are noticeable in [model] galaxies g15807_Irr, g15784_Irr, g1536_MW and g5664_MW ... These results represent evidence that the different mass components affect each other throughout the disc region as they co-evolve within a ΛCDM Universe.

But inspection of the model rotation curves reveals that – as really must be the case – features in the rotation curve only reflect features in the (normal) matter distribution when the latter dominates over the dark matter. What makes Renzo's rule so surprising, of course, is the fact that in observed galaxies, it holds even in regions where (according to a standard-model cosmologist) the baryons contribute negligibly to the density.

§

I conclude that (claims by some standard-model cosmologists notwithstanding) Milgrom's confirmed prediction of a functional relation between $|a|$ and $|g_N|$ represents an anomaly for the standard cosmological model, and that this is true with regard to whichever of the three forms of the prediction one considers. Musgrave's condition for novelty of the Milgromian prediction is therefore satisfied. These conclusions are summarized in Table 4.1.

§§

The gravitational acceleration[19] g_N that appears in Milgrom's postulates is understood to be the total acceleration due to gravity. In the case of a system, S, that is freely falling in an external gravitational field, there will be two contributions

---

[19] Recall that 'gravitational acceleration' here means 'force per unit mass due to gravity'.

to g_N: from the matter associated with S, and from the matter that is responsible for the external field in which S is moving. In this case, Postulate I ("Newton's second law ... breaks down when the acceleration is sufficiently small") implies that the internal dynamics of S will deviate from Newtonian only if the *summed* gravitational fields has amplitude less than a_0. In Milgromian dynamics, even a spatially constant external field can affect the internal dynamics of a system. This is different, of course, from Newtonian dynamics, according to which the internal dynamics of a system is independent of any external field that is constant across the system. It also conflicts with the strong equivalence principle of general relativity.

The existence of this 'external field effect' (EFE) follows from the three foundational postulates and it was discussed by Milgrom already in his first paper from 1983. I will argue that testable predictions of the EFE are only feasible under theory variant T_1. The basis of the EFE will nevertheless be introduced here, following the discussion in Milgrom (1983a).

Let g_N^i and g_N^e be the gravitational accelerations due to the matter internal and external to S respectively, and define a^e as the acceleration due to the external matter alone, so that

$$\mu \left(a^e / a_0 \right) a^e = g_N^e. \tag{4.17}$$

The total acceleration, $a = a^e + a^i$, satisfies

$$\mu \left(a / a_0 \right) a = g_N^e + g_N^i. \tag{4.18}$$

Consider the limiting case $a^i \ll a^e$. Expanding the total acceleration in terms of the small quantity a^i / a^e yields

$$\frac{a}{a_0} \approx \frac{a^e}{a_0} - e \cdot \frac{a^i}{a_0} \tag{4.19}$$

and

$$\mu \left(\frac{a}{a_0} \right) \approx \mu \left(\frac{a^e}{a_0} \right) \left[1 - L \left(e \cdot a^i \right) / a^e \right], \tag{4.20}$$

where e is a unit vector in the direction of g_N^e and $L \equiv d \ln[\mu(x)]/d \ln x$ at $x = a^e/a_0$. Combining Equations (4.17)–(4.20) yields for the acceleration due to matter in S:

$$a^i - Le \left(e \cdot a^i \right) = \left[\mu \left(a^e / a_0 \right) \right]^{-1} g_N^i. \tag{4.21}$$

In theory variant T_0, the transition function $\mu(x)$ is not specified, but its limiting forms are known (Equation 4.8). In the case of a strong external field, $a^e \gg a_0$, $\mu = 1$, $L = 0$, and Equation (4.20) implies $a^i = g_N^i$: the internal motion is Newtonian regardless of whether a^i is large or small compared with a_0. In the case $a^i < a^e < a_0$, reasonable assumptions about the smoothness of the function $\mu(x)$

imply that L is at most of order unity, and Equation (4.20) implies that $\boldsymbol{a}^i$ is approximately proportional to $\boldsymbol{g}_N^i$ but with a potentially large prefactor $\left[\mu\left(a^e/a_0\right)\right)\right]^{-1}$. The motion is then essentially Newtonian but with a larger effective gravitational constant.

Recall that theory variant T_0 is restricted in its application to systems that are static and symmetric. Testable predictions that follow from the EFE – for instance, about the internal motion of a satellite galaxy orbiting around the Milky Way – will become possible under theory variant T_1, which is applicable to time-dependent systems with any geometry.

§§

In summary: Theory variant T_0 is theoretically progressive in that it predicts a number of facts that are novel according to at least one of the criteria established in Chapter 2. These novel predictions include the asymptotic constancy of disk galaxy rotation speeds, the baryonic Tully–Fisher relation (BTFR), and the mass discrepancy–acceleration relation (MDAR).

The Milgromian research program at this stage is also empirically progressive, since each of these novel predictions has been confirmed through observation.

Theory variant T_0 also predicts the existence of a single dimensioned parameter, a_0 (Milgrom's constant), which appears, in different ways, in each of its three novel predictions. The prediction of a universal 'acceleration scale' has also been observationally corroborated; a fuller discussion of this result is deferred until Chapter 8.

The 'external field effect' is also a novel prediction of T_0. However, it was argued here that this prediction only becomes testable under theory variant T_1.

Since it is both theoretically and empirically progressive, theory variant T_0 represents, according to Lakatos's criterion, a 'progressive problemshift.'

5

Theory Variant T_1: A Non-relativistic Lagrangian

In his discussion of the heuristic of the Newtonian research program, Lakatos notes that Newton

first worked out his programme for a planetary system with a fixed point-like sun and one single point-like planet. It was in this model that he derived his inverse-square law for Kepler's ellipse. But this model was forbidden by Newton's own third law of dynamics, therefore the model had to be replaced by one in which both sun and planet revolved round their common centre of gravity. This change was not motivated by any observation (the data did not suggest an 'anomaly' here) but by a theoretical difficulty in developing the programme (Lakatos, 1970, p. 50).

Theory variant T_1 in the Milgromian research program came about for a similar set of reasons – and, as in Newton's case, as an expression of the program's heuristic, not in response to any observational anomaly.

Applying the force law in the form of Milgrom's second postulate to a system of massive bodies implies a net acceleration of the system's center of mass (Milgrom, 1983a; Felten, 1984). Only for highly symmetric systems, e.g. two equal-mass particles, is momentum conserved. Theory variant T_0 side-stepped this problem by – as we have seen – restricting application of the postulates to idealized systems, e.g. a test mass moving in the fixed gravitational field of a spherical or disklike galaxy. Theory variant T_1 solves the problem by including a scheme for computing the particle equations of motion that respects the usual conservation laws, regardless of the system's geometry. In addition, T_1 addresses, and solves, another "theoretical difficulty" associated with T_0: it yields a value for the center-of-mass motion of a bound system that respects the modified dynamics even if the system's *internal* gravitational accelerations are large compared with a_0.

Theory variant T_1 achieves both aims by deriving the equations of motion as stationary points of an action. Conservation of energy, linear momentum and angular momentum for a system of particles is guaranteed by the invariance of the action under translations and rotations.

Recall from classical mechanics that the Lagrangian for a single point particle is $L = K - V$, where K is the kinetic energy and V the potential energy. The action,

S, is the integral with respect to time of $L(x, \dot{x})$, and the equations of motion can be derived by finding critical (stationary) points of the action as a function of the particle trajectory. The critical points of the action turn out to be those that satisfy the Euler–Lagrange equations:

$$\frac{\partial L}{\partial x_i} - \frac{d}{dt}\left(\frac{\partial L}{\partial \dot{x}_i}\right) = 0, \quad i = 1, \dots, 3. \tag{5.1}$$

In the case of a collection of particles, the Lagrangian is replaced by the Lagrange density, $\mathcal{L}$, defined in the Newtonian case as

$$\mathcal{L}_N = \frac{\rho v^2}{2} - \rho \Phi_N - \frac{|\nabla \Phi_N|^2}{8\pi G}. \tag{5.2}$$

In this equation, $\rho(x, t)$ is the matter density, $v(x, t)$ is the rms particle velocity, and $\Phi_N(x, t)$ is the (Newtonian) gravitational potential. The Newtonian action is then:

$$S_N = \int \frac{\rho v^2}{2} d^3x \, dt - \int \rho \Phi_N d^3x \, dt - \int \frac{|\nabla \Phi_N|^2}{8\pi G} d^3x \, dt. \tag{5.3}$$

Varying (minimizing) the first two terms with respect to particle positions yields the Newtonian equations of motion, $d^2x/dt^2 = -\nabla \Phi_N$. Varying the full action with respect to Φ yields Poisson's equation, $\nabla^2 \Phi_N = 4\pi G\rho$.

Bekenstein and Milgrom (1984) sought a version of Equation (5.3) that preserves the particle equations of motion, i.e. $a = d^2x/dt^2 = -\nabla \Phi_N$, but which modifies Poisson's equation by changing the gravitational action. In other words: they sought a 'modified gravity,' as opposed to a 'modified inertia,' theory (see Chapter 2). They showed that replacing the last two terms in Equation (5.3) by

$$-\int \rho \Phi d^3x \, dt - \frac{a_0^2}{8\pi G} \int \mathcal{F}\left[\frac{(\nabla \Phi)^2}{a_0^2}\right] d^3x \, dt \tag{5.4}$$

with $\mathcal{F}$ an arbitrary function, yields, after variation with respect to Φ, the modified Poisson equation:

$$\nabla \cdot \left[\mu\left(\frac{|\nabla \Phi|}{a_0}\right) \nabla \Phi\right] = 4\pi G\rho, \tag{5.5}$$

where $\mu(x) \equiv d\mathcal{F}/dz$ and $z = x^2$. The density can be eliminated from this equation using Poisson's equation for the Newtonian potential, $\nabla^2 \Phi_N = 4\pi G\rho$, so that

$$\nabla \cdot \left[\mu\left(\frac{|\nabla \Phi|}{a_0}\right) \nabla \Phi - \nabla \Phi_N\right] = 0. \tag{5.6}$$

The expression in brackets is the curl of some field, h. Writing $\nabla \Phi = -a$ and $\nabla \Phi_N = -g_N$,

$$\mu(a/a_0)a = g_N + \nabla \times h. \tag{5.7}$$

Disregarding the curl term, the function $\mu(x)$ appears in this equation in precisely the same way as the 'transition function' $\mu(x)$ that was defined above in the discussion of rotation-curve analyses, and it is reasonable to identify the two functions. The asymptotic forms of $\mathcal{F}$ are then

$$\mathcal{F}(z) \to z \text{ for } z \gg 1 \text{ and } \mathcal{F}(z) \to \frac{2}{3}z^{3/2} \text{ for } z \ll 1. \tag{5.8}$$

Bekenstein and Milgrom showed in fact that the curl term in Equation (5.7) falls off with distance from a bound system more rapidly than the other two terms, and in certain highly symmetric (e.g. spherical) systems the curl term is identically zero.

Bekenstein and Milgrom (1984) demonstrated that their hypothesis resolves another "theoretical difficulty" (Lakatos's term) with theory variant T_0. As Milgrom pointed out in his first paper from 1983, the internal gravitational acceleration of a system such as a star or gas cloud may greatly exceed a_0; but in applying his postulates to the motion of stars or gas clouds in the weak gravitational field of the Galaxy, he assumed that the center-of-mass acceleration of the system is described by his second postulate, without regard to the (large) internal accelerations experienced by the component particles. Bekenstein and Milgrom showed that a consequence of their auxiliary hypothesis is that

A composite particle (say a star or a cluster of stars) moving in an external field, say of a galaxy, moves like a test particle according to the MOND rules [i.e. the three postulates of T_0], even if within it the relative accelerations are large (Bekenstein and Milgrom, 1984, p. 8).

The fact that any isolated object is predicted to fall in exactly the same way in a constant external gravitational field implies that the weak equivalence principle is satisfied by theory variant T_1.[1]

Bekenstein and Milgrom's theory variant is called AQUAL (for 'AQUAdratic Lagrangian'). Their particular modification of the gravitational action is not the most general that can be imagined; for instance, the action could contain a sum of terms each having a different $\mathcal{F}$. Another variation, discussed by them, would be to write the acceleration of a test particle in terms of *two* potentials as $a = -\nabla(\Phi_1 + \Phi_2)$, where

$$\nabla^2 \Phi_1 = 4\pi G (1 - \lambda) \rho(x),$$
$$\nabla \cdot \left[\overline{\mu} (|\nabla \Phi_2| \times \lambda/a_0) \nabla \Phi_2 \right] = 4\pi G \lambda \rho(x), \tag{5.9}$$

where the function $\overline{\mu}(x)$ has the asymptotic forms

$$\overline{\mu}(x) \approx 1, \, x \gg 1, \quad \overline{\mu}(x) \approx x, \, x \ll 1. \tag{5.10}$$

---

[1] The weak equivalence principle is discussed in more detail in Chapter 6.

Setting $\lambda = 1$ recovers Equation (5.5); setting $\lambda = 0$ yields Newtonian gravity. The corresponding Lagrange density is

$$-\rho\,(\Phi_1 + \Phi_2) - \frac{1}{8\pi G}\left\{(1-\lambda)^{-1}\,(\nabla\Phi_1)^2 + \lambda^{-3}a_0^2\,\mathcal{F}\left[(\nabla\Phi_2)^2\,\lambda^2/a_0^2\right]\right\}. \quad (5.11)$$

Bekenstein and Milgrom (1984) noted that this more general version of the theory implies the same conservation laws as AQUAL, and retains the property that the center-of-mass acceleration of a system is independent of the internal properties of the system.

§

Yet another way to modify the gravitational action, while leaving the conservation laws intact, was proposed by Milgrom (2010) and is called QUMOND, for 'QUasi-linear formulation of MOND.' One starts by writing the gravitational action in the form

$$\frac{1}{8\pi G}\int\left(2\nabla\Phi\cdot\mathbf{g}_N + g_N^2\right)d^3x\,dt, \quad (5.12)$$

which is the same as the Newtonian expression if $\Phi = \Phi_N$. One then replaces the quantity g_N^2 in the integrand by some nonlinear function of g_N, and writes $g_N = -\nabla\Phi_N$, so that expression (5.12) becomes

$$-\frac{1}{8\pi G}\int\left[2\nabla\Phi\cdot\nabla\Phi_N - a_0^2 Q\left(|\nabla\Phi_N|^2/a_0^2\right)\right]d^3x\,dt. \quad (5.13)$$

Variation of the action then gives the two relations

$$\nabla^2\Phi_N = 4\pi G\rho, \quad (5.14a)$$

$$\nabla^2\Phi = \nabla\cdot\left[\nu\left(\frac{|\nabla\Phi_N|}{a_0}\right)\nabla\Phi_N\right], \quad (5.14b)$$

where $\nu(y) = dQ/dz$ and $z = y^2$. The first relation is just Poisson's equation for the Newtonian potential; having computed Φ_N, this result can be substituted into the second relation to yield the potential from which the equations of motion follow as $\mathbf{a} = -\nabla\Phi$. The function $\nu(y)$ appears in this equation in the same way as the function $\nu(y)$ that was defined above in the discussion of rotation-curve analyses, and it is reasonable to identify the two functions. The asymptotic forms of Q are then

$$Q(z) \to z \text{ for } z \gg 1 \text{ and } Q(z) \to \frac{4}{3}z^{3/4} \text{ for } z \ll 1. \quad (5.15)$$

As in the case of AQUAL, the equations of motion are the same as in theory variant T_0 excluding a curl-field correction. However, this correction is different from the one that obtains in the case of AQUAL. Except in highly symmetrical systems, AQUAL and QUMOND will therefore not make identical predictions.

The functions $\mathcal{F}(z)$ or $\mathcal{Q}(z)$ that appear in theory variant T_1 take the form of auxiliary hypotheses. It was noted in Chapter 4 that – under theory variant T_0 – the functions $\mu(x)$ or $\nu(y)$, which roughly speaking are equivalent to $\mathcal{F}(z)$ or $\mathcal{Q}(z)$, can in principle be 'read off' from a plot of the radial acceleration relation or the mass discrepancy–acceleration relation. It would be reasonable, therefore, to argue that these functions, and perhaps also $\mathcal{F}(z)$ and $\mathcal{Q}(z)$, comprise part of the 'background knowledge' incorporated by theory variant T_1.

A nice feature of the QUMOND hypothesis is that it allows one to easily construct the 'effective density' of matter that would generate the observed dynamics in the Newtonian (i.e. the standard model) framework. Setting the left hand side of Equation (5.14a) equal to $4\pi G$ times this effective density, ρ_{eff}, yields

$$\rho_{\text{eff}} = -\frac{1}{4\pi G}\nabla \cdot \left[\nu\left(\frac{|\nabla\Phi_N|}{a_0}\right)g_N\right]. \tag{5.16}$$

Milgrom (2010) defines the 'phantom mass density' as $\rho_{\text{ph}} = \rho_{\text{eff}} - \rho$. As he points out, a standard-model cosmologist, who assumes the correctness of Newton's law of gravity (in the non-relativistic limit), would interpret ρ_{ph} as the 'dark matter density.' Under AQUAL, the phantom density is also computable, as $\rho_{\text{ph}} = (4\pi G)^{-1}\nabla^2\Phi - \rho$.

Like AQUAL, QUMOND predicts that the center-of-mass motion of a bound system in an external field is independent of the internal accelerations (Milgrom, 2010).

In spite of the formal differences between the AQUAL and QUMOND hypotheses, the contents of testable statements derivable from them are often similar (as discussed in more detail below) and it is reasonable to consider either, or both, as comprising theory variant T_1.

§§

In Chapter 4, a distinction was made between 'modified gravity' theories and 'modified inertia' theories. Both AQUAL and QUMOND are modified gravity theories. Alternatively, one can modify the kinetic action (the first term in Equation 5.3), which leaves the relation between gravitational potential and density unchanged but changes the equations of motion. Auxiliary hypotheses of this form were first investigated by Milgrom (1994a, 2011b), who showed that the single-particle equation of motion becomes

$$A\left[\{x(t)\}, a_0\right] = -\nabla\Phi_N, \tag{5.17}$$

where A is a functional of the entire trajectory $\{x(t)\}$. The 'nonlocality' of modified inertia theories can greatly complicate the mathematical work of deriving predictions from the theory. It implies, for instance, that deviations from Newtonian motion occur not only when the *instantaneous* acceleration falls below $\sim a_0$, but much more generally. Partly because of these mathematical complications, much less work has been devoted by Milgromian dynamicists to modified inertia theories than to modified gravity theories, and the former will not be discussed in the remainder of this chapter. Nonlocal theories will however appear again in Chapter 6 when relativistic generalizations are discussed.

§§

Theory variant T_1 allows one to compute the acceleration due to gravity that is generated by an arbitrary (non-symmetric, time-dependent) distribution of matter in the absence of relativistic corrections. Any number of potentially testable predictions follow. For instance: given the detailed distribution of mass in a (disk or non-disk) galaxy or star cluster, one could compute the gravitational acceleration as a function of position and compare it with the observed accelerations of its component stars. At least at the present time, most such tests are not feasible. The three-dimensional distribution of mass in a galaxy is essentially impossible to reconstruct from the observed, two-dimensional surface density on the plane of the sky (disk galaxies are an exception because disks are two-dimensional objects). The vector accelerations of individual stars are also generally inaccessible: stellar orbits in non-disk galaxies need not be even approximately circular, nor is the location of a star within the body of a galaxy generally inferable from its apparent position projected against the image of the galaxy.[2]

In spite of these practical difficulties, a number of testable predictions from theory variant T_1 have been identified. These include:

1. A relation between the central surface densities of normal and 'dark' matter in galaxies;
2. A predicted dependence of the rms, vertical velocity of stars on distance above or below the plane of the Milky Way galaxy;
3. A relation between the mass of a gravitating system and the rms velocity of its components;
4. A relation between the internal velocity dispersions of two, otherwise identical spherical systems that are located in different external fields.

Theory variant T_1 also allows one to predict the behavior of two galaxies, or a group of galaxies, as they interact, by integrating the particle equations of motion forward in time (e.g. Nipoti et al., 2007a). Here there is potential for a number of novel predictions. For instance: Galaxies that collide with impact parameter less

---

[2] As usual, there are exceptions, at least partial, to these statements. For nearby stellar systems, roughly speaking those within the Local Group, proper motion measurements combined with Doppler shifts can yield three-dimensional velocities. For sufficiently large systems, e.g. clusters of galaxies, use of 'standard candles' can sometimes locate a single galaxy along the line of sight with sufficient accuracy to assign it a three-dimensional position within the cluster.

than a few times the galaxies' half-mass radii will experience dynamical friction, a transfer of kinetic energy from bulk to internal motions, and in favorable cases the transfer can be so effective that the galaxies would be expected to merge into a single system. In the absence of dark matter, such mergers are expected to be extremely rare (Toomre, 1977): typical encounters would be too distant, and/or too rapid, to transfer much of the galaxies' bulk motion into internal kinetic energy. In the standard cosmological model, on the other hand, galaxies are postulated to be embedded in spatially extended 'dark matter haloes,' and that theory predicts that mergers should be common: almost every large galaxy would have experienced at least one 'major merger' event since its formation.[3] Under Milgromian dynamics the predicted rate of mergers would be drastically lower.

The latter is a bona fide prediction, and it contributes to the 'theoretical progressivity' of the Milgromian research program. That fact is worth recording. But ideally, one would hope to find some *corroboration* of the low-merger-rate prediction. Unfortunately, in spite of much effort by observational cosmologists, it has proven extremely difficult to place convincing constraints on the rate at which galaxy populations evolve due to mergers. A standard approach (e.g. Patton et al., 2000) is to conduct a census of close galaxy pairs, then to *assume* that the galaxies are embedded in dark matter haloes that extend far beyond the luminous parts of the galaxies. The statistical merger rate is then computed by dividing the number of such systems by the time for the *dark matter haloes* to merge – the latter obtained from computer simulations. As Pavel Kroupa (2015a, p. 176) notes, "the profuse merging behavior of galaxies in the [standard model of cosmology] is entirely the result of the assumed existence of exotic DM particles, which form the dark halo potentials within which the galaxies exist." Indeed there is a view – even among some standard-model cosmologists – that the observations essentially rule out galaxy mergers at rates comparable with standard-model predictions (Matteucci, 2003; Weinzirl et al., 2009; Kormendy et al., 2010). While this is hardly a *majority* view, few cosmologists would be so bold, perhaps, as to maintain that the galaxy merger rate is an observationally well-determined quantity.[4]

The lack of observational corroboration means that the prediction of a low merger rate, which might be stated as:

5. A rate of galaxy mergers low enough that it does not contribute significantly to the evolution of most galaxies

---

[3] A 'major merger' is defined as one in which the mass ratio between the merging galaxies is no more extreme than 1:3. Such mergers are predicted to induce global changes in the morphology of the larger galaxy; for instance, converting a disk galaxy into an elliptical galaxy. Under the standard model, this is the way that elliptical galaxies are assumed to form.

[4] A great deal more can be said on the subject of galaxy mergers and the prospects for distinguishing Milgromian from standard-model predictions. An excellent starting point for the interested reader would be the insightful review articles by Pavel Kroupa (2012; 2015a; 2015b).

Table 5.1 *Corroborated excess content, theory variants $T_0 - T_1$*

Theory variant	Prediction	Status	Novelty			
			P	Z	M	C
T_0	$V(R) \to V_\infty$	confirmed	no	no	–	yes
	$V_\infty = \left(a_0 G M_{\text{gal}}\right)^{1/4}$	confirmed	yes	yes	yes	–
	$a = f\left(g_N/a_0\right) g_N$	confirmed	yes	yes	yes	–
	Renzo's rule	corroborated	yes	yes	yes	–
T_1	$\Sigma_{\text{ph}}(0) \lesssim a_0/\left(2\pi G\right)$	corroborated	yes	yes	yes	–
	Central surface density relation	confirmed	yes	yes	yes	–
	Vertical kinematics in Milky Way	corroborated	yes	yes	yes	–
	$V_{\text{rms}} \approx \left(a_0 G M\right)^{1/4}$	partially corroborated[a]	yes	yes	yes	–
	External field effect	possibly corroborated[a]	yes	yes	yes	–
	Low merger rate	neither confirmed nor refuted	no	–	yes	–

[a] See text

can, at best, contribute to the theoretical (and not the empirical) progressivity of T_1. Following the rules set down in Chapter 2 for evaluating novelty in such a case, I conclude that prediction number five is not novel according to criterion P (since one would have made a similar prediction based on T_0), but that it is novel according to criterion M (since the standard cosmological model makes a very different prediction). These conclusions are summarized in Table 5.1.

We now turn to predictions 1-4, assessing their novelty and degree of corroboration in turn. In assessing Popper's criterion (P) for novelty ("A predicted fact is novel if it was unknown prior to the formulation of the theory") we need to assign a date of 'theory construction.' I will take that date to be 1984, the year that Bekenstein and Milgrom published their 'AQUAL' Lagrangian.

§§

1. *A relation between the central surface densities of normal and 'dark' matter in galaxies*: This prediction has been formulated in two versions: an essentially exact, or functional, relation that is valid only for a class of idealized (disklike) stellar systems; and a more approximate relation that applies to a more general class of galaxies. As we will see, both predictions satisfy at least one of the conditions for novelty, and both have been observationally corroborated.

The more approximate of the two relations specifies a maximum value for the surface density of phantom dark matter associated with galaxies. The prediction was motivated in the following way by Milgrom:

> The excess acceleration that MOND produces over Newtonian dynamics, for a given mass distribution, cannot much exceed a_0 ... This simply follows from the fact that MOND differs from Newtonian dynamics only when the accelerations are around or below a_0. Put in terms of DM [dark matter] this MOND prediction would imply that the acceleration produced by a DM halo alone can never much exceed a_0, according to MOND. There is no known reason for this to hold in the context of the DM paradigm (Milgrom, 2008, p. 6).

There are two reasons for expressing this result in terms of surface density, rather than acceleration. First, when standard-model cosmologists infer the parameters that describe the 'dark matter haloes' of galaxies, the most robust way to express their results is typically in terms of the 'dark matter density' projected onto the plane of the sky – i.e. in terms of its surface density. (This is particularly true in the case of lensing data.) Second, in the idealized case of a thin planar mass distribution, the perpendicular acceleration g_N is independent of position and given by the simple relation $g_N = 2\pi G \Sigma$. Equating g_N with the (approximate) upper limit claimed by Milgrom, or a_0, would then imply an upper limit to the surface density of $\sim a_0/(2\pi G) \equiv \Sigma_0$. How closely this value corresponds to the upper limit in non-disk geometries can only be determined by applying T_1 to different mass distributions; but from dimensional considerations, the result must be some constant times Σ_0.

Figure 5.1 shows the results of such calculations. Plotted there is the surface density (i.e. the projection along the line of sight of the mass per unit volume) of the phantom mass generated by a set of spherical galaxy models: a point mass, and four models having constant density (real, not phantom) within some outer radius. The QUMOND version of T_1 was used to compute the phantom mass densities. All five galaxy models have the same total mass.[5]

For the three more compact models, the Newtonian gravitational acceleration g_N even at the model's outer edge is greater than a_0; these galaxies are in the 'Newtonian regime' and can be thought of as simple models of high surface brightness galaxies. The other two models have low enough densities that $g_N < a_0$ in their outer regions; they are simple models of low surface brightness galaxies. The projected central densities of the phantom dark matter around the high surface brightness galaxy models never greatly exceed a value of $\Sigma_0 \equiv a_0/(2\pi G)$.

Brada and Milgrom (1999) were the first to note the result illustrated in Figure 5.1, i.e. that the central surface density of the phantom dark matter associated

---

[5] Spherically symmetric galaxy models are simple enough that many predictions based on them – including the predictions illustrated in this figure – are identical under T_0 and T_1.

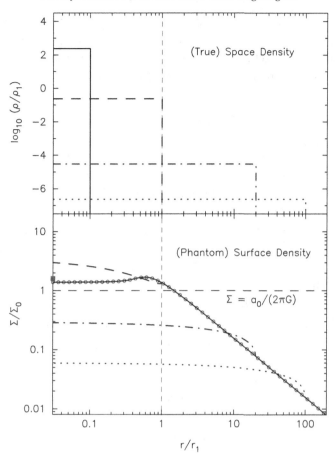

Figure 5.1 The top panel shows four, highly idealized galaxy models, each having constant density, ρ, within some outer radius; all models have the same total mass. The two, more compact models are fully within the Newtonian regime, i.e. $g_N > a_0$. The bottom panel shows the projected (surface) density of the 'phantom dark matter' for each model, as well as for a point-mass model (open circles), computed using QUMOND with the transition function described in the text. The more compact galaxy models have roughly the same value, $a_0/(2\pi G)$, of the central phantom density. A standard-model cosmologist would interpret the phantom density as the density of dark matter. The filled square symbol on the vertical axis shows the determination of the central dark matter density in high surface brightness galaxies by Donato et al. (2009), confirming the Milgromian prediction. The parameters r_1 and ρ_1 are defined as $r_1 = (GM_{\text{gal}}/a_0)^{1/2}$, $\rho_1 = M_{\text{gal}}/r_1^3$.

with high density galaxies exhibits a maximum value given by $\sim\Sigma_0$. Expressed in terms of characteristic physical values,

$$\Sigma_0 \approx 138 \left(\frac{a_0}{1.2 \times 10^{-10}\,\text{cm s}^{-2}} \right) M_\odot\,\text{pc}^{-2}, \qquad (5.18)$$

where $M_\odot$ is the solar mass. The approximate condition for this result to hold is $\overline{\Sigma} > \Sigma_0$, where $\overline{\Sigma}$ is the mean surface density of the (normal) matter in the galaxy. Milgrom (2009b) showed that in galaxies with $\overline{\Sigma} < \Sigma_0$ the relation between central phantom density, $\Sigma_{ph}(0)$, and $\overline{\Sigma}$ is given approximately by

$$\Sigma_{ph}(0) \approx 2.4 \sqrt{\frac{\overline{\Sigma}}{\Sigma_0}} \, \Sigma_0, \qquad (5.19)$$

i.e. the phantom density scales as the square root of the matter density. Thus, the characteristic value of $\Sigma_{ph}(0)$ in high surface brightness galaxies can also be described as a maximum value (Brada and Milgrom, 1999). Figure 5.1 is a concrete illustration of this result.

The predicted value of $\Sigma_{ph}(0)$ depends weakly on the form assumed for the functions $\mathcal{F}$ or $\mathcal{Q}$, i.e. of μ or ν (Milgrom and Sanders, 2008). The results shown in Figure 5.1 were computed using QUMOND and a transition function from the family of Equation (4.12), with $n = 2$.

The prediction of a characteristic, or maximum, projected central density of 'dark matter' around galaxies has been well corroborated observationally (Milgrom and Sanders, 2005; Donato et al., 2009). Both of the cited studies were carried out in a Newtonian framework, in which a parameterized model of a 'dark matter halo' was fit to data for a sample of galaxies. Milgrom and Sanders assumed for the three-dimensional density of the 'dark matter'

$$\rho_{MS}(r) = \frac{\rho_0 r_0^2}{r^2 + r_0^2}, \qquad (5.20)$$

a functional form which closely mimics, after projection, the phantom density profiles of Figure 5.1; the projected central density corresponding to ρ_{MS} is $\Sigma(0) = \pi \rho_0 r_0$, and $\Sigma \propto r^{-1}$ for $r \gg r_0$. Donato et al. assumed a different functional form,

$$\rho_D(r) = \frac{\rho_0 r_0^3}{(r + r_0)(r^2 + r_0^2)}, \qquad (5.21)$$

which projects to a central density $\Sigma(0) = \pi \rho_0 r_0 / 2$. Equation (5.21) implies $\Sigma \propto r^{-2}$ at large radii, different from the behavior predicted by the modified dynamics. Milgrom and Sanders analyzed rotation-curve data for a sample of 17 disk galaxies in the Ursa Major group; they plotted the inferred ρ_0 versus r_0 and found that these parameters lie near a line of constant $\Sigma(0) = 10^2 \, M_\odot \, pc^{-2}$. Donato et al. (2009) fit their density law to data from a larger sample of galaxies including both late type (i.e. disk) and early type (elliptical) and spanning a range of almost 10^6 in luminosity. Data of various sorts relating to the mass distribution were used, including rotation curves and gravitational lensing data. Donato et al. found a remarkably constant value of $\Sigma(0) \approx 220 \, M_\odot \, pc^{-2}$ with little dependence on galaxy luminosity or type. This value is consistent with the predicted central density for high surface brightest galaxies (Figure 5.1), confirming the Milgromian prediction. The Donato et al. sample contained only a few galaxies of low enough surface brightness to test the $\Sigma_{ph}(0) \propto \overline{\Sigma}^{1/2}$ prediction; Milgrom (2009b) subsequently argued that the data for these (mostly dwarf) galaxies do not allow a robust determination of their mass distribution.

The second, essentially exact version of the prediction applies only to pure disk galaxies. Milgrom (2016c) showed that for precisely axisymmetric and thin disks, T_1 implies a functional relation between $\Sigma_{ph}(0)$ and $\Sigma(0)$. The latter quantity is the surface mass density at the disk center, and the former quantity is defined in terms of a projection parallel to the disk symmetry axis. Using version QUMOND of T_1, Milgrom derived this relation as

$$\Sigma_{ph}(0) + \Sigma(0) = \Sigma_0 S\left[\Sigma(0)/\Sigma_0\right], \quad S(y) \equiv \int_0^y \nu(y)dy. \quad (5.22)$$

Adopting for $\nu(y)$ the form

$$\nu(y) = \frac{1 + \left(1 + 4y^{-1}\right)^{1/2}}{2}, \quad (5.23)$$

which is often used in disk galaxy rotation-curve studies, one finds for $S(y)$

$$S(y) = \frac{y}{2} + \sqrt{y\left(1 + \frac{y}{4}\right)} + 2\sinh^{-1}\sqrt{\frac{y}{4}}. \quad (5.24)$$

Milgrom showed that precisely the same result holds under theory sub-variant AQUAL, if the function $\mu(x)$ that appears in that theory is related to $\nu(y)$ as in Equation (4.9), i.e. $\mu(x) = x/(1+x)$. Milgrom (2016c) called the prediction (5.22) of T_1 the 'central surface density relation,' or CSDR.

In the high surface brightness limit, i.e. $\Sigma(0) \gg \Sigma_0$, the function $S(y) \to 1 + y + \log y + \cdots$, and the relation (5.22) becomes

$$\Sigma_{ph}(0) \approx \Sigma_0 \left[1 + \log\left(\frac{\Sigma(0)}{\Sigma_0}\right)\right], \quad \Sigma(0) \gg \Sigma_0, \quad (5.25)$$

showing again the existence of an approximate upper limit to $\Sigma_{ph}(0)$ of $\sim \Sigma_0$. In the low surface brightness limit, $S(y) \to 2\sqrt{y}$, and the prediction becomes

$$\Sigma_{ph}(0) \approx \sqrt{4\Sigma_0\Sigma(0)} - \frac{1}{2}\Sigma(0), \quad \Sigma(0) \ll \Sigma_0. \quad (5.26)$$

The existence of a tight correlation between the true and 'dark' central surface densities of disk galaxies was first established by Lelli et al. (2016b), shortly before the prediction of the CSDR by Milgrom.[6] Figure 5.2 plots the data tabulated by

---

[6] Here we have an example of the situation described in Chapter 2, where the novel prediction was made a long time after the theory was formulated – in this case, about 32 years (in the case of AQUAL) or six years (in the case of QUMOND) later. Thus, the Lelli et al. confirmation can not plausibly be described as having been made during the course of an attempted refutation.

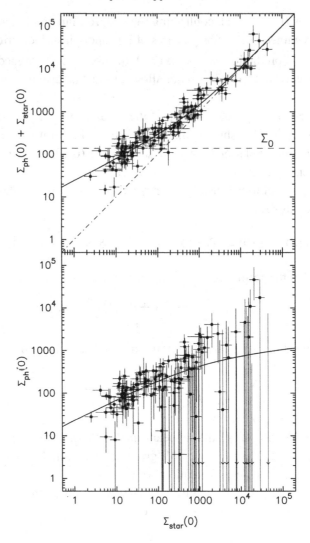

Figure 5.2 The central surface density relation (CSDR) for disk galaxies, in two representations. The top panel plots the 'dynamical' central surface density, i.e. an estimate of the the actual plus phantom densities, versus the central surface density in stars alone. The bottom panel plots just the excess of the 'dynamical' surface density over the density in stars. Data points are from the observational study of Lelli et al. (2016b), with a correction factor due to Milgrom (2016c) that is discussed in Chapter 8. The horizontal dashed line in the top panel is $\Sigma_0 = a_0/(2\pi G)$, assuming $a_0 = 1.27 \times 10^{-8}$ cm s^{-2}. Solid curves in both panels were computed using Equations (5.22–5.24) and represent the prediction of theory variant T_1.

Lelli et al. for a sample of 135 disk galaxies, spanning Hubble types from S0 to Irr and total stellar masses from $\sim 10^7 M_\odot$ to $\sim 10^{11} M_\odot$. Those authors did *not* use Milgromian dynamics to compute the phantom mass distribution associated with their observed galaxies. Instead, they relied on an approximate (Newtonian) formula due to Alar Toomre (1963) that relates the central surface density to the full rotation curve, $V(R)$, for precisely thin disks:

$$\Sigma_{\text{dyn}}(0) = \frac{1}{2\pi G} \int_0^\infty \frac{V^2(R)}{R^2} dR. \tag{5.27}$$

Equation (5.27) takes no account of 'dark matter,' phantom or otherwise. Lelli et al. nevertheless argued that Σ_{dyn} as given by Equation (5.27) is a reasonable estimate of the surface density that is required, dynamically, to yield the observed $V(R)$ under Newtonian gravity, hence that it is a reasonable proxy for $\Sigma_{\text{ph}}(0) + \Sigma(0)$.

The lower panel of Figure 5.2 plots the quantity $\Sigma_{\text{dyn}}(0) - \Sigma(0)$ along the vertical axis; this quantity can be taken as a proxy for $\Sigma_{\text{ph}}(0)$. As in Figure 5.1, one sees evidence here for a characteristic, or maximum, central surface density of magnitude $\sim\Sigma_0$.

Based on Figures 5.1 and 5.2, it is reasonable to conclude that these two predictions of Milgrom have been corroborated, and perhaps (modulo the limitations in the modeling of the data) confirmed.

Both the prediction of a maximum surface density of 'dark matter haloes,' and the prediction of the central surface density relation, are clearly novel according to criteria P and Z: neither fact was known prior to the construction of T_1 (which, as noted above, is assumed to have occurred in 1984) and the theory could not have been designed to accommodate an unknown fact. Criterion C is inapplicable.

Criterion M requires that the predicted fact be 'prohibited or improbable' as seen from the vantage point of the standard cosmological model. Consider first the case of the predicted maximum value of the central 'dark matter' surface density. Quoting from Donato et al. (2009) (whose standard-model commitments are obvious[7]):

Considering that DM [dark matter] haloes are (almost) spherical systems, it is surprising that their central surface density plays a role in galaxy structure. One could wonder whether the physics we witness in μ_{0D} [i.e. $\Sigma_{\text{ph}}(0)$] is instead stored separately in the quantities r_0 and ρ_0. This reasonable interpretation has, however, a problem: r_0 and ρ_0 do correlate with the luminous counterparts (the disc length-scale and stellar central surface density) while μ_{0D} does not... it is difficult to envisage how such a relation can be achieved across galaxies which range from DM-dominated to baryon-dominated in the inner regions. In addition, these galaxies have experienced significantly different evolutionary histories (e.g. number of mergers, significance of baryon cooling, stellar feedback, etc.) (Donato et al., 2009, p. 1174).

Is the existence of the CSDR surprising from the standpoint of the standard cosmological model? Milgrom notes:

---

[7] As is often the case in the standard-model literature, Donato et al. (2009) do not acknowledge that they are testing a prediction of the Milgromian research program, nor do they cite the Milgrom and Sanders paper from 2005 which first tested the prediction using rotation-curve data.

Even with schemes, such as 'feedback', 'abundance matching', and the like – put in by hand to save this paradigm from various embarrassments – one would expect large scatter in any relation between the 'dynamical' and baryonic properties, which, to boot, one would expect to depend on galaxy properties.

As of this writing, standard-model cosmologists appear not to have taken notice of the CSDR. I conclude that this prediction of T_1 is novel according to Musgrave's criterion M.

These results are summarized in lines five and six of Table 5.1.

§§

2. *A predicted dependence of the rms, vertical velocity of stars on distance above or below the plane of the Milky Way galaxy*: A galaxy like the Milky Way contains, in first approximation, a compact, central spheroidal 'bulge' and a thin axisymmetric disk; the Sun is located near the disk midplane, at a distance $R_0 \approx 8$ kpc from the Galactic center and well outside of the bulge. The distribution of mass (stars, gas) in the Milky Way is reasonably well known, especially in the vicinity of the Sun. Theory variant T_1 contains an algorithm for computing the gravitational acceleration of a galaxy having any specified distribution of mass. T_1 therefore makes a definite prediction about the gravitational acceleration near the Sun, and this prediction is testable using the observed motions of nearby stars (Milgrom, 1983b).

The 'phantom mass density' produced by a disk was first described by Milgrom and collaborators (Brada and Milgrom, 1995; Milgrom, 2001b). Assuming that the disk is thin, Gauss's theorem gives for the *Newtonian* acceleration just above or below the disk plane $g_N = 2\pi G\Sigma$, with Σ the disk surface mass density; this acceleration is normal to the disk plane. In the modified dynamics, g_N is set equal to $\mu(g_z/a_0)g_z$ with g_z the actual vertical acceleration.[8] The latter would be attributed by a standard-model cosmologist to a disk of surface density $(2\pi G)^{-1}g_z = \Sigma/\mu$, with $\mu = \mu(g_z/a_0)$ evaluated just outside the disk. (In this approximation, g_z is independent of z.) It follows that the phantom, or 'dark matter,' surface density of the disk is related to its true density by

$$\Sigma_{ph}(R) \approx \left[\frac{1}{\mu(g_z/a_0)} - 1\right]\Sigma(R) \equiv \eta(R)\Sigma(R). \tag{5.28}$$

A more careful calculation (Milgrom, 2001b) takes into account the observed fact that the disks of galaxies like the Milky Way have finite vertical extent; for instance, in the case of the Milky Way, the 'vertical scale height' near the solar circle is

---

[8] As in the case of a precisely spherical galaxy, the gravitational acceleration produced by infinite and by finite but infinitely thin disks are identical under T_1 or T_0.

about 300 pc.[9] In a finite-thickness disk, the vertical acceleration in the modified dynamics increases with z at a given R, from zero at the midplane ($z = 0$), to a nearly constant value at large z. At values of R such that the radial acceleration exceeds the vertical acceleration (which is the case near the Sun), the phantom density, ρ_{ph}, is approximately proportional to the true density, $\rho_{ph}(z) \propto \rho(z)$; nearer to the center of a galaxy, the phantom density is more concentrated to the midplane than the true density.

Calculation of the gravitational acceleration predicted by T_1 using a full mass-model of the Milky Way (Nipoti et al., 2007b; Bienaymé et al., 2009) yields a value $\eta(R_0) \approx 0.6$ in Equation (5.28), if z is set equal to 1.1 kpc and Σ is defined as the surface density obtained by integrating the (true space) density over the vertical integral -1.1 kpc $< z < 1.1$ kpc. One reason for expressing the prediction in this way – rather than as a curve of vertical acceleration versus z – is that the data sets used in carrying out the test are generally so small that the full z-dependence of the potential is difficult to infer from them.[10] If the Milgromian prediction is correct, then, from the point of view of a Newtonian (i.e. standard-model) dynamicist, the Galactic disk would appear to contain 'dark matter,' with a density that is comparable to the density of the normal matter.

Assuming a steady state, the gravitational acceleration computed via T_1 from the observed distribution of stars and gas will be reflected in the velocity distribution of stars moving in response to that acceleration. The starting point is the so-called 'Jeans equation,'[11] the stellar-dynamical analog of the hydrostatic equilibrium equation for a fluid. Define a reference (x, y) plane coincident with the galactic disk, and a reference (z) axis that passes normally through the disk center. Cylindrical-polar coordinates are then $P = (R, z, \phi)$ where R is measured from the z-axis and ϕ is an azimuthal angle. In these coordinates, the equation of equilibrium with respect to z-motions is

$$\frac{\partial}{\partial z}\left(n\sigma_z^2\right) + \frac{1}{R}\frac{\partial\left(nR\sigma_{Rz}\right)}{\partial R} = -n\frac{\partial\Phi}{\partial z}. \tag{5.29}$$

Here, $n(R, z)$ is the number density of some tracer population (e.g. stars), $\sigma_z(R, z)$ is the rms velocity of the stars in the z-direction, and $\sigma_{Rz}(R, z) = \langle v_R v_z \rangle$. The first and last terms in Equation (5.29) resemble two terms in the equation of hydrostatic equilibrium for a gas, with $n\sigma_z^2$ playing the role of the gaseous pressure. The middle, or 'tilt,' term is due to the collisioness nature of the stellar 'fluid.' By considerations of symmetry, this term must be negligible near the disk plane, and direct evaluation

---

[9] In fact, number counts of stars near the Sun are typically modeled using two 'populations' with different vertical scale heights: a 'thin disk' population with scale height ∼200–300 pc, and a 'thick disk' population with scale height ∼800–1100 pc (Siegel et al., 2002; Chang et al., 2011; Polido et al., 2013).

[10] The choice of 1.1 kpc as a fiducial vertical distance has been standard since the work of Kuijken and Gilmore (1989a, b, c; 1991), who based their analysis on a sample of stars that was complete up to 1.1 kpc above the disk plane.

[11] Actually first derived by J. C. Maxwell.

of the term using observed stellar motions near the Sun (e.g. Siebert et al., 2008; Moni Bidin et al., 2012) reveals that its magnitude is less than 10% that of the other terms for distances up to 1 kpc from the disk plane.[12] If this term is ignored, the resulting equation has only one independent variable, z, and functions such as $n(R, z)$ can be written $n(z) \equiv n(z; R_0)$. The test then consists of (*i*) measuring the positions (i.e. heliocentric distances) and radial velocities of a sample of stars in a direction perpendicular to the disk plane; (*ii*) estimating $n(z)$ and $\sigma_z(z)$ for this sample via regression; (*iii*) comparing the gradient of $n\sigma_z^2$ with the right hand side of Equation (5.29), adopting for $\partial \Phi / \partial z$ the gravitational acceleration as computed from the observed Milky Way mass distribution via AQUAL or QUMOND.

The majority of such studies (as reviewed by Einasto, 2005 and Read, 2014) have been carried out by standard-model cosmologists assuming Newtonian gravity. Results from such studies are typically presented in a different way: as a comparison between $\Sigma (R_0)$ and $\Sigma_{\mathrm{dyn}} (R_0)$: the former is the actual disk surface density near the solar circle, based on a census (and excluding 'dark matter'); the latter is the surface density that would generate the measured velocity gradients in Equation (5.29) assuming Newtonian dynamics. (Some studies, particularly the earliest ones, instead compare ρ and ρ_{dyn}, defined as the midplane space densities.) These quantities (especially the space densities) have considerable uncertainties, but all studies since the early 1980s (and some before that date) have concluded that $\Sigma_{\mathrm{dyn}} > \Sigma$ and/or $\rho_{\mathrm{dyn}} > \rho$, at least qualitatively in agreement with Milgrom's prediction.

Nevertheless, these same studies often include a statement that the kinematical data are consistent with 'no dark matter in the disk.'[13] That would appear, at first blush, to *contradict* the Milgromian prediction of phantom dark matter.

In fact there is no contradiction. Since the early 1980s, standard-model cosmologists have *taken for granted* that the Milky Way is embedded in an approximately spherical dark matter 'halo,' having whatever radial mass distribution is needed to explain the anomalously (from a standard-model perspective) rapid Galactic rotation.[14] Such a component would necessarily contribute also to the vertical force above and below the disk at any R, and this assumption is built into the modeling. When a standard-model cosmologist says that the kinematical data are consistent with 'no dark matter in the disk,' what she means is that no *additional* dark matter is needed; i.e. that the dark matter *assumed to be present* in the form of a halo is

---

[12] The magnitude of the 'tilt' term has itself been treated as a testable prediction of the modified dynamics. However, relating the expected magnitude of this term to the observed mass distribution requires additional assumptions about the global dynamics of the Galaxy, assumptions that are not required in carrying out the test described in this section; see Cuddeford and Amendt (1991), Bienaymé (2009).

[13] Kuijken and Gilmore (1991, p. L11): "There remains no evidence for any significant unidentified mass in the Galactic disk"; Crézé et al. (1998, p. 936): "There is no room left for any disk shaped component of dark matter"; Holmberg and Flynn (2000, p. 209): "We conclude that there is no compelling evidence for significant amounts of dark matter in the disc."

[14] The *assumption* by these authors that dark matter is present reflects the rapid evolution of the dark matter hypothesis – in the decade after 1980 – into a component of the standard model's unchallengable hard core.

sufficient (or can be made sufficient) to increase the vertical force near the Sun to the observed value.

Now, in recent studies of this type (Smith et al., 2012; Bienaymé et al., 2014; Piffl et al., 2014), the local density assigned to the dark matter halo is about $0.015 \, M_\odot$ pc^{-3}. The density of such a halo varies weakly (by assumption) with z at fixed R; the implied surface density of the dark matter halo, integrated from $-|z|$ to $+|z|$ at $R = R_0$, is therefore

$$\Sigma_{\mathrm{halo}}(<z) \approx 2 \int_0^z \rho_{\mathrm{DM}}(z')dz'$$

$$\approx 2z \times \rho_{\mathrm{DM}}$$

$$\approx 30 \left(\frac{z}{\mathrm{kpc}} \right) \frac{M_\odot}{\mathrm{pc}^2} .$$

For comparison, Equation (5.28) (with $\eta = 0.6, z = 1.1$ kpc and $\Sigma = 45 \, M_\odot \, \mathrm{pc}^{-2}$) predicts for the Milgromian phantom surface density $\sim 29 \, M_\odot \, \mathrm{pc}^{-2}$ – essentially the same value. This quantitive agreement is (at least in part) fortuitous. But it means that a finding, by a standard-model researcher, that no dark matter is required to reproduce the vertical motions in the disk is essentially a confirmation of the Milgromian prediction.[15]

§

Attempts to determine ρ_{dyn} were published as early as 1922 (Kapteyn, 1922; Jeans, 1922), long before either the standard cosmological model, or the Milgromian research program, had come into existence. From then, until the 1990s, there was considerable study-to-study variation in determinations of ρ_{dyn} or Σ_{dyn}. The first, widely credited claims of a dynamically inferred density in excess of the expected value were by E. R. Hill (1960) and Jan Oort (1960). Oort concluded that his value for ρ_{dyn} in the solar neighborhood

is somewhat higher than what might have been expected from an extrapolation of the known stars... The recent data concerning the large optical thickness of some interstellar clouds... may bring about an increase in the estimates of the average density of interstellar material by a factor of perhaps 1.5... About one third of the total density would thus remain unexplained (Oort, 1960, p. 50).

Of course, Oort's work preceded the dark matter hypothesis by two decades.

Starting in the early 1980s, studies of the Galaxy's vertical force law began to incorporate the assumption that the Galaxy is embedded in a 'dark-matter halo.'

---

[15] Only "essentially," because (*i*) all of these quantities have uncertainties, and (*ii*) the standard-model cosmologist may be (and generally is) assuming a vertical dependence for the dark matter density that differs from that of the phantom density predicted by T_1. In addition, the contribution of the halo dark matter to the vertical force is reduced by an amount that depends on the assumed Galaxy rotation curve; however for a flat or nearly-flat rotation curve this correction is negligible (e.g. Garbari et al, 2012, Eq. 12).

Table 5.2 *Estimates of the surface density of the Milky Way disk*

Reference	Σ_{dyn} ($|z| < 1.1$ kpc)[a]	Σ ($|z| < 1.1$ kpc)	Σ_{dyn}/Σ
Kuijken and Gilmore (1991)	71 ± 6	48 ± 9	1.48 ± 0.30
Siebert et al. (2003)	85 ± 23	–	$[1.88]^b$
Holmberg and Flynn (2004)	74 ± 6	53	1.39 ± 0.11
Bienaymé et al. (2006)	$57 - 79$	–	$[1.26 - 1.76]$
Flynn et al. (2006)	–	48.7^c	–
Smith et al. (2012)	66	–	[1.47]
Bovy and Trelayne $(2012)^d$	64	–	[1.42]
Zhang et al. $(2013)^e$	67 ± 6	55 ± 6	1.22 ± 0.17
Bovy and Rix (2013)	68 ± 4	38 ± 4	1.79 ± 0.22
Bienaymé et al. (2014)	68.5 ± 1	–	[1.52]
Piffl et al. (2014)	69 ± 15^f	–	[1.53]

[a] Units: $M_\odot$ pc^{-2}
[b] Quantities in brackets assume $\Sigma = 45\,M_\odot$ pc^{-2}
[c] Upper limit on $|z|$ not specified
[d] Corrects Moni Bidin et al. (2012)
[e] $|z| < 1.0$ kpc
[f] $|z| < 0.9$ kpc

Bahcall (1984b,a) concluded that the kinematics of stars near the Sun imply a significantly larger density than can be accounted for in the stars and gas, and that a discrepancy may exist even if the 'halo dark matter' is accounted for:

The unseen material that is inferred from galaxy rotation curves at large galactocentric distances [i.e. the dark matter halo] ... may not be the same as the unobserved disk matter discussed in the present paper (Bahcall, 1984a, p. 287).

Since about 2000, estimates of Σ_{dyn} ($|z| < 1.1$ kpc) based on Newtonian modeling of the kinematical data have converged on a value of $\sim 70\,M_\odot$ pc^{-2}, with an error that is typically estimated to be about $\pm(5\text{–}10)M_\odot$ pc^{-2} (Table 5.2). Estimates of the surface density in 'known' components, Σ ($|z| < 1.1$ kpc), are typically $\sim 45\,M_\odot$ pc^{-2}; uncertainties in this quantity are difficult to judge since Σ contains significant contributions from components that are not directly observed (stellar remnants, brown dwarves, cold gas). Taking both numbers at face value, the effective density is $\sim 55\%$ higher than the actual density. This result is quite consistent with the Milgromian prediction of $\Sigma_{dyn}/\Sigma \approx 1.6$ (Nipoti et al., 2007b; Bienaymé et al., 2009). On this basis, it is reasonable to state that the Milgromian prediction has been corroborated, and possibly (modulo the observational uncertainties) confirmed.

§

I now consider the novelty of this corroborated prediction. Consider first criterion *P* ("A predicted fact is novel if it was unknown prior to the formulation of the theory"). As noted above, the date associated with 'formulation of the theory' is taken here to be 1984, the year that Bekenstein and Milgrom published their AQUAL Lagrangian. Assuming that the predicted quantity is Σ_{dyn}: On what date was consensus reached about the observationally determined value of this quantity?

As Read (2014) points out, data-based estimates of ρ_{dyn} or Σ_{dyn} published prior to the 1990s showed considerable variation, and it was not until roughly 2000 that a consensus began to emerge; furthermore – according to that consensus – the early determinations by Oort and Bahcall were substantially in error (in the sense of claiming too large a discrepancy). I will therefore adopt 2000 as the date on which intersubjective agreement was reached about the value of Σ_{dyn}. On this basis, the Milgromian prediction of the value of Σ_{dyn} clearly meets Popper's criterion (*P*) for novelty. Zahar's criterion (*Z*) is also satisfied.[16]

I proposed in Chapter 2 that a Milgromian prediction is not novel according to Musgrave's criterion (*M*) if the then-current version of the standard model predicts the same fact. Here we must delve a bit more deeply into the history; in part because the 'standard cosmological model' was in a rapid state of evolution in the period just before and just after Milgrom's 1983 publications.

Prior to about 1980, the universal expectation was consistency of the kinematical data with the Newtonian prediction, the latter based on the observed (or inferred) mass in stars and gas. For instance, Oort (1960) wrote:

In the present note an attempt is made to combine Hill's data with the condition that the gravitational force must be due to stars and interstellar matter, the approximate distribution of which may, at least partly, be supposed to be known. In other words, a solution for Kz [i.e. g_z] is sought which satisfies Poisson's law, and at the same time give a reasonable agreement with the observed data for the K giants (Oort, 1960, p. 45).

The theoretical studies of Woolley and Stewart (1967) and Vandervoort (1970) likewise assumed that the vertical motions should be explainable in terms of the gravitational acceleration produced by the observed mass components.

As discussed above, by about 2000, a consensus had been reached that these Newtonian predictions were inconsistent with the measured ρ_{dyn} or Σ_{dyn}. By this time, standard-model cosmologists had been incorporating a 'dark matter halo' in their modeling of the vertical force for about two decades. In principle, these studies *could* have been designed to test the predictions of the 'dark halo' models for the vertical force. But, remarkably, this appears never to have happened: apparently

---

[16] Here we have a second example of the situation mentioned in Chapter 2, where a prediction is first made long after the theory on which it is based was published. The first Milgromian predictions of the vertical force law were by Nipoti et al. (2007b) and Bienaymé et al. (2009), some 25 years after the publication of T_1. This delay does not affect the novelty of the prediction according to the version of Popper's criterion that is adopted here, but it would be relevant if one insisted (as Popper sometimes did) that a novel fact be discovered during an attempted falsification.

because the dark matter hypothesis has never been considered well enough defined to generate testable predictions about the vertical force law in the Milky Way.

Bahcall's two papers from 1984 – the first to appear after 'dark matter' was postulated – are interesting transitional cases. Bahcall was clearly aware at this time of the dark matter hypothesis (i.e. DM-1); already in 1980, he and Raymond Soneira (Bahcall and Soneira, 1980) had published a mass model for the Milky Way that included a spherical 'dark matter halo,' having a mass distribution designed to explain the Galaxy's anomalous rotation. The Bahcall and Soneira model makes a definite prediction for the space density of dark matter at the Sun's location ($\rho_{\text{dyn}} \approx 0.011 \, M_{\odot} \, \text{pc}^{-3}$) and also for the z-dependence of the dark matter density at $R = R_0$. In principle, Bahcall could have chosen to test this prediction via his analysis of the vertical motions. But *he did not do so*, either in 1984, or in his 1992 paper with Flynn and Gould. As Bahcall explained in a review article from 1987:

Since we haven't yet observed the unseen material, we don't know how it is distributed. Therefore we have to try different models for the unseen material to see how the results depend upon our assumptions (Bahcall, 1987, p. 19).

And indeed, Bahcall (1984b,a) treats the dark matter density and its dependence on z as unknown quantities. One set of models assigns to $\rho_{\text{DM}}(z)$ a constant (but undetermined) value; another assumes that the "unseen material" has a density that is a fixed multiple P of the density of observed matter at all z. To the extent that Bahcall can be said to have been testing a prediction, that test clearly failed, since his models require a dark component in addition to what he assigns to the halo; thus his conclusion that "*The unseen material must be mostly in a disk*" (Bahcall, 1984a, p. 942).

Interestingly, although Bahcall never claims to be testing a *standard model* prediction, he *does* note that his results are explainable via Milgrom's theory:

The modification that Milgrom has proposed implies that there should be missing matter in the z-direction, as well as in the plane of the Galaxy... In fact, the modified dynamics model predicts that P ∼ 1, in agreement with the results summarized [here] (Bahcall, 1987, p. 25).

Holmberg and Flynn (2000; 2004) followed essentially the same procedure as Bahcall (1984a,b).

All other studies of the vertical force problem by standard-model cosmologists (including all of those listed in Table 5.2) have likewise shied away from casting their analyses as *tests* of a standard-model prediction – apparently because any such prediction has always been considered too uncertain to be worth testing. For instance, Smith et al. (2012) write (italics added):

If we assume our background mass represents the dark halo, it corresponds to a local dark matter density of 0.57 GeV cm^{-3}, which is noticeably larger than the canonical value of 0.30 GeV cm^{-3} typically assumed (e.g., Jungman et al. 1996). As pointed out by various authors (e.g., Gates et al. 1995; Weber & de Boer 2010; Garbari et al. 2011), *the local dark matter density is uncertain by a factor of at least two.* Our analysis adds still more weight to the argument that the local halo density may be substantially underestimated by the canonical value of 0.30 GeV cm^{-3} (Smith et al., 2012, p. 11).[17]

Most of these studies acknowledge the need for enforcing consistency with the Galaxy's rotation curve, and some studies incorporate that information into the analysis as a constraint or a 'prior' (in the Bayesian sense) (Bienaymé et al., 1987; Kuijken and Gilmore, 1991; Holmberg and Flynn, 2000, 2004). But the *local* value of ρ_{DM} – the predicted quantity that could be falsified via an analysis of the the vertical motions – is typically decoupled from the rotation-curve constraint in these studies by allowing the dark matter halo to be nonspherical (e.g. Garbari et al., 2012; Bienaymé et al., 2014). By treating the halo axis ratio as an extra, freely adjustable parameter, the local dark matter density can be varied by large factors without changing the predicted rotation curve, thus effectively nullifying any predictive power of the rotation data.[18]

We are now in position to evaluate the novelty of the Milgromiam prediction according to the criterion of Musgrave (a corroborated fact provides evidential support for a theory if the best rival theory fails to predict the fact). As I documented, standard-model cosmologists have never claimed to be able to make useful, quantitative predictions of the values of ρ_{dyn} or Σ_{dyn}, nor are their studies of the vertical force law ever cast in the form of predictions or tests. I conclude that the Milgromian prediction satisfies Musgrave's criterion for novelty. These conclusions are summarized in the seventh line of Table 5.1.

$$\S\S$$

3. *A relation between the mass of a gravitating system and the rms velocity of its components*: As in the case of the central surface density relation discussed above, this prediction exists in two versions: an essentially exact relation that applies only to a class of idealized systems, and a more approximate relation that applies more generally. As we will see, corroboration – of a partial nature – exists only for the second, weaker version of the prediction.

Both variants of T_1 (AQUAL, QUMOND) imply a theorem similar to the virial theorem of classical mechanics. In the case of AQUAL, Milgrom (1994b, 1997) showed that the virial relation can be written $2K + W_A = 0$, where $K = M\langle V^2\rangle/2$ is the total kinetic energy of the system, M is its total mass, $\langle V^2\rangle$ is the mean-square velocity of its components (e.g. stars or gas molecules), and W_A is a generalized potential energy given by

---

[17] 0.30 GeV cm$^{-3} \approx 0.008\, M_\odot$ pc^{-3}.

[18] Read (2014) notes that these studies sometimes require the putative dark matter halo to have a 'negative elongation,' i.e. to be elongated perpendicular to the Milky Way disk. Such a geometry is quite unnatural according to the standard-model paradigm of galaxy formation.

$$W_A = -\frac{2}{3}\sqrt{GM^3a_0} \tag{5.30}$$

$$-\frac{1}{4\pi G}\int\left[\frac{3}{2}a_0^2 F(|\nabla\Phi|^2/a_0^2) - \mu(|\nabla\Phi|/a_0)|\nabla\Phi|^2\right]d^3x.$$

In the case of QUMOND, the virial relation can also be written as $2K + W_Q = 0$, now with

$$W_Q = -\frac{2}{3}\sqrt{GM^3a_0} \tag{5.31}$$

$$-\frac{1}{4\pi G}\int\left[-\frac{3}{2}a_0^2 Q(|\nabla\Phi_N|^2/a_0^2) + 2\nu(|\nabla\Phi_N|/a_0)|\nabla\Phi_N|^2\right]d^3x$$

(Milgrom, 2010). These two versions of the virial theorem assume – as does the classical virial theorem – that the system under consideration is in a state of dynamical equilibrium: a steady state.

In the high acceleration, or classical, limit, both quantities W become the Newtonian total gravitational potential energy, and the two virial relations reduce to the classical virial theorem.

In the opposite limit of low accelerations ($g_N \ll a_0$), both expressions for W take the form

$$W_Q = W_A = -\frac{2}{3}\sqrt{GM^3a_0}\,. \tag{5.32}$$

This expression for W is appropriate when applying the virial theorem to a system in which the internal accelerations satisfy $g_N \ll a_0$ *everywhere*. For such a system, the Milgromian virial theorem, using either AQUAL or QUMOND, then implies

$$\langle V^2\rangle^{1/2} \equiv V_{\text{rms}} = \sqrt{\frac{2}{3}}(GMa_0)^{1/4}, \tag{5.33}$$

which is a unique relation between total mass M and V_{rms}. This essentially exact version of the prediction applies only to an idealized class of systems: those in which the internal dynamics are strongly in the Milgromian regime everywhere.[19]

Now, it turns out that – in the modified dynamics – a relation very similar to Equation (5.33) can be shown to apply to a *second* class of ideal systems, including some that are *not* fully in the low-acceleration regime. Milgrom (1984) considered the properties of so-called 'isothermal spheres' in the modified dynamics; these are systems in which the internal motions are everywhere isotropic (with respect to direction) and the rms velocity is independent of distance from the center, conditions that are believed to be reasonably well satisfied by at least some galaxies and star clusters. Under Newtonian dynamics, isothermal spheres

---

[19] The prediction (5.33) is independent of the shape and internal velocity distribution of the system. However, if the system is nonspherical in shape, or if its velocity distribution is anisotropic, extracting the quantity V_{rms} from an image projected onto the plane of the sky will generally not be possible without assumptions about the system geometry or internal kinematics.

constitute a two-parameter family; the parameters can be taken as σ, the rms velocity in any direction, and ρ_0, the central mass density.[20] These parameters can take any value, however the total mass is always divergent, which limits the usefulness of the models for describing real systems. In the modified dynamics (with specified a_0), there is again a two-parameter family of solutions (the 'Milgrom spheres'), but their total mass is now finite – certainly a desirable property for models of real galaxies.

Regardless of the choice of the two parameters, Milgrom spheres always extend to radii where the gravitational acceleration falls below a_0, hence a large part (and sometimes all) of any such model is in the low-acceleration regime. Partly as a consequence, certain global (i.e. integrated) properties of the solutions are bounded or highly restricted. Milgrom showed that the mean projected density – averaged within a circle containing one-half of the total mass – is always less than $\sim \Sigma_0 \equiv a_0/(2\pi G)$; the limiting value occurs when the central space density is high.[21] Furthermore, near the half-mass radius, the local gravitational acceleration in the Milgrom spheres is always $g_N \approx 0.7a_0$, implying that a fraction $\gtrsim 1/2$ of the mass must be in the low-acceleration regime. But most important in the present context is the fact that Milgrom spheres obey a relation of the form

$$V_{rms} = \alpha \, (GMa_0)^{1/4} \tag{5.34}$$

between total mass and rms velocity; here α is a quantity that is fixed by the two parameters that define the solution, and it turns out that α always lies in the narrow range $1 \leq \alpha \leq \sqrt{2/3}$ – never far from the asymptotic value $\alpha = \sqrt{2/3}$ of Equation (5.33). This is a genuinely surprising result, with no parallel in Newtonian dynamics.

If real self-gravitating systems satisfied the conditions that were assumed in deriving either Equation (5.33) or (5.34), these results of Milgrom would constitute well-defined, testable predictions of T_1. However, it turns out (under both Milgromian and Newtonian dynamics) that allowing the internal velocity distribution of a stellar system to be even slightly anisotropic or non-isothermal implies a steady-state structure that can differ very substantially from the precisely isothermal case. Rms velocities are in fact observed to decline with radius in most early-type (elliptical) galaxies, implying non-isothermality, and the generically nonspherical shapes of these galaxies implies a degree of anisotropy as well. And indeed it turns out that observed elliptical galaxies are *not* well described, as a class, by Milgrom spheres: they are too compact, i.e. too dense, at a given σ. One way to see this is to note that g_N/a_0 in observed elliptical galaxies is roughly 6 at the

---

[20] There are in fact other independent families of solution to the differential equation describing Newtonian isothermal spheres, but only one family is physically relevant (Chandrasekhar, 1967). No doubt the same is true under the modified dynamics; apparently this question has never been investigated.

[21] This occurs for essentially the same reasons as the maximum surface density of phantom haloes discussed above.

half-light radius (Sanders, 2010), an order of magnitude greater than in Milgrom's spheres. This fact is not inconsistent with the modified dynamics: modeling that starts from the observed density profiles, rather than from Milgrom's isothermal solutions, and that assumes Milgromian dynamics is quite capable of reproducing the observed kinematics of early-type galaxies (Sanders, 2000, 2010; Sanders and Land, 2008; Cardone et al., 2011; Chiu et al., 2017). (Newtonian modeling is similarly successful, but only if adjustable 'dark matter haloes' are included.) The fact that real galaxies often have much higher average surface densities than allowed by Milgrom spheres must reflect some aspect of their formation, about which T_1 makes no prediction one way or the other.

But the fact that the Milgrom spheres – unlike Newtonian isothermal spheres – define an effectively one-parameter sequence in the (V_{rms}, M) plane is a novel enough result that it warrants comparison with real systems, even if the latter are known *not* to be precisely isothermal. And in fact the prediction turns out to be reasonably well corroborated, over the entire range of masses that characterize known equilibrium systems – from roughly 10^4 to 10^{15} solar masses (Sanders, 1994, 2010). This remarkable result is illustrated in Figure 5.3.

As Figure 5.3 shows, systems more massive than dwarf galaxies are typically offset from the line that defines the asymptotic (low-acceleration) relation, Equation (5.34). That in itself is not surprising; an offset is expected, by an amount that is predicted to depend, in a complicated but potentially computable way, on the degree to which an observed system fails to satisfy the conditions for which the asymptotic relation is satisfied. The offset of these points in Figure 5.3 reflects the previously mentioned fact that early-type galaxies, as a class, are not isothermal and are more compact than Milgrom spheres.

The least massive systems plotted on Figure 5.3 include the so-called 'dwarf spheroidal galaxies,'[22] systems of very low surface brightness. Many of these dwarf systems (particularly at the lowest masses) have Newtonian internal accelerations as small as $\sim 0.1a_0$, and so would be expected to fall close to the asymptotic line on Figure 5.3 – as indeed they do. This fact constitutes a successful prediction of the modified dynamics. In his second paper from 1983, Milgrom wrote:

we predict that when velocity dispersion data is available for the dwarfs [i.e dwarf galaxies], a large mass discrepancy will result when the conventional dynamics is used to determine the masses. The dynamically determined mass is predicted to be larger by a factor of order 10 or more than that which can be accounted for by stars (Milgrom, 1983b, p. 381).

---

[22] Dwarf galaxies lie outside the standard (Hubble, de Vaucouleurs) classification systems and the terminology used to describe them is fluid. Dwarf *spheroidal* galaxies are typically defined, ostensively, as galaxies that are structurally similar to the low-luminosity, low-surface density satellite galaxies of the Milky Way and the Andromeda Galaxy. Many of these were discovered only recently; their luminosities are sometimes no greater than those of star clusters, which has motivated astronomers to ask: How, exactly, should 'galaxy' be defined? One proposed requirement (Gilmore et al., 2007; Willman and Strader, 2012), which has been gaining traction recently (Forbes and Kroupa, 2011), is the presence of dark matter. Such a definition must strike Milgromian researchers as truly bizarre: rather like defining a human being as a creature with a soul. See Lazutkina (2017) for an interesting discussion of the use by cosmologists of "intellectual artefacts" in defining physical entities.

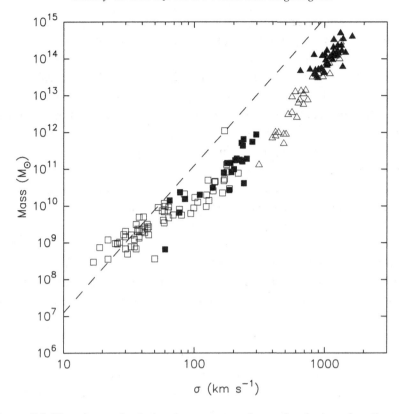

Figure 5.3 The observed relation between rms internal velocity of stellar and galactic systems and their total mass. The horizontal axis plots σ, the line-of-sight velocity dispersion; for a spherical and isotropic system, $\sigma = V_{rms}/\sqrt{3}$. The vertical axis is the total known mass (stars and gas). Different symbols refer to different types of system, as follows: open squares are dwarf galaxies (data from Kourkchi et al., 2012); filled squares are elliptical galaxies (Cappellari et al., 2006); open triangles are groups of galaxies (Angus et al., 2008); filled triangles are galaxy clusters (Sanders, 2003). The dashed line is the asymptotic Milgromian prediction, Equation (5.33); as discussed in the text, observed systems are only expected to fall on this line if they satisfy certain conditions. Nevertheless, the fact that gravitating systems with such a wide range of masses fall close to the Milgromian relation is impressive; no such dependence is predicted by Newtonian theory absent additional hypotheses.

Milgrom noted further that many of these dwarf galaxies would be subject to the 'external field effect' by virtue of their orbiting in the outer parts of the Milky Way, and that this fact complicates the *a priori* prediction of their internal dynamics. Nevertheless the prediction that dwarf spheroidal galaxies should be strongly 'dark matter dominated' has been corroborated; the first confirmation (based on the measured velocities of only three stars!) was published almost simultaneously with Milgrom's prediction (Aaronson, 1983).

However, the same can not be said for the *most* massive systems for which points are plotted in Figure 5.3: the galaxy clusters. Clusters of galaxies are gravitationally bound systems containing from a few to a few hundred bright galaxies and a much larger number of dwarf galaxies. In principle, the motion of galaxies in a cluster could be used in the same way as motion of stars in a galaxy to infer the internal gravitational accelerations. It turns out that the number of bright galaxies is too small, even in a rich cluster, for this program to be feasible (Merritt, 1987). But galaxy clusters also contain hot intracluster gas. Gas molecules move isotropically, and while the gas is not isothermal, spectral data allow the extraction of the gas temperature as a function of position in at least some galaxy clusters, thus removing the main degeneracies that plague dynamical modeling of stellar systems. Analysis of data like these reveals that galaxy clusters are typically *not* describable via Milgromian dynamics (isothermal or otherwise), in the following sense: at a given V_{rms}, the mass required by the Milgromian equations of equilibrium is roughly a factor of two greater than the mass inferred from the stars and the intracluster gas (Gerbal et al., 1992; Sanders, 1999, 2003, 2007; Aguirre et al., 2001). Stated differently: if galaxy clusters were well described by Milgromian dynamics, the corresponding points on Figure 5.3 would be displaced upwards, closer to the asymptotic line.

Does this discrepancy constitute a failure of prediction for T_1? Possibly; but not necessarily. In the case of systems such as (single) galaxies and star clusters, there are good reasons to believe that almost all of the normal (i.e. non-dark) matter has been detected at one wavelength or another, and the contribution to the total mass from 'dark baryons' such as stellar-mass black holes and cold gas can at least be estimated. But there is much less justification for believing that we have a complete census of matter in galaxy clusters. Indeed, until the 1980s, it was assumed that essentially all of the matter in galaxy clusters was contained in the stars and gas belonging to their component galaxies. Following the advent of space-based X-ray observatories, it was discovered that the mass in intracluster gas is typically an order of magnitude greater than the mass in stars – an enormously larger gas fraction than in isolated galaxies (Sarazin, 1988). That is to say: until the 1980s, the dominant component of galaxy clusters, by mass, was undetected at the wavelengths that were accessible to astronomers. That may still be the case; there may still be undetected components – for instance, gravitationally bound clumps of cold gas, or massive neutrinos – and this additional mass could conceivably account for the apparent discrepancy with the predictions of Milgromian dynamics in these systems.

To sum up this story so far: T_1 predicts a relation between V_{rms} and total mass M for self-gravitating systems, a relation which takes on a particularly simple form ($V_{rms} \propto M^{1/4}$) for systems that are isothermal, or that are fully in the low-acceleration regime. Observed systems (with the likely exception of low-density dwarf galaxies) probably do not fall completely in either of these two regimes, but they are nevertheless found to approximately respect the $V_{rms} \propto M^{1/4}$ relation over a very wide range in mass (Figure 5.3). The poorly known internal dynamics of most elliptical galaxies makes it difficult to judge whether their offset from that relation (which is expected) is consistent with the modified dynamics; there is currently no compelling reason to believe that it is not. However, the availability of X-ray data for the most massive bound systems (clusters of galaxies) allows one to state unequivocally that the internal velocities of these systems are typically higher than predicted by the modified dynamics – at least, if it is assumed that there are no unaccounted-for forms of matter.

The possibility that theory variant T_1 is contradicted by the galaxy cluster data is important, of course, and Milgromian cosmologists have taken the possibility seriously. As we will see, this failure of prediction is a principal motivation for theory variant T_3.

At the same time, the fact that gravitationally bound systems with such a wide range of masses obey, even approximately, a $V_{rms} = $ const. $\times M^{1/4}$ relation (Figure 5.3), with the constant of proportionality approximately equal to the (asymptotic) Milgromian value, is also impressive. As Robert Sanders notes:

The amplitude of the relation is even more significant than its slope because this is related directly to the magnitude of a_0. It is easily demonstrated that pure Newtonian systems (obeying the Newtonian virial theorem) with a constant mean surface density will also fall on an $M \propto \sigma^4$ relation. The essential point is that a_0 sets this characteristic value of the surface density (a_0/G) for all near-isothermal systems ... The fact that the magnitude of this constant ($\sim 10^{-8}$ cm s^{-2}) is the same as that required by the scale of spiral galaxy rotation curves (the Tully–Fisher relation) is one more powerful indication that such a fundamental acceleration scale exists in the Universe and is operative in gravitational physics (Sanders, 2010, p. 1133).

Sanders alludes here to the (baryonic) Tully–Fisher relation for disk galaxies, which has a similar functional form, $V_\infty = $ const. $\times M^{1/4}$, as the $V_{rms} \approx $ const. $\times M^{1/4}$ relation of Figure 5.3. As discussed above, the former relation is predicted (by T_0) to be exact, in part because the quantity V_∞ is (by definition) measured in the asymptotic, low-acceleration regime. The relation discussed here is not predicted to be exact, but the variety of systems to which it is observed to apply is extremely broad, in total mass and in type. It is primarily in respect to the latter property that the novelty of Milgrom's $V_{rms} \approx $ const. $\times M^{1/4}$ prediction resides.

Based on this discussion, I conclude that prediction number three is 'partially corroborated.' It is *well* corroborated in the case of dwarf spheroidal galaxies, which are precisely the systems for which the assumptions underlying the prediction are most likely to be correct. But in the case of more massive bound structures, testable predictions are either difficult to make (elliptical galaxies) or they are contradicted by the data (clusters of galaxies). Nevertheless, the overall data do show a well-defined trend of V_{rms} with total mass that has approximately the same functional dependence ($V_{rms} \sim M^{1/4}$) as predicted.

§

In assessing the novelty of the Milgromian prediction, a first question, as always, is: What was known about the predicted fact at the time the theory was constructed? (Recall that we decided to identify 1984 as the date of construction of theory variant T_1.)

That *elliptical galaxies* define an approximate relation of the form $L \propto \sigma^m$, $3 \lesssim m \lesssim 5$ was known already a few years prior to 1984 (Faber and Jackson, 1976); here L is the (optical) luminosity and σ is a velocity dispersion, typically defined as the rms, line-of-sight velocity measured in an aperture centered on the galaxy nucleus. The two dozen galaxies in the Faber and Jackson sample extended over a range of only $\sim 10^2$ in mass, and all were of the same type (luminous elliptical galaxies). Roughly speaking, these data from 1976 correspond to the filled squares in Figure 5.3 (the latter are based on more recent observations and cover a wider range in mass).

Here a judgment is called for. The Faber–Jackson relation was clearly part of the 'background knowledge' that existed in 1983, and one could reasonably argue that that relation (after a conversion of L to M via some prescription) is close enough to the relation predicted by Milgrom to violate Popper's criterion P ("A predicted fact is novel if it was unknown prior to the formulation of the theory") – particularly given that the Milgromian prediction is not quantitatively very definite in this case.

However, I choose to rule in the opposite sense. As noted above, two aspects of the Milgromian prediction that are most surprising, and most clearly corroborated – the fact that dwarf galaxies should fall accurately on the asymptotic relation, and the fact that systems of essentially *any* mass should lie *close* to the relation – are not confronted by the Faber–Jackson sample.

If Popper's condition P for novelty is satisfied, so necessarily is Zahar's condition Z.

We are left with Musgrave's criterion M (a corroborated fact provides evidential support for a theory if the best rival theory fails to explain the fact, or if the fact is prohibited or improbable from the standpoint of that theory).

Under the standard cosmological model, a relation between V_{rms} and (baryonic) mass is necessarily a relation between dark and non-dark matter. For instance, standard-model cosmologists Desmond and Wechsler (2017, p. 820) write (referring to the Faber–Jackson and baryonic Tully–Fisher relations) "Since the internal motions of galaxies are set largely by their dark matter (DM) mass, these relationships provide key insight into the connection between galaxies and their host haloes."

Recall from Chapter 4 that accommodation of the BTFR under the standard model is primarily a task of selectively removing baryons from dark matter haloes, so that a halo of specified V_∞ contains the correct M_{gal}. Exactly the same requirement faces a standard-model theorist who seeks to explain a relation like that of Figure 5.3; but in addition, she must relate V_{rms} of the baryons to V_∞ of the dark halo. Indeed, in their 2017 paper, Desmond and Wechsler "re-deploy and expand the framework" of their 2015 paper on the BTFR to explain the Faber–Jackson relation. By "expand," they mean: adding an algorithm that computes V_{rms} for a putative elliptical galaxy, given the galaxy's mass, and given the rotation speed determined by the dark halo in which it resides. To a first approximation, $V_{rms} \approx V_\infty$, but the exact relation depends on a number of additional factors, including the dependence on radius of the density and the velocity anisotropy of the baryons. The former is reasonably well constrained by observations; the latter, not so much. A common practice (Courteau et al., 2007; Dutton et al., 2011) is to assume a simple relation between the two velocities, e.g. to assume that the line-of-sight velocity dispersion within a galaxy's half-light radius (roughly speaking, the observer's proxy for V_{rms}) is a fixed multiple of the peak circular velocity of the dark halo (the theorist's proxy for V_∞). Desmond and Wechsler (2017) adopt a slightly more complicated algorithm; but for the current discussion, the essential point is that accommodating the M_{gal}–V_{rms} relation under the standard model requires more assumptions than accommodating the BTFR.

This work has been reasonably successful (Ciotti et al., 1996; Trujillo-Gomez et al., 2011; Dutton et al., 2013; Desmond and Wechsler, 2017). But all of these papers, as well as the papers cited in the previous paragraph, address only the $M - V_{rms}$ relation as it applies to (luminous) elliptical galaxies. Standard-model cosmologists seem not yet to have appreciated that a relation such as the one plotted in Figure 5.3 – extending from dwarf galaxies to galaxy clusters – exists, and that it requires a unified explanation.[23] Instead, they have addressed the different types of system in Figure 5.3 in separate theoretical studies, applying different auxiliary hypotheses to each subclass of objects and achieving differing degrees of 'success,' that is, agreement with the data.

Galaxy clusters, the most massive systems plotted on Figure 5.3, are large enough that one expects them to reflect the 'cosmological' ratio of normal to dark matter, a ratio that is specified as part of the standard model.[24] The observed mass of galaxy clusters (in gas and stars) falls systematically short of this prediction, by a factor of $\sim$1.5–2 (Vikhlinin et al., 2006; Gonzalez et al., 2007).[25] Standard-model cosmologists view this as a 'problem to be solved' rather than as a falsification of the standard model. Proposed solutions generally invoke 'feedback' during the formation of the observed stars, the same sort of feedback invoked by them when accommodating galaxy rotation curves, as discussed in Chapter 4. But one finds that additional, ad hoc sources of feedback must be included in order to bring the baryonic (i.e. intracluster gas) properties in accord with observations. For instance, Bode et al. (2009, p. 993) conclude:

---

[23] For instance, standard-model cosmologists have never assigned a name to the relation. Were they to do so, a likely choice might be the 'baryonic Faber–Jackson relation.'

[24] This is the 'concordance' value of Ω_b / Ω_{DM}. Chapter 6 describes how standard-model cosmologists assign a number to this quantity.

[25] This discrepancy contributes to the 'missing baryons problem' of the standard model; see Chapter 6.

Core collapse SNe [supernovae] could only add 0.4 keV per particle ... so in this case there must be some additional source such as AGN [active galactic nucleus] energy from accreting black holes. If the mass in black holes is $10^{-3}M_\star$, and 2% of the rest-mass energy of the accreted material is converted into thermal energy (Allen et al. 2006), then an additional 1.0 keV per particle could be added to the gas.

This study, and others (Nagai et al., 2007; Fabjan et al., 2010), include additional sources of star-formation-related feedback, but without addressing the corresponding observational constraints, e.g. that the stellar mass produced in the model is consistent with what is observed (Andreon, 2010).

Moving downward in mass, the next class of objects plotted on Figure 5.3 are the (luminous) elliptical galaxies. In accommodating these data, standard-model cosmologists typically relate a galaxy's dark mass to its non-dark mass through an algorithm called 'abundance matching';[26] for instance, this was the hypothesis adopted in the 2017 study by Desmond and Wechsler cited above. Then (as just discussed) additional assumptions are made about the distribution of normal matter within the dark halo, and the velocity anisotropy of the stellar motions, in order to bring the predicted V_{rms} in accord with data at each M_{gal}.

Finally, the objects of lowest mass on Figure 5.3 are the dwarf galaxies. In the eyes of a standard-model scientist, these are 'dark matter dominated' systems, and standard-model predictions of both the total dark mass associated with dwarf galaxies, and its spatial distribution, fail dramatically; the latter failure was discussed in Chapter 4, the former is described in Chapter 7.

In summary: Standard-model cosmologists have not attempted to find a unified explanation for the existence of a relation between M and V_{rms} like that plotted in Figure 5.3, and attempts to explain limited regimes of this relation have been ad hoc, unsuccessful or unconvincing. I conclude that the relation constitutes an anomaly for standard-model researchers and therefore that Musgrave's condition for novelty of the Milgromian prediction is satisfied. These conclusions are summarized in the eighth line of Table 5.1.

§§

4. *A relation between the internal velocity dispersions of two, otherwise identical systems that are located in different external fields*: A star cluster or dwarf galaxy orbiting in the outskirts of a large galaxy like the Milky Way will experience that galaxy's gravity as a nearly homogeneous, external field with acceleration $g_N^e < a_0$. Furthermore, as just discussed, internal accelerations in dwarf spheroidal satellite galaxies are of order a_0 or less. These are conditions under which the 'external field effect' (EFE) that was discussed briefly in Chapter 4 is predicted to become important. Figure 5.4, which is based on that discussion, plots three

---

[26] See Chapter 7 for a detailed discussion of abundance matching.

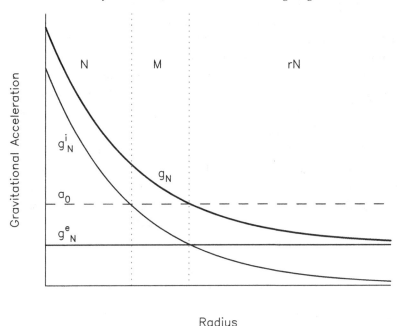

Figure 5.4 This figure illustrates the external field effect (EFE). A spherical stellar system (e.g. a dwarf satellite galaxy) is assumed to be orbiting in the gravitational field of a much larger system (e.g. the Milky Way). The gravitational acceleration experienced by a star inside the dwarf galaxy has two contributions: from all the other stars in the dwarf galaxy (g_N^i), and a component due to the gravitational field of larger galaxy, approximated here as having a constant value g_N^e throughout the dwarf galaxy. This figure assumes that $g_N^e < a_0$. The horizontal axis is the distance measured from the center of the dwarf galaxy. Motion inside the dwarf galaxy falls into one of three regimes. (*i*) Near the center, $g_N^i \gg a_0$ and the motion is Newtonian (N). (*ii*) Farther out, $g_N = g_N^i + g_N^e$ is comparable to a_0, and motion obeys the modified dynamics (M). (*iii*) In the outskirts of the dwarf galaxy, where g_N^i falls below a_0, the EFE is important; as showed in Chapter 4, the motion in this regime is approximately Newtonian but with a larger effective constant of gravitation (rN). McGaugh and Milgrom (2013a,b) argue that the dwarf satellite galaxy And XVII is mostly in the rN regime, while the satellite galaxy And XXVIII, due to its much greater separation from its host galaxy, is mostly in the M regime. (McGaugh and Milgrom call these two regimes 'EFE' and 'isolated,' respectively.)

regimes of motion of a star that orbits inside such a satellite system. (*i*) Near the satellite galaxy's center (assuming it is sufficiently dense), its internal acceleration, g_N^i, satisfies $g_N^i > a_0 > g_N^e$ and the motion is Newtonian (N). (*ii*) Farther from the center, where g_N^i is comparable to or less than a_0, the motion is described by the modified dynamics (M). (*iii*) Farther out still, where the total acceleration (internal plus external) falls below a_0, the motion is 'rescaled Newtonian' (rN), i.e. Newtonian with a larger effective constant of gravitation.

A testable consequence of the EFE using dwarf satellite galaxies was described by McGaugh and Milgrom (2013a,b). Suppose one identifies two satellite systems that are morphologically similar (same mass and size) and which are both in the low-acceleration regime in terms of their internal accelerations, i.e. $g_N^i \lesssim a_0$ everywhere. Suppose in addition that the two satellites differ in terms of their location in the external field, such that one is effectively 'isolated' ($g_N^i > g_N^e$), while in the other, the external field dominates the internal acceleration ($g_N^e > g_N^i$). In terms of Figure 5.4, the first satellite is in the 'M' (Milgromian) regime and the second is in the 'rN' (rescaled-Newtonian) regime. The rms velocity of stars should be different in the two systems, by a factor that is computable via the formalism of T_1.

The isolated satellite (call it S_1) satisfies the conditions under which Equation (5.33) was derived, hence its internal velocity dispersion is given uniquely in terms of its mass by that equation, i.e.

$$V_{\text{rms}} = \sqrt{\frac{2}{3}} (GMa_0)^{1/4}.$$

The second satellite (S_2) obeys the Newtonian virial theorem, but with a rescaled (larger) value of the effective gravitational constant. Writing

$$K = \frac{1}{2}MV_{\text{rms}}^2, \quad W = -C\frac{G_{\text{eff}}M^2}{R_{\frac{1}{2}}}$$

and setting $2K + W = 0$ yields

$$V_{\text{rms}} = \sqrt{C\frac{G_{\text{eff}}M}{R_{\frac{1}{2}}}}. \tag{5.35}$$

In these expressions, $R_{1/2}$ is the radius containing one-half of the satellite's mass, and C is a constant, of order unity, that depends on the detailed distribution of mass inside the satellite. The effective gravitational constant is given by

$$G_{\text{eff}} = \frac{G}{\mu(g_N^e/a_0)}. \tag{5.36}$$

Of course, Equations (5.33), (5.35) and (5.36) constitute predictions for any single satellite galaxy that is known to be in one of the two specified regimes. One reason for casting the test as a comparison between *two* systems is to maximize the novelty of a positive result: if S_1 and S_2 are morphologically similar, then the quantity C (which is never known precisely) would be the same for both, and under standard (Newtonian) dynamics, the two systems would be predicted to have the same V_{rms}.

This prediction was tested by McGaugh and Milgrom (2013b) using a pair of dwarf spheroidal galaxies orbiting in the outskirts of the nearby Andromeda Galaxy (NGC 224). The two satellites (Andromeda XXVIII $\equiv S_1$; Andromeda XVII $\equiv S_2$) have similar total luminosities, $L \approx 2 \times 10^5 L_\odot$, and similar sizes, $R_{1/2} \approx 300$ pc. Due to the relative nearness of NGC 224 to our Galaxy ($\sim$780 kpc), it is possible to determine accurate line-of-sight distances to individual satellites, which allows their separation, D, from the center of Andromeda to be determined. And XVII is estimated to be 67^{+20}_{-24} kpc from the center of its host galaxy (Conn et al., 2012), while And XXVIII sits currently at a distance of 365^{+17}_{-1} kpc (Slater et al., 2011).

McGaugh and Milgrom (2013a) estimated the amplitude of the external field using

$$g_N^e \approx \frac{V^2_{\text{And}}(D)}{D}, \tag{5.37}$$

with $V_{\text{And}}(D)$ the (known) galactic rotation speed at D. In this way, S_1 was found to be securely in the isolated (M) regime, and S_2 in the rescaled-Newtonian (rN) regime. McGaugh and Milgrom predicted line-of-sight velocity dispersions, $\sigma = V_{\text{rms}}/\sqrt{3}$, of 4.5 km s^{-1} for S_1 and 2.3 km s^{-1} for S_2. This difference is large in a relative sense, but not large compared with the precision with which σ can be measured. The measured values are

$S_1 : \sigma = 4.9 \pm 1.6$ km s^{-1} (Tollerud et al., 2013, based on 18 stars)
$\quad\quad = 6.6(+2.9, -2.1)$ km s^{-1} (Collins et al., 2013, based on 17 stars),

$S_2 : \sigma = 2.9(+1.0, -0.7)$ km s^{-1} (Collins et al., 2013, based on 8 stars).

Based on the F-test (Alder and Roessler, 1968), the two satellites have velocity dispersions that differ with probability 92% (based on the first measured σ for S_1) or 98% (based on the second measured σ for S_1). On this basis, it is reasonable to conclude that the Newtonian prediction for the velocity dispersion ratio (i.e. that they are equal) has been falsified. The difference in the measured velocity dispersions is either 2.0 km s^{-1} or 3.7 km s^{-1}, compared with the Milgromian prediction of 2.2 km s^{-1}; their ratio is either 1.69 or 2.28, compared with the Milgromian prediction of 1.96. Modulo the uncertainties, it is reasonable to conclude that the Milgromian prediction has been corroborated (McGaugh and Milgrom, 2013b).

The EFE is clearly a novel prediction with respect to criteria P and Z. An effect like the EFE is precluded by the so-called strong equivalence principle, which is generally assumed by standard-model cosmologists to be valid, hence criterion M is also satisfied. Finally, criterion C is not applicable, since the EFE was not put into the theory in order to explain an observed fact. These results are summarized in the final line of Table 5.1.

§§

We are now in a position to judge the progressivity of the Milgromian research program including theory variant T_1.

The requirement of heuristic progress ("successive modifications of the protective belt must be in the spirit of the heuristic") is clearly satisfied. As discussed in the opening paragraphs of this chapter, the auxiliary hypotheses that define T_1 were added to T_0 in order to resolve theoretical difficulties that had been recognized, from the start, by Milgrom and collaborators. They were not added in response to any unexpected or anomalous observations.

Theory variant T_1 also satisfies Lakatos's condition for a theoretically progressive problemshift: as summarized in Table 5.1, six of its predictions satisfy one or more conditions for novelty. And finally, T_1 constitutes an empirically progressive problemshift since at least three, and possibly as many as five, of its novel predictions have been corroborated or confirmed.

At the same time, T_1 fails in prediction in one important way. As discussed above, its predictions for the internal rms velocities of galaxy clusters (or equivalently, for the mean temperature of the intracluster gas) are systematically in error. As we will see in Chapter 7, theory variant T_3 contains an auxiliary hypothesis that successfully accounts for this anomaly.

6

Theory Variant T_2: A Relativistic Theory

As discussed in Chapter 3, the heuristic of the Milgromian research program directs the researcher to construct theories first as extensions of Newtonian dynamics, then as extensions of general relativity, and ultimately in a form that has no pre-ordained connection with Newton's or Einstein's theories. In addition, of course, all variants $T_0, T_1, T_2, \ldots$ of the theory must embody the three hard-core postulates, including a departure from Newtonian dynamics for accelerations $a \lesssim a_0$ and spacetime invariance in the deep-MOND ($a \ll a_0$) limit.

According to Bekenstein and Milgrom (1984, p. 8), there are two principle motivations for developing a relativistic version of Milgrom's theory: "(a) to help incorporate the principles of MOND into the framework of modern theoretical physics; (b) to provide tools for investigating cosmology in light of MOND." They add: "This last is particularly pressing since some of the arguments adduced in support of the MOND from the empirical viewpoint [in Milgrom (1983c)] have a cosmological aspect which cannot be treated self-consistently without a relativistic cosmology."

That last sentence refers to the observed dynamics of galaxy clusters as interpreted in the light of Milgromian dynamics. But in the three decades following Bekenstein and Milgrom's 1984 paper, MOND's rival theory, the standard cosmological model, expanded its explanatory scope to accommodate a much wider range of data relating to the large-scale structure of the universe, including the statistics of gravitational lensing and fluctuations in the cosmic microwave background (CMB) radiation. When standard-model cosmologists make statements like the statement of Olive et al. that opens Chapter 1 ("The concordance model is now well established, and there seems little room left for any dramatic revision of this paradigm") it is to these successes that they are primarily referring. And when standard-model cosmologists *dismiss* MOND, as they very often do, it is typically on the ground that no relativistic generalization of Milgromian dynamics has yet been found that is as successful as the standard model at accommodating those data. In the words of standard-model cosmologist Katherine Freese (2017, p. 2), "While these [Milgromian] models have been shown to fail, particularly by CMB observations, they may provide an interesting phenomenological fit on small scales."

At the same time, it is important to recognize just how the successes of the standard cosmological model were achieved. That model failed, again and again, to predict in advance (or to "anticipate," as Lakatos would have said) the observational facts that are now adduced in its support (Kroupa, 2012). Its eventual 'success' was attained via a series of auxiliary hypotheses and parameter adjustments. Indeed, the acronym commonly associated with the current standard model – the 'ΛCDM' model – encodes two of the most important of those auxiliary hypotheses: cold dark matter (CDM) and dark energy (Λ). Both postulates (or rather, both sets of postulates) were responses to unexpected observational facts. Furthermore, standard-model cosmologists have not hesitated to adjust their postulates again and again, as needed in order to accommodate new data. For instance, determining the 'equation of state of dark energy' – that is, adjusting the assumed properties of 'dark energy' so as to maintain consistency of the observations with Einstein's equations – is an active (and at least in the eyes of standard-model scientists, legitimate) field of research (e.g. Daly and Djorgovski, 2005).

Nothing in the preceding paragraph is intended as a criticism of the standard cosmological model. But it is important, when interpreting the claims of standard-model cosmologists, to understand that they rarely view the development of their own theory in Popperian or Lakatosian terms. Rather than conceive of dark matter or dark energy as postulates invoked in response to falsifying observations, standard-model cosmologists typically interpret those same observations as tantamount to the *discovery* of dark matter or dark energy (Merritt, 2017). And so when a *Milgromian* researcher successfully invokes an auxiliary hypothesis to reproduce the CMB temperature fluctuation spectrum – as Garry Angus did, in 2009 – standard-model cosmologists like Freese tend to be unimpressed, forgetting (or so it would appear) that the standard cosmological model, itself, only managed to accommodate those data with the help of a carefully crafted set of auxiliary hypotheses. Indeed, as discussed later in this chapter, when adjusting the parameters of their theory to accommodate the CMB data, standard-model cosmologists were forced to bring their theory *out* of agreement with some well-established facts. The resulting anomalies have persisted until the present day.[1] Angus's Milgromian solution, whatever its other perceived shortcomings, does not suffer from that epistemic defect.

As interesting as these comparative judgments are, we are not obliged to make them. The essential, Lakatosian criterion for progressivity of a scientific research program is 'incorporation with corroborated excess content.' Both 'incorporation' and 'excess content' are defined with respect to previous theory variants in the

---

[1] The 'lithium problem' and the 'missing baryons problem' were both created in this way. Neither anomaly exists from the standpoint of a Milgromian researcher.

same – that is, the Milgromian – research program: in the present case, that means with respect to theory variants T_0 and T_1. The failure of theory variant T_2 to explain *some* known facts – including facts that may be successfully explained by some theory in a rival research program – is taken for granted; in Lakatos's words, "All theories . . . are born refuted and die refuted." Such failures are certainly worth noting, but it is assumed that they will be dealt with in some future variant $(T_3, T_4, \ldots)$ of the theory.

At the same time, there *is* a role to be played here by the 'rival research program.' As discussed in Chapter 2, Alan Musgrave proposed that a relevant factor in deciding whether a corroborated fact should be considered 'novel' – that is, capable of providing evidential support for a theory – is whether that fact is prohibited or improbable from the standpoint of the rival theory. And so, as in Chapters 4 and 5, we will apply Musgrave's criterion together with the other three criteria that were discussed in Chapter 2 when evaluating the novelty of predictions from theory variant T_2. That means that we will be particularly interested in successful predictions of T_2 which are *not* naturally explained from the standpoint of the standard cosmological model. As we will see, such predictions do exist.

Most of the work on theory variant T_2 has taken place relatively recently; since, roughly, the late 1990s. This has also been a period of rapid advancement in the state of our knowledge about the universe. These are mundane facts, but they have two consequences that are relevant for judging progressivity according to Lakatos's criteria. First: the scope for *novel* predictions by theory variant T_2 has necessarily narrowed over time, at least if novelty is judged by Popper's criterion ("A predicted fact is novel if it was unknown prior to the formulation of the theory.") Second: Milgromian theorists have sometimes been able to evaluate and discard their proposed relativistic generalizations in 'real time,' so to speak, by checking the predictions of their theories against the most recent observations. One example is gravitational lensing. The first relativistic version of Milgrom's theory ('RAQUAL') was proposed before much data were available that were relevant to lensing, but it took only a few years to recognize that RAQUAL was inconsistent with observations of weak gravitational lensing and so could be ruled out.

This interplay between theory development and observation has continued, and as a result, it probably does not make sense to pick a *single* of the many, proposed relativistic generalizations as uniquely defining 'theory variant T_2.' As we will see, it is still possible to identify novel predictions that would be entailed by virtually any acceptable relativistic version of the theory. But the fact that theory development has often been driven by unanticipated observational results constitutes a violation of Lakatos's condition of 'heuristic progress': that is, the requirement that research programs should evolve autonomously and not simply in response to anomalies.

§§

As an expression of the research program's heuristic, a directive such as 'Construct a relativistic version of the theory' is rather vague. There is no shortage of possible ways to modify or generalize Einstein's gravitational theory, and even summary discussions of the currently viable alternatives to the general theory of relativity (GR) – most of which were not designed with Milgrom's postulates in mind – can run to hundreds of pages (e.g. Clifton et al., 2012).

Writing in 1992, after some progress had already been made at constructing relativistic versions of Milgrom's theory, Jacob Bekenstein (1992, p. 910) elaborated on the "guiding principles" (that is, heuristic) that a Milgromian researcher should follow:

1. Action principle: The theory must be derivable from an action principle from which flow the usual conservation laws.
2. Covariance: The action should be a scalar so that all equations of the theory are relativistically invariant.
3. Causality: The equations deriving from this action should not permit superluminal propagation of any measurable field, thus precluding acausality.
4. Positivity of energy: Fields in the theory should not carry negative energy (otherwise an instability of the vacuum would ensue).
5. Universality of free fall: In harmony with the weak equivalence principle, free particle trajectories should be geodesics in some metric $\tilde{g}_{\alpha\beta}$ special to the theory. This will insure that $a = -\nabla\phi$ for a certain ϕ independent of particle internal structure ... [and] will also insure that the gravitational redshift is governed by the same ϕ.
6. Departures from Newtonian gravity: The theory should exhibit a preferred scale of acceleration below which departures from Newtonian gravity should set in.

The weak equivalence principle (WEP) mentioned in point 5 (sometimes defined as 'equivalence of inertial and gravitational mass') states that all objects fall with the same acceleration in an external gravitational field. The WEP has been tested by laboratory experiments and no violations have been observed; gravitational and inertial masses appear to differ at most by about one part in 10^{13}. At the same time, Bekenstein was quite aware of another property that an acceptable relativistic theory should have: it should *violate* the *strong* equivalence principle (SEP), for the reasons discussed in Chapter 5.

By "action principle," Bekenstein meant a generalization of the Newtonian principle of least action for a single particle (as discussed in Chapter 5) to a relativistically invariant action principle for fields. In classical (non-relativistic) field theory, the coordinates are replaced by a set of fields, $\Phi^i(x,t)$, which themselves depend on position and time; i is an index labeling an individual field. The field-theory Lagrangian in *flat* spacetime can be defined as an integral over space of a Lagrange density $\mathcal{L}$, or

$$L = \int d^3x \, \mathcal{L}\left(\Phi^i, \partial_u \Phi^i\right), \tag{6.1}$$

where ∂_μ indicates a partial derivative. The action is

$$S = \int dt\, L = \int d^4x\, \mathcal{L}\left(\Phi^i, \partial_\mu \Phi^i\right), \tag{6.2}$$

where $d^4x = dt\, dx\, dy\, dz$. (Henceforth 'Lagrangian' will be taken to mean 'Lagrange density.') The Euler–Lagrange equations are derived by assuming that the action is unchanged under small changes in the fields. The result can be shown to be

$$\frac{\partial \mathcal{L}}{\partial \Phi^i} - \partial_\mu \left(\frac{\partial \mathcal{L}}{\partial \left(\partial_\mu \Phi^i\right)}\right) = 0, \quad \mu = 1, \ldots, 4, \tag{6.3}$$

similar to Equation (5.1).

In *curved* spacetime, the Lagrangian is a function of the fields and their covariant derivatives, which we denote by ∇_μ:

$$S = \int d^4x\, \mathcal{L}\left(\Phi^i, \nabla_\mu \Phi^i\right). \tag{6.4}$$

The Euler–Lagrange equations become

$$\frac{\partial \hat{\mathcal{L}}}{\partial \Phi^i} - \nabla_\mu \left(\frac{\partial \hat{\mathcal{L}}}{\partial \left(\nabla_\mu \Phi^i\right)}\right) = 0, \tag{6.5}$$

where $\mathcal{L} = \sqrt{-g}\, \hat{\mathcal{L}}$ and g is the determinant of the metric tensor.

The action that yields Einstein's equations, $S = S_{\text{GR}}$, is well known to be

$$S_{\text{GR}} = S_{\text{gravity}} + S_{\text{matter}}, \tag{6.6a}$$

$$S_{\text{gravity}} = S_{\text{H}} = \frac{c^4}{16\pi G} \int d^4x\, \sqrt{-g}\, R[g_{\mu\nu}], \tag{6.6b}$$

the sum of the Hilbert action (S_{H}) and the matter action; here $R = R_{\mu\nu} g^{\mu\nu}$ is the curvature scalar and $R^{\mu\nu}$ is the Ricci curvature tensor. The matter action can be written in terms of the energy-momentum tensor $T_{\mu\nu}$; in the simplest case of a free point particle, of mass m and velocity v,

$$S_{\text{matter}} = -mc \int dt\, \sqrt{-g_{\mu\nu}(x) v^\mu v^\nu}. \tag{6.7}$$

The metric $g_{\mu\nu}$ plays the role of a dynamical variable; varying the total action with respect to it yields Einstein's field equations:

$$R_{\mu\nu} - \frac{1}{2} R g_{\mu\nu} = \frac{8\pi G}{c^4} T_{\mu\nu}, \tag{6.8}$$

and varying the matter action with respect to the matter fields degrees of freedom yields the geodesic equation for a point particle:

$$\frac{d^2 x^\mu}{d\tau^2} = -\Gamma^\mu_{\alpha\beta} \frac{dx^\alpha}{d\tau} \frac{dx^\beta}{d\tau}, \tag{6.9}$$

where τ is the proper time, i.e. $c\,d\tau = \sqrt{g_{\mu\nu}\,dx^\mu dx^\nu} \approx c\,dt$ for slowly moving particles, and $\Gamma^\mu_{\alpha\beta}$ is the Christoffel symbol, expressible in terms of first derivatives of the metric.

In the weak-field limit, the single-particle equation of motion implied by Equation (6.9) becomes

$$\frac{d^2 x^i}{dt^2} = -\Gamma^i_{00} = -\frac{\partial \Phi}{\partial x^i}, \tag{6.10}$$

with Φ the (Newtonian) potential. Evidently, the Christoffel symbol plays the role of the acceleration. But since the Christoffel symbol is not a tensor, one can not simply use it to insert an 'acceleration scale' into a general relativistic theory.

Bekenstein and Milgrom (1984) took the following alternative approach. In the same paper where they first described their AQUAL ('AQUAdratic Lagrangian') theory variant of Milgromian dynamics, these authors included a short appendix entitled "A relativistic theory which satisfies the requirements of MOND." Their goal was to find a relativistically invariant Lagrangian that reduced to the AQUAL Lagrangian, Equation (5.4), in the non-relativistic limit. Recall that the term responsible for the modified dynamics in that Lagrangian was $\mathcal{F}\left(|\nabla\Phi|^2 / a_0^2\right)$ with $\mathcal{F}$ an arbitrary function. It is clear that writing $\mathcal{F} = \mathcal{F}(R)$, a function of the curvature scalar, will not achieve this, since R contains second derivatives of the metric; for a general $\mathcal{F}(R)$ the result will be a Lagrangian that contains second derivatives of Φ in the non-relativistic limit. Bekenstein and Milgrom chose instead to assume that the quantity Φ that appears as an argument of $\mathcal{F}$ may not correspond only to the g_{00} metric component but also to some extra field in the theory. The simplest such choice would be a *scalar* field; call it ψ. The covariant generalization of $|\nabla\Phi|^2$ is $g^{\mu\nu}\partial_\mu\psi\,\partial_\nu\psi$. Bekenstein and Milgrom therefore proposed an action of the form

$$S = \frac{c^4}{16\pi G} \int d^4 x \sqrt{-g}\left[R - \frac{a_0^2}{c^4}F(\xi)\right] + S_{\text{matter}}, \tag{6.11}$$

with $\xi \equiv (c^4/a_0^2)g^{\mu\nu}\partial_\mu\psi\,\partial_\nu\psi$, and F a dimensionless function that is chosen to yield the correct Milgromian behavior in the low-acceleration limit.

But this can not be the whole story: unless the matter action also depends on ψ, the ψ equation will be sourceless, and its simplest solution is $\psi = $ constant. At the same time, satisfying the weak equivalence principle demands that ψ couple to all matter fields, $\Psi_1, \Psi_2, \ldots$, in the same way. Stated differently: the matter action must be expressible in covariant form in terms of a single metric, i.e.

$$S_{\text{matter}} = \int d^4 x\, \mathcal{L}_{\text{matter}}\left(\Psi_1, \Psi_2, \ldots, \tilde{g}_{\mu,\nu}\right)(-\tilde{g})^{1/2}. \tag{6.12}$$

Einstein's equations have $g_{\mu\nu} = \tilde{g}_{\mu\nu}$, and as a consequence, Einstein's equations satisfy the *strong* equivalence principle. Bekenstein and Milgrom chose instead to write

$$\tilde{g}_{\mu\nu} = e^{2\psi/c^2} g_{\mu\nu}, \tag{6.13}$$

where $\tilde{g}_{\mu\nu}$ is the "physical" or "Jordan" metric and $g_{\mu\nu}$ is the "gravitational" or "Einstein" metric. Equation (6.13) is a conformal transformation.[2]

In the particle equations of motion, the scalar field plays the role of an additional potential. Bekenstein and Milgrom derived the expression for $F(\xi)$ that corresponds, in the non-relativistic limit, to any specified $\mathcal{F}$ in Equation (5.4); for instance, such that the gravitational potential due to a point mass becomes $\sim \sqrt{GMa_0}\log r$ for $a \ll a_0$. In the high-acceleration regime the theory becomes a weakly coupled Jordan–Brans–Dicke theory. This version of theory variant T_2 is called RAQUAL, for "Relativistic AQUAL."

It was soon recognized that RAQUAL has a serious flaw (Romatka, 1992; Bekenstein, 1992). If the stress-energy of the scalar field is negligible compared with that of the matter, then the curvature of the Einstein metric, $g_{\mu\nu}$, is generated by the 'baryonic' mass alone. But photons move along null geodesics of $g_{\mu\nu}$; in other words, bending of light in the low-acceleration regime (or 'weak lensing' as it is called by astrophysicists) in RAQUAL will be determined entirely by the baryonic matter. But observations since the 1980s have shown that weak lensing is well accommodated by assuming Einsteinian gravity plus 'dark matter,' with the latter having roughly the same spatial distribution that is required to fit galaxy rotation curves under the standard cosmological model.

The undesirable behavior of photon trajectories under RAQUAL can be traced to the fact that Maxwell's equations are conformally invariant, and $\tilde{g}_{\mu\nu}$ and $g_{\mu\nu}$ are related by a conformal transformation. Bekenstein (1992, 1993) considered the possibility that the relation between the two metrics is *dis*formal; for instance,

$$\tilde{g}_{\mu\nu} = e^{2\psi/c^2} \left[A\,(\xi)\,g_{\mu\nu} + B\,(\xi)\,\partial_\mu\psi\,\partial_\nu\psi \right]. \tag{6.14}$$

Unlike a conformal transformation, which 'stretches' spacetime isotropically, the $\partial_\mu\psi\,\partial_\nu\psi$ term now selects out a particular direction, namely the radial direction in spherically symmetrical systems. The extra contribution remains negligible in the equations of motion of test particles but photon trajectories can be made sensitive to the presence of the scalar field. In a static, spherically symmetric situation, the radial metric component can be shown to have the additive term $(B/A^2)(\partial_r\psi)^2$; setting $B > 0$ then implies a greater bending of light. For instance, choosing $(B/A^2)(\partial_r\psi)^2 = 4\sqrt{GMa_0}/c^2$ with M the total baryonic mass (admittedly a fine-tuning) yields a deflection

$$\Delta\theta = \Delta\theta_{\mathrm{GR}} + \frac{2\pi\sqrt{GMa_0}}{c^2} + \mathcal{O}\left(\frac{1}{c^4}\right). \tag{6.15}$$

---

[2] A conformal transformation is one in which the physical metric is multiplied by a spacetime-dependent function $\omega^2(x^\mu)$, i.e. $g_{\mu\nu} \to \omega^2(x^\mu)g_{\mu\nu}$. Equivalently, the line element transforms as $ds^2 \to \omega^2(x^\mu)ds^2$.

The second term reproduces exactly the prediction of GR in the presence of 'dark matter' if the latter has a spatial distribution that implies (according to Newton) the asymptotic Milgromian acceleration.

Unfortunately, setting $B > 0$ also has one undesirable consequence: gravitons (i.e. perturbations of the Einstein metric) propagate superluminally (Bekenstein, 1993). Bekenstein, Sanders and coworkers considered this to be a fatal flaw and rejected the theory on that ground.[3] Sanders (1997) suggested an alternate form for the disformal relation between $\tilde{g}_{\mu\nu}$ and $g_{\mu\nu}$:

$$\tilde{g}_{\mu\nu} = A^2\,(\psi)\,g_{\mu\nu} + B\,(\psi,\xi)\,\mathcal{U}_\mu\mathcal{U}_\nu, \tag{6.16}$$

replacing the gradient $\partial_\mu\psi$ by a vector field $\mathcal{U}_\mu$. The latter is chosen to be a non-dynamical, timelike vector of unit norm, $\mathcal{U}_\mu = (-1,0,0,0)$. The second term in (6.16) affects only the $\tilde{g}_{00}$ component of the metric; provided the scalar field comes from a RAQUAL-type equation, Milgromian dynamics are recovered and the lensing is augmented by the correct amount. However, because $\mathcal{U}_\mu$ is a constant vector (pointed in the time direction), the resulting theory is not covariant.

Bekenstein (2004) achieved covariance by converting $\mathcal{U}_\mu$ into a *dynamical* vector field $\mathcal{A}_\mu$ and defining the disformal relation

$$\tilde{g}_{\mu\nu} = e^{-2\psi/c^2} g_{\mu\nu} - 2\sinh(2\psi)\,\mathcal{A}_\mu\mathcal{A}_\nu. \tag{6.17}$$

The action that determines the evolution of $\mathcal{A}_\mu$ is

$$S_{\mathcal{A}} = -\frac{c^4}{16\pi G}\int d^4x\sqrt{-g}\left[\frac{K}{2}g^{\alpha\beta}g^{\mu\nu}\mathcal{F}_{[\alpha,\mu]}\mathcal{F}_{[\beta,\nu]} - \lambda(g^{\mu\nu}\mathcal{A}_\mu\mathcal{A}_\nu + 1)\right], \tag{6.18}$$

where $\mathcal{F}_{\mu\nu} = \partial_\mu\mathcal{A}_\nu - \partial_\nu\mathcal{A}_\mu$ and K is a positive dimensionless number. The quantity λ is a Lagrange multiplier that is chosen to force the unit norm $\mathcal{A}^\mu\mathcal{A}_\nu = g^{\mu\nu}\mathcal{A}_\mu\mathcal{A}_\nu = -1$; if this were not done the magnitude of scalar-field effects would depend on the background norm of the vector field. The action for the scalar field is defined as in RAQUAL, Equation (6.11), but with $\xi \equiv (c^4/a_0^2)\,(g^{\mu\nu} - \mathcal{A}^\mu\mathcal{A}^\nu)\,\partial_\mu\psi\partial_\nu\psi$.

Equations (6.17) and (6.18) are the basis of Bekenstein's 'Tensor–Vector–Scalar' or 'TeVeS' theory (Bekenstein, 2004). TeVeS satisfies all of the Bekensteinian "guiding principles" listed above: in particular, it reduces to AQUAL in the non-relativistic limit. In addition, it yields the desired result for bending of light trajectories, that is, it predicts the same pattern of gravitational lensing that Einstein's theory would predict, if the latter is supplemented with whatever distribution of 'dark matter' is required to fit the observed rotation curve. Bekenstein (2004) showed as well that cosmological models based on TeVeS are quite similar to

---

[3] Bruneton and Esposito-Farèse (2007) argue that this property is not actually deadly since it does not imply causality violation.

those based on Einstein's theory, and it has been shown that TeVeS implies no cosmological evolution of the Newtonian gravitational constant G, and only minor evolution of Milgrom's constant a_0 (Famaey et al., 2007; Bekenstein and Sagi, 2008). In the words of Famaey and McGaugh (2012, p. 91), TeVeS "has played a true historical role as a *proof of concept* that it was possible to construct a fully relativistic theory both enhancing dynamics and lensing in a coherent way and reproducing the MOND phenomenology for static configurations."

Contaldi et al. (2008) criticized TeVeS on the ground that $\mathcal{A}_\mu$ tends to form 'caustic singularities,' i.e. the integral curves of the vector are timelike geodesics that meet each other when falling into gravity potential wells. Similar behavior is exhibited in some so-called 'Einstein-æther' theories, and Contaldi et al. (2008) suggested a solution similar to the one adopted for those theories: adding an extra term to the Lagrange density for the vector field. This suggestion was followed by Skordis (2008) and Sagi (2009). The formulation of Skordis (2008) is currently the 'standard' version of TeVeS; it includes an action corresponding to the vector field of

$$S_{\mathcal{A}} \equiv -\frac{c^4}{16\pi G} \int d^4x \sqrt{-g}\, \left[K^{\alpha\beta\mu\nu} \mathcal{A}_{\beta,\alpha} \mathcal{A}_{\nu,\mu} - \lambda(g^{\mu\nu} \mathcal{A}_\mu \mathcal{A}_\nu + 1) \right], \qquad (6.19)$$

where the new term in the Lagrange density has the form

$$K^{\alpha\beta\mu\nu} = c_1 g^{\alpha\mu} g^{\beta\nu} + c_2 g^{\alpha\beta} g^{\mu\nu} + c_3 g^{\alpha\nu} g^{\beta\mu} + c_4 \mathcal{A}^\alpha \mathcal{A}^\mu g^{\beta\nu} \qquad (6.20)$$

and c_1, c_2, c_3, c_4 are a set of adjustable constants, a certain combination of which yields the original TeVeS theory (Skordis and Zlosnik, 2012).

§

Recall from Chapter 5 that *non*-relativistic variants of Milgromian dynamics can be grouped into two broad classes: 'modified gravity' theories and 'modified inertia' theories. As discussed in that chapter, most of the work on the part of Milgromian researchers has focussed on modified gravity theories. Modified *inertia* theories can be constructed, in the non-relativistic case, by modifying the kinetic part of the action, leaving the relation between gravitational potential and density unchanged. In such theories, the equation of motion of a particle is defined in terms of functionals that depend on the particle's entire trajectory. Another way of describing this is to say that modified inertia theories are *nonlocal* (Milgrom, 1994a, 1999). One consequence is that modified inertia theories are mathematically much less tractable than modified gravity theories, and this is why so little attention has been given to them.

The *relativistic* theories described so far in this chapter are all 'local' and they can be thought of as relativistic generalizations of modified-gravity theories. The same is true for a number of other relativistic variants of Milgrom's theory (none of which will be discussed in detail here): so-called Einstein-æther theories (Zlosnik et al., 2007), bi-metric models (Milgrom, 2009), and Moffat's 'STVG' theory (Moffat, 2006). In all of these theories, the 'MOND' contribution to the gravitational force is carried by an extra field or fields.

But one is also free to construct relativistic versions of modified inertia theories. Such theories would modify the 'matter action,' S_{matter}, in Equation (6.6), which corresponds to the 'kinetic' part of the action in the non-relativistic case. These relativistic theories are, like their non-relativistic counterparts, necessarily nonlocal.[4]

It is interesting to ask why Bekenstein chose *not* to do something like this when developing TeVeS. In his words:

> Whenever it becomes necessary to formulate a new theory of gravity, a conservative way to proceed in order to avoid immediate conflict with the tests of GR is to invoke a Riemannian metric $g_{\alpha\beta}$, build the Einstein–Hilbert action for the geometry's dynamics out of it, and effect the departure from standard GR by prescribing the relation between $g_{\alpha\beta}$ and the physical geometry on which matter propagates (Bekenstein, 1993, p. 3641).

This is essentially an argument from mathematical convenience (rather than from some deeper physical principle), and so we should not be too surprised, perhaps, if a theory like TeVeS turns out not to be the correct one.

In nonlocal formulations of gravity, the gravitational interactions are described by the dynamics of a single metric tensor without introducing any extra (scalar or vector) fields. Deffayet et al. (2011) established some of the properties that such a 'pure-metric' theory would need to have in order to exhibit Milgromian behavior in the low-acceleration limit, and to give the 'correct' expression for gravitational lensing, in the restricted case of a static, spherically symmetric geometry. They then demonstrated the existence of Lagrangians which yielded the correct limiting behavior. In a subsequent paper, Deffayet et al. (2014) derived a set of full field equations for such a theory and specialized them to the case of a homogeneous, isotropic and spatially flat universe. Their model contained two scalar functions (not fields!), or 'nonlocal invariants,' which could be adjusted to yield Milgromian behavior in the weak field limit. A natural property of these models is that Milgrom's constant, a_0, is some functional of the metric so that it changes as the universe evolves, and in such a way that the 'MOND force' only becomes effective at recent times. Kim et al. (2016) showed that such models were capable of reproducing the same expansion history of the universe as predicted by the ΛCDM model over almost the full range of cosmic evolution.

Other discussions of nonlocal gravitational theories in the context of Milgromian dynamics include Barvinsky (2003), Soussa and Woodard (2004), Deser and

---

[4] It is probably not obvious to most readers *why* theories constructed in this way must be nonlocal. This conclusion is usually justified with reference to to an 1850 (!) theorem of M. V. Ostrogradski which states that the Hamiltonian must be unbounded from below if the matter action depends on a *finite* number of time derivatives. See e.g. Soussa and Woodard (2004).

Woodard (2007), Hehl and Mashhoon (2009b), Hehl and Mashhoon (2009a), Blome et al. (2010) and Arraut (2014).

§

In theories such as TeVeS, which relate $g_{\mu\nu}$ and $\tilde{g}_{\mu\nu}$ via a disformal transformation, null geodesics of the physical metric do not correspond to null geodesics of the gravitational metric. Photons track null geodesics of the physical metric while gravitational waves follow null geodesics of the Einstein metric and are not affected by the scalar field. One implication is that the arrival time of gravitational waves from a distant source will be very different, in general, from the arrival time of photons (or neutrinos) that were generated in the same energetic event (Kahya and Woodard, 2007). The difference can be very substantial; in the case of a source located at the distance of the nearby Andromeda galaxy ($D \approx 1$ Mpc), the photons could arrive years after the gravitational waves (Kahya, 2008).[5]

Very recently, a claim has been made of near-simultaneous ($\Delta t \approx 2$ s) detection of bursts of gravitational waves and photons from a source located at about 40 Mpc from the Earth (Abbott et al., 2017; Goldstein et al., 2017; Savchenko et al., 2017). Assuming that this result withstands critical scrutiny (and that other, similar events are observed), the consequences for the Milgromian research program would be substantial, since one would be able to rule out all theories in which gravitational waves couple to a different metric than matter. A critical rationalist, whether Milgromian or otherwise, could hardly hope for a more felicitous circumstance.

Of the theory variants discussed so far in this chapter, the 'local' theories have the property that the Milgromian 'force' is carried by fields other than the metric, and such theories would be ruled out by a requirement that photons and gravitons propagate at the same rate. However the 'nonlocal,' or pure-metric, theories would remain viable.

§§

Some ideas about how a Milgromian universe might evolve over cosmological timescales can be obtained from a simple *non*-relativistic model, similar in spirit to the so-called 'Newtonian cosmology' that reproduces the basic features of a Friedmann universe under general relativity (McCrea and Milne, 1934). As we will see, a natural approach (Felten, 1984; Sanders, 1998, 2001) leads to the conclusion that a Milgromian universe would expand uniformly on scales of the cosmological horizon, as described by the Friedmann equation, while on sub-horizon scales

---

[5] Amusingly (from the point of view of a Milgromian researcher), these authors label theories like TeVeS 'dark matter emulators.' Given that dark matter is a hypothetical entity, it would make precisely as much sense to describe the dark matter postulates in the standard cosmological model as 'MOND emulators.' As far as I know, no Milgromian researcher has been so presumptuous.

density inhomogeneities would inevitably grow, and at a rate that exceeds the Newtonian prediction.

Consider an isolated, spherical, homogenous (i.e. constant-density) region of radius r. Invoking Milgrom's second postulate, the equation of motion for r in the low-acceleration regime is

$$\ddot{r} = -\left[\frac{4\pi G a_0}{3}\rho r\right]^{1/2},$$ (6.21)

with $\rho = \rho_0 (r/r_0)^{-3}$ the density of the sphere. Here r_0 and ρ_0 are the sphere's radius and density at some fiducial time, which we choose to be the present, so that r_0 plays the role of the 'comoving radius' of the sphere.

Anticipating the result that the large-scale expansion will be Friedmann-like, we define a dimensionless density parameter Ω_0 in the usual way as

$$\Omega_0 \equiv \frac{8\pi G\rho_0}{3H_0^2},$$ (6.22)

with H_0 the current Hubble parameter, i.e. expansion rate $|\dot{r}/r|$. Then Equation (6.21) becomes

$$\ddot{r} = -\left[\frac{\Omega_0}{2}H_0^2 r_0^3 a_0\right]^{1/2} r^{-1}.$$ (6.23)

Integrating once with respect to time, we obtain the Milgromian analog of the Friedmann equation:

$$\dot{r}^2 = u_i^2 - \left(2\Omega_0 H_0^2 r_0^3 a_0\right)^{1/2} \ln\left(r/r_i\right),$$ (6.24)

where r_i is the initial radius of the sphere and u_i is its initial velocity of expansion or contraction.

The minus sign in Equation (6.24) guarantees that any initial expansion will stop at some maximum radius, r_m, and the region will recollapse. The maximum radius is given by

$$r_m = r_i e^{q^2}, \quad q^2 = \frac{u_i^2}{\left(2\Omega_0 H_0^2 r_0^3 a_0\right)^{1/2}}.$$ (6.25)

So far we have *assumed* that the sphere is entirely in the low-acceleration regime. In our homogenous (constant-density) universe, the gravitational acceleration inside a spherical region increases linearly with distance from the center, and so there will be a critical radius, r_c, below which the (Newtonian) acceleration inside is everywhere less than a_0. This radius is the solution to

$$r_c = \sqrt{\frac{GM(r_c)}{a_0}} = \left[\frac{4\pi G\rho r_c^3}{3a_0}\right]^{1/2}.$$ (6.26)

Writing $\rho = \rho_0/x^3$, with $x = r_c/r_0$ the 'scale factor' of the region (equal to one at the current time), we find

$$r_c = \frac{2a_0}{H_0^2 \Omega_0} x^3. \tag{6.27}$$

Thus, in a matter-dominated universe, Milgromian dynamics is only expected to apply in regions smaller than r_c, and this radius increases with time. It is straightforward to show (Felten, 1984; Sanders, 1998) that r_c increases with time faster than both the scale factor and the horizon radius in both the radiation-dominated and matter-dominated regimes. There will therefore come a time after which the entire universe is in the Milgromian regime and the expansion is governed by a different equation than Friedmann's. At earlier times, it is reasonable to expect that Friedmann's equation still applies to regions greater in size than r_c at that time. But there will always be regions smaller than r_c in which Milgromian dynamics applies. These regions will expand more slowly than the universe at large, causing their density to increase above the mean value ρ.

We can estimate when the *entire* universe (within the horizon) will have entered the Milgromian regime by adopting values for the constants H_0 and Ω_0. The current consensus would have $H_0 \approx 75$ km s^{-1} Mpc^{-1}. In the case of Ω_0, a Milgromian researcher would likely identify it with Ω_b, the 'baryon' (i.e. normal matter) density parameter:

$$\Omega_b \equiv \frac{8\pi G \rho_b}{3H^2}. \tag{6.28}$$

As discussed in some detail below, a robust estimate for $\rho_{b,0}$ exists based on nucleosynthesis arguments, and on a direct census of baryons in the nearby universe. For the adopted value of H_0, the current baryon density parameter is then $\Omega_{b,0} \approx 0.024$. (Standard-model cosmologists have adopted a larger value for this parameter, as discussed below.) For these parameter choices, the entire, causally-connected universe (aside from any over-dense regions) will have entered into the Milgromian regime at a scale factor of ~ 0.23, corresponding to a redshift of ~ 3.3 (Sanders, 1998).

These arguments suggest that the appropriate choice for the initial radius r_i in Equation (6.24) is just r_c. If we consider a region with the (baryonic) mass of a galaxy – say, $10^{11} M_\odot$ – its 'initial' radius would then be ~ 14 kpc and the region would enter into the Milgromian regime at the large redshift of ~ 140. The 'initial' expansion velocity would be set by the universal expansion on a scale of r_i, or $u_i \approx 320$ km s^{-1}. Equation (6.25) then implies that $r_m/r_i \approx 3.5$: in other words, a galaxy-sized region would expand by a factor of about four before recollapsing. The elapsed time, from turnaround to 'virialization,'[6] would be about 10^8 years.

At first sight, the existence of tiny regions that depart from the overall expansion would seem to play havoc with the usual assumption of homogeneity at early times – an assumption that is standardly made, for instance, by particle physicists who compute the rates of formation of light nuclides via nucleosynthesis. But Sanders (1998) argues that, in the radiation-dominated universe, pressure gradients would strongly inhibit the growth of matter perturbations, with the consequence that no significant deviation from homogeneity would be expected. The thermal history of the early universe would then be identical to that of the standard model, and the rates of nucleosynthesis would also be essentially the same. This conclusion will be important in what follows.

The assumption of standard-model cosmologists that most of the matter in the universe is 'dark' would imply that the universe becomes 'matter-dominated' – that is, that the energy density of (mostly dark) matter exceeds that of radiation – at very early times, when the mean temperature is about 10^4 K. This is before the era of 'recombination,' when electrons combine with protons to form neutral hydrogen. The matter density in a Milgromian universe is much smaller, and as a result, the universe would remain in the radiation-dominated regime until the scale factor has increased by about a factor of ten above its value at recombination, i.e. $z \approx 10^2$. But once this happens, rapid collapse can occur for regions within a certain range of masses: masses smaller than about $4 \times 10^9 \, M_\odot$, which is the mass contained within a region of size r_c; and masses greater than about $300 \, M_\odot$, which is the 'Jeans mass' at this epoch – the latter being the mass of a region for which gravitational forces exceed pressure forces. Thus, the first objects to form would have masses between those of current-day star clusters and dwarf galaxies.

Recall our estimate that the entire, causally-connected universe will have entered into the Milgromian regime at a scale factor of about 0.23. At the *current* time, recollapse would be occurring for regions up to about 30 Mpc in size. Such regions should be observed to depart from the universal expansion, and their densities should decline with distance from their 'centers,' in a manner that can be computed (Sanders, 1998). The result of this simple picture would be a 'fractal' universe in which the distribution of matter is clumped in a complicated, non-analytic way, a conclusion for which there is some observational support (Coleman and Pietronero, 1992). But such a universe would be difficult to reconcile with the idea of a uniform expansion characterized by a single scale factor at a given cosmological time.

On the other hand, recall that in a number of relativistic generalizations of MOND, including TeVeS, the Friedmann equation in its usual form remains

---

[6] This ugly word is standardly used by astrophysicists to mean the complicated process by which a self-gravitating system reaches a state of equilibrium, starting from some strongly non-equilibrium state. Galaxies in the current universe are usually assumed to be 'virialized.'

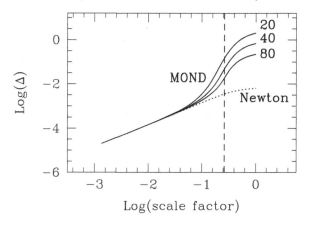

Figure 6.1 The growth of density fluctuations ($\Delta = \delta\rho/\rho$) in a Milgromian, baryons-only universe as computed in a two-field Langrangian theory of MOND. The initial over-density is assumed to be $\Delta = 2 \times 10^{-5}$, similar to the value assumed by standard-model cosmologists. The solid curves correspond to regions with comoving radii of 20, 40 and 80 Mpc. The dotted line is the standard Newtonian evolution. The assumed, background cosmological model has $\Omega_m = \Omega_b = 0.034$, $\Omega_\Lambda = 1.01$ and $h = 0.75$; the vertical dashed line indicates the epoch at which the cosmological constant begins to dominate the universal expansion. Calculations like this one suggest that primordial density perturbations in a Milgromian universe would reach over-densities at the current epoch that are similar to those predicted in the standard (dark matter dominated) cosmological model. Figure reprinted with permission from Robert Sanders, "The formation of cosmic structure with modified Newtonian dynamics," *The Astrophysical Journal*, 560, p. 1–6, 2001. Reproduced by permission of the AAS.

approximately valid, apart from small contributions due to source terms resulting from the energy density of the scalar and vector fields (if present). Sanders (2001) made a first attempt to compute the growth of perturbations in such a universe, under the assumption that Milgromian dynamics applies only to regions of *perturbed* density and that the 'background' cosmology is Friedmannian. His approach was based on the non-relativistic theory variant AQUAL. Recall from Chapter 5 that AQUAL can be formulated in terms of two potentials, one Milgromian and one Newtonian. Sanders (2001) considered a similar, two-field theory in which one field represents normal gravity while the other, the 'MOND' force, is assumed to act only on over- or under-dense regions. Following the same procedure outlined above, Sanders derived the equation describing the growth of initially small density fluctuations. Figure 6.1, from Sanders (2001), shows that the growth is nonlinear even in the regime where the density fluctuations are small; the growth becomes dramatically rapid where the background acceleration vanishes, i.e. when the vacuum energy density as described by a 'cosmological constant' becomes comparable to the matter energy density.

Figure 6.1 suggests that primordial density perturbations corresponding to galaxy-cluster-sized objects would reach over-densities in a Milgromian universe that are similar to those predicted in the standard cosmological model – a surprising result, given that standard-model cosmologists depend on the presence of dark matter to achieve the desired growth factor. The growth of over-densities on smaller scales would be even more enhanced since smaller regions enter the MOND regime earlier. Thus, Milgromian theory would appear capable of accommodating the observed large-scale structure, while at the same time predicting that smaller objects (galaxies, star clusters) would form at much earlier times than in the standard cosmological model.

§§

Some predictions of theory variant T_2 must depend on precisely which relativistic generalization is adopted. Indeed, as we have seen, Milgromian researchers have already made use of this fact to rule out some relativistic theories, through a process analogous to the 'differential diagnoses' used by physicians to rule out certain diseases. But calculations like those of Sanders just described suggest that there ought be *some* predictions that follow from essentially any version of theory variant T_2, and which, furthermore, differ from the corresponding prediction of the standard cosmological model. Two such predictions will be discussed in this chapter. Both predictions appear to be robust; both have been experimentally corroborated; and both are 'novel' according to at least some of the four criteria discussed in Chapter 2. The two predictions are:

1. An early epoch of cosmic reionization, at a redshift of $z \approx 15$ or greater;
2. A predicted ratio of approximately 2.4 between the amplitudes of the first and second peaks in the power spectrum of the cosmic microwave background.

As we will see, the absence of 'dark matter' in Milgromian theory is an essential factor in the success of these two predictions.

§§ .

1. *An early epoch of cosmic reionization, at a redshift of $z \approx 15$ or greater.* A generic prediction of big bang cosmologies, including the standard cosmological model, is an epoch in the early universe that is sometimes called the 'dark ages': an interval of time that begins with 'recombination,' the formation of neutral hydrogen when the plasma temperature has dropped below about 3000 K, and that ends when the first structures capable of producing ionizing radiation are formed. During the dark ages, there were (by definition) no light-producing sources. However, the universe would have been essentially transparent to light due to the low

scattering cross section of photons on neutral atoms,[7] and radiation produced at earlier epochs – notably, the photons produced during recombination itself – could have propagated through the neutral hydrogen gas with almost no attenuation.[8]

The dark ages came to an end with the formation of the first generation of stars – so-called 'Population III' stars. Presumably, star formation occurred (as it does in the local universe) inside of structures with masses of order $10^3 M_\odot$ or greater, the masses corresponding to star clusters or dwarf galaxies. Young, massive stars shine mostly at ultraviolet wavelengths and so are able to photoionize the surrounding hydrogen gas. While it is not obvious, it can be argued that this first generation of stars would have been very efficient at photoionizing the surrounding gas, even in the face of opposing factors such as a clumpy intergalactic medium or absorption of the ionizing photons within the galaxies (e.g. Barkana and Loeb, 2001).

A brief account was given above of cosmogony in a Milgromian universe. Since the matter density is relatively low (in comparison with the standard model, with its 'dark matter'), the era of 'matter dominance' occurs later than in the standard cosmological model, beginning roughly when the cosmological scale factor has increased by a factor of ten above its value at recombination. But once this happens, perturbations in the matter density can grow very quickly, since density fluctuations are in the low-acceleration regime. Thus, structure formation in a Milgromian universe first lags behind that of a dark-matter-dominated universe, then suddenly, at a redshift of $z \approx 200$, speeds up. The first objects to form, perhaps at a redshift of $z \approx 100$, would have the masses of current-day globular clusters or dwarf galaxies. Structure would grow 'from the bottom up,' with objects having the (baryonic) masses of galaxies like the Milky Way forming at redshifts $25 \gtrsim z \gtrsim 10$ (Nusser, 2002; Llinares et al., 2008; Sanders, 2008).

Formation of the first generation of *stars* would mark the end of the dark ages. Milgromian cosmologists have attempted to predict, approximately, when that would happen. In his 1991 paper, Sanders argued that stars would form before galaxies; galaxy formation would then be 'dissipationless,' i.e. relatively little of the matter would remain in gaseous form. Sanders's calculations did not include any of the gaseous physics that would set the timescale of star formation. Stachniewicz and Kutschera (2001, 2005) were the first to attempt this. They adopted essentially the same framework as Sanders for computing the growth of density fluctuations in a Milgromian universe. For reasons of computational convenience, they assumed (as had Sanders) spherical symmetry for the evolving gas cloud. But they improved on his treatment by representing the gas in terms of eight atomic and molecular species $(H, H^-, H^+, He, He^+, He^{++}, H_2, H_2^+)$ and they incorporated a comprehensive

---

[7] With the important exception of light blueward of the Lyman-α line at 1216 Å, which is energetic enough to ionize neutral hydrogen.

[8] These are the photons identified with the 'cosmic microwave background,' or CMB, so-called because their redshifted wavelengths as observed at the Earth fall mostly into the microwave band of the electromagnetic spectrum.

set of equations describing how the abundance of the various species would change with time due to chemical reactions, ionization and dissociation. Stachniewicz and Kutschera assumed that initial over-densities in the 'baryons' would only begin to grow after the start of matter dominance, which occurs, in a Milgromian cosmogony, at a redshift $z \approx 500$. They found that clouds of mass $\sim 3 \times 10^3 M_\odot$ would collapse by a redshift of $z \approx 30$ and suggested that the central parts, or 'cores,' of these collapsing structures would comprise the first generation of stars. McGaugh (2004) argued, based on these results, that a reasonable estimate for the redshift of reionization is $z_{\text{reion}} \approx 15$ or greater.

Star formation is undoubtedly a more complex process than modeled by Stachniewicz and Kutschera. But those authors provided a useful point of comparison. In a 2003 paper, they applied the same description of gaseous physics to estimate the time of formation of the first stars in the standard cosmological model. In the standard model (as discussed in a bit more detail below), star formation involves an extra step: growth of over-densities in the dark matter must occur first, before 'baryons,' in the form of gas, can fall into the dark matter potential wells and form stars. Using the same approximations adopted by standard-model cosmologists to describe the formation of the dark matter structures, Stachniewicz and Kutschera concluded that formation of baryonic objects that could possibly reionize the universe was very unlikely before a redshift of $z \approx 8$.[9] They argued, in addition, that relaxing their assumption of spherical symmetry was unlikely to speed up the formation process appreciably due to the timescale set by cooling of the gas. These conclusions were at least approximately consistent with those of standard-model cosmologists, as discussed in more detail below.

§

Stachniewicz and Kutschera's prediction of an early epoch of reionization in a Milgromian cosmogony is unambiguously 'novel' according to at least two of the criteria defined in Chapter 2: criterion P, which requires that the theory on which the prediction is based pre-date the prediction's experimental confirmation; and criterion Z, which requires that the theory (in this case, T_2) not have been designed in such a way as to yield a known result. In this respect, theory variant T_2 can be said to be 'theoretically progressive' as Lakatos defined that term. But has the prediction been *confirmed*? – which is a necessary requirement for *empirical* progressivity. Here, unfortunately, the situation is not so clear.

---

[9] Redshift is a well-defined quantity from an observational point of view, but the relation between z and cosmic time (measured since the big bang) will be different in different cosmological models. It is perfectly acceptable to directly compare Milgromian and standard-model predictions for z_{reion} but it should be kept in mind that the corresponding *times* in the two models may be different even if z_{reion} is the same.

Until about 2003, the only observational constraints on the epoch of reionization consisted of lower limits on z_{reion} (that is, upper limits on the corresponding time). The technique was based on observations of the spectra of quasi-stellar objects, or 'quasars,' which are known to emit some fraction of their radiation in the UV, at wavelengths shorter than the wavelength corresponding to the Lyman-alpha (Lyα) transition at 1215 Å. Such photons are efficiently absorbed by neutral hydrogen, and so the fact that the spectra of some quasars with redshifts as large as $z \approx 6$–7 do not exhibit Lyα absorption indicates that the intergalactic medium was highly ionized already at these redshifts. There is general agreement that such observations imply that reionization occurred at some redshift later than $z \approx 1000$ (i.e. recombination) and earlier than $z \approx 6$.

Starting around 2003, another technique came into use, based on observations of the cosmic microwave background (CMB). Photons from the CMB travel unimpeded through neutral hydrogen, but once they encounter free electrons they are subject to Thomson scattering. One consequence is a reduction in the observed amplitude of fluctuations in the CMB, by a factor $e^{-\tau}$, where τ is the optical depth with respect to Thomson scattering. The quantity τ is evidently a measure of the state of the intergalactic medium *after* reionization; in fact it measures the integrated effect of electron scattering from z_{reion} all the way until $z = 0$. One could imagine measuring τ directly from the amplitude of the CMB temperature fluctuation spectrum, but there is an obvious degeneracy between the inferred amplitude of those fluctuations (at recombination) and the scattering optical depth.

But the scattering of photons by free electrons has another observable consequence: it results in polarization of the scattered radiation (Kaplinghat et al., 2003). One might think that the net polarization of all the scattered photons would be zero, but this is only the case if the radiation field is precisely isotropic. In fact, the CMB radiation is anisotropic (as discussed in a little more detail below) and this results in a small net polarization as observed from the Earth. The degree of polarization depends on the fraction of CMB photons that undergo scattering (among other factors), which allows one to measure the scattering optical depth in the post-reionization universe. Determination of τ by this technique is challenging: the signal is intrinsically small, and calibration of the instruments is difficult since there is no 'standard' source in the sky for polarization measurements. In addition, the amplitude of the electron scattering signal is comparable to that produced by 'foreground' sources such as polarized dust emission. Sources of systematic uncertainty like these continue to be grappled with.

Given a measured value of τ, the simplest way to estimate z_{reion} is to assume that the intergalactic hydrogen transitioned from a fully neutral to a fully ionized state in zero time.[10] In reality, of course, the ionization process was gradual, and relating

---

[10] 'Fully ionized' is typically defined as a volume-weighted neutral hydrogen fraction that is less than 10^{-3}.

τ to z_{reion} requires some assumption about the cosmic time at which the first ionizing sources formed and the timescale over which their ionizing effects were felt. For a given redshift, z_{reion}, of complete ionization, a more gradual onset of ionization implies a larger value of τ.

In 2003, polarization data from the *WMAP* satellite observatory[11] were used to estimate τ. Bennett et al. (2003, p. 1) wrote: "The optical depth of reionization is $\tau = 0.17 \pm 0.04$, which implies a reionization epoch of $t_r = 180^{+220}_{-80}$ Myr (95% CL) after the big bang at a redshift of $z_r = 20^{+10}_{-9}$ (95% CL) for a range of ionization scenarios." This early result came as a considerable surprise to standard-model cosmologists, whose expectations had been set by the quasar observations described above. For instance, Bromm and Larson (2004, p. 26) wrote: "Such a high value [of τ] is surprising. We know that the IGM [i.e. intergalactic medium] was completely ionized again at redshifts $z \lesssim 6$ ... The observed excess must have arisen from the scattering of CMB photons off free electrons ionized by star formation at higher redshifts," while Milgromian cosmologist Stacy McGaugh noted that "180 Myr [is] a very early time by the standard of any prior expectation in the context of CDM" (McGaugh, 2004, p. 35).

Of course, a reionization redshift of ~ 20 would be quite consistent with the *Milgromian* prediction of Stachniewicz and Kutschera. But after 2003, estimates of τ and z_{reion} based on CMB polarization data trended gradually downwards (that is, the epoch of reionization was shifted toward later times). Analysis of the accumulated nine years of data from the *WMAP* satellite led to the estimate $\tau = 0.089 \pm 0.014$, corresponding to an instantaneous reionization redshift $z_{\text{reion}} \approx 10.6 \pm 1.1$ (Hinshaw et al., 2013).

Data from the *WMAP* satellite ceased after 2010. *WMAP* was followed by *Planck*.[12] The initial data releases from *Planck* did not include polarization data, which were considered too uncertain due to poorly understood sources of systematic error. But in 2014 an estimate of $\tau = 0.089 \pm 0.032$ (Planck Collaboration XVI, 2014) was published that did not depend on polarization measurements; this estimate was based on the measured amplitude of CMB temperature fluctuations, and the degeneracy mentioned above was removed by using an independent method to estimate the actual amplitude of the fluctuations at recombination. Subsequent papers from the Planck Collaboration began to include polarization-based estimates of τ. The first was $\tau = 0.078 \pm 0.019$, implying reionization at $z_{\text{reion}} \approx 9.9^{+1.8}_{-1.6}$ (Planck Collaboration XIII, 2016), followed by an even lower $\tau = 0.055 \pm 0.009$ (Planck Collaboration XLVI, 2016). Thus, while the early measurements of τ were quite consistent with Stachniewicz and

---

[11] *WMAP*, the *Wilkinson Microwave Anisotropy Probe* (originally known as the *Microwave Anisotropy Probe*, or *MAP*), is a satellite observatory launched from Florida on 30 June 2001. It remained in operation until 2010.

[12] The *Planck* satellite observatory (originally named *COBRAS/SAMBA*, the *Cosmic Background Radiation Anisotropy Satellite/Satellite for Measurement of Background Anisotropies*) was launched on 14 May 2009 from Guiana. It remained active until October 2013.

Kutschera's Milgromian prediction ($z_{reion} \gtrsim 15$), later measurements of τ seemed more consistent with standard-model expectations.

§

What were those expectations, exactly? In reading the standard-model literature, one often comes across statements about the 'predicted' redshift of reionization or the electron-scattering optical depth. For instance, in the paper Planck Collaboration XLVI (2016) just cited, which gave the low measured value of $\tau \approx 0.055$, the authors wrote (italics mine):

Reionization history models based on astrophysical observations of high redshift sources *predict* asymptotic values of τ at high redshift in the range $0.048 - 0.055$ (figure 7 of Mashian et al. 2016) or $0.05 - 0.07$ (figure 2 of Robertson et al., 2015). Our results are fully consistent with these *expectations*.

The two cited papers – Robertson et al. (2015) and Mashian et al. (2016) – are theoretical studies. On what basis do these two studies claim to 'predict' the redshift of ionization?

In fact, neither paper makes any such claim. What the authors of both papers do is to *assume* the correctness of the most recent *measurement* of $\tau = 0.066 \pm 0.012$ – a measurement *published by the Planck Collaboration itself*, just prior to their paper XLVI – and then ask: What must we *postulate* about the cosmic star formation history if we are to reconcile this measured value of τ with the standard cosmological model?

For instance, Robertson et al. (2015, p. A17) write (italics added):

The Planck Collaboration et al. (2015) [Paper XIII, published in 2016] has recently reported a significantly lower value of the optical depth, $\tau = 0.066 \pm 0.012$, consistent with a reduced redshift of instantaneous reionization, $z_{reion} = 8.8^{+1.2}_{-1.1}$. *Here we determine the extent to which the Planck result reduces the need for significant star formation in the uncharted epoch at $z > 10$.*

And Mashian et al. (2016, p. 2101) write (italics added):

the inclusion of galaxies with SFRs [star formation rates] well below the current detection limit . . . leads to a fully reionized universe by $z \sim 6.5$ and an optical depth of $\tau \simeq 0.054$, consistent with the recently derived Planck value at the 1σ level. . . . These findings, in line with previous analyses, indicate that *a significant population of low-mass star-forming galaxies is necessary for cosmic reionization* and strengthen the conclusion that the bulk of photons responsible for reionizing the early universe emerged from ultrafaint galaxies.

In both of these studies (and many similar ones), the authors are explicitly invoking auxiliary hypotheses in order to *accommodate* the latest measurements of τ to the standard cosmological model – even if, in so doing, they are forced to make assumptions that are not observationally testable ("uncharted epoch," "below the current detection limit"). There are no predictions here about τ or z_{reion}. That the authors of Planck Collaboration XLVI (2016) describe the conclusions of these two papers as 'predictions' is not surprising to anyone familiar with the standard-model literature: the distinction between 'prediction' and 'accommodation' – a distinction that is so important from the point of view of a critical rationalist – is often blurred, if not ignored completely, by standard-model cosmologists.

As always, our goal here is not to critique the standard model of cosmology. But recall that in judging the novelty of a predicted fact according to Musgrave's criterion:

M: A corroborated fact provides evidential support for a theory if the best rival theory fails to explain the fact, or if the fact is prohibited or improbable from the standpoint of that theory

we decided that the *rival* theory was not allowed to 'explain' the fact via modifications made purely in response to the fact itself, especially if those modifications were fine-tuned, manifestly ad hoc or otherwise unreasonable. All, or essentially all, of the standard-model studies of cosmic reionization published after about 2002 fail to meet this condition. Almost always, they frame the theoretical question in the same way as in the studies cited above by Mashian et al. (2016) and Robertson et al. (2015): namely, what *assumptions* can we make that allow us to accommodate the (most recently) measured value of the scattering optical depth to the standard cosmological model?

If we are to find a bona-fide *prediction* of τ or z_{reion} by standard-model cosmologists, we must return to the pre-*WMAP* literature, since essentially all papers published after the first CMB polarization measurements incorporate a measured value of τ as a constraint or a 'prior' into their modeling.[13] Fortunately, a comprehensive review article was published in 2001 with the title "The first sources of light and the reionization of the universe." The authors wrote that, according to the standard cosmological model, the first generation of stars "reionized most of the hydrogen in the universe by $z = 7$" (Barkana and Loeb, 2001, p. 125).

This prediction for z_{reion} is essentially the same as the (standard-model) prediction of Stachniewicz and Kutschera cited above. Recall that those authors predicted values of z_{reion} under both Milgromian and standard-model assumptions. One would like to see, in the *standard-model* literature, a comparative prediction of this sort. No such study appears to exist; in fact, in the voluminous standard-model literature on this subject, Stachniewicz and Kutschera's papers do not receive a single citation. One possible interpretation of this remarkable fact is that standard-model cosmologists have no interest in making *a priori* predictions of quantities like z_{reion}.[14]

---

[13] The first determinations of τ from CMB data actually preceded *WMAP* by a few years, e.g. Schmalzing et al. (2000); Wang et al. (2003).

[14] Another possibility, of course, is that standard-model cosmologists reflexively disregard any results published by Milgromian researchers.

Table 6.1 *Corroborated excess content, theory variants $T_0 - T_2$*

Theory variant	Prediction	Status	Novelty			
			P	Z	M	C
T_0	$V(R) \to V_\infty$	confirmed	no	no	–	yes
	$V_\infty = (a_0 G M_{\text{gal}})^{1/4}$	confirmed	yes	yes	yes	–
	$a = f(g_N/a_0)\, g_N$	confirmed	yes	yes	yes	–
	Renzo's rule	corroborated	yes	yes	yes	–
T_1	$\Sigma_{\text{ph}}(0) \lesssim a_0/(2\pi G)$	corroborated	yes	yes	yes	–
	Central surface density relation	confirmed	yes	yes	yes	–
	Vertical kinematics in Milky Way	corroborated	yes	yes	yes	–
	$V_{\text{rms}} \approx (a_0 G M)^{1/4}$	partially corroborated	yes	yes	yes	–
	External field effect	possibly corroborated	yes	yes	yes	–
	Low merger rate	neither confirmed nor refuted	no	–	yes	–
T_2	$z_{\text{reion}} \gtrsim 15$	possibly corroborated	yes	yes	yes	–
	$A_{1:2} \approx 2.4$	confirmed	yes	yes	yes	–

In any case: I conclude that the standard-model prediction for z_{reion} is ~ 7–8. This value is lower than most, if not all, estimates of z_{reion} derived from polarization studies of the CMB.

Table 6.1 summarizes these conclusions. Given the evolution over time of the measured value of z_{reion}, I consider the Milgromian prediction of this quantity to be 'possibly corroborated' at best. The reader is encouraged to consult the post-2017 literature for the current consensus on the measured value of this quantity.

§§

2. *A ratio of approximately 2.4 between the amplitudes of the first and second peaks in the power spectrum of the cosmic microwave background.* In 1999, Stacy McGaugh proposed a test based on a prediction which, he argued, should follow from essentially any version of theory variant T_2. McGaugh began from the argument of Sanders (1998, 2001) that was presented above: that the early thermal history of a Milgromian universe should be identical to that of the standard cosmological model. One implication is that calculations under the standard model of so-called 'big bang nucleosynthesis' (BBN) – the formation of light elements in the first few minutes after the big bang – should carry over, without significant

change, to a Milgromian universe. This component of the standard cosmological model was in a mature and stable state by the end of the twentieth century: there was intersubjective agreement that the relevant nuclear reactions had been identified, that the rate constants of those reactions were (for the most part) well established, and that the dependence of the predicted abundances of elements such as helium, lithium and deuterium on the cosmological parameters were well determined and unlikely to change. These conclusions were unaffected by the addition of the dark matter postulate DM-2 to the model (Chapter 1), for two reasons: the energy density at the time of BBN would have been dominated by photons, whether or not dark matter was present, so that the predicted expansion history of the universe would be unchanged; and the dark matter particles were, by assumption, weakly interactive with normal matter, so they would not appear as source terms in the nuclear reaction equations.

A second consequence of the Sanders argument was that the *mathematical formalism* used by standard-model cosmologists to compute the early growth of primordial density fluctuations could also be carried over to a Milgromian cosmology – with the important understanding that in a Milgromian universe there was no need to postulate the existence of dark matter. That is, the equations describing the early growth of perturbations in a Milgromian universe (roughly speaking, until the time when the universe was matter-dominated and the density perturbations were in the low-acceleration region) would be the same as in the standard model, even though the *results* of those calculations would differ due to the absence of dark matter.

Given these two *Ansätze*, McGaugh was in a position, in 1999, to carry out a calculation which – remarkably – had not, at that time, been done by any standard-model cosmologist. McGaugh asked what a Milgromian cosmologist and a standard-model cosmologist would each predict for the form of the power spectrum of temperature fluctuations in the cosmic microwave background (CMB). In a Milgromian cosmology, the critical parameter influencing the form of the predicted spectrum is the mean 'baryon' density (that is, the density of normal matter), which was well determined at the time based on BBN calculations combined with observed light-element abundances. Whereas in the standard cosmology, the mass density of the universe would have been dominated by dark matter, implying a different form for the CMB fluctuation spectrum (for reasons that are discussed in more detail below).

At the time of McGaugh's prediction, observations of the CMB were just beginning to approach the resolution and sensitivity needed to make useful statements about the statistics of the temperature fluctuations. McGaugh chose to frame his prediction in terms of a quantity which, he expected, would be among the first to be determined by future observations, and which furthermore was strongly dependent on the assumed matter (i.e. 'baryon') density: the amplitude ratio of the first two 'peaks' in the power spectrum.

McGaugh's Milgromian prediction was confirmed via observations just three years later, in 2002. The same observations that *confirmed* the Milgromian prediction, *refuted* the standard-model prediction. Thus, McGaugh's exercise fits the definition given by Karl Popper of a 'crucial experiment': an experiment that confirms a prediction of a new theory, while simultaneously refuting the old theory.

Most of the remainder of this chapter will be devoted to this remarkable story. As we will see, Musgrave's condition for evidential support – a corroborated fact provides evidential support for a theory if that fact is prohibited or improbable from the standpoint of the rival theory – is particularly difficult to apply in this case, due to the manner in which standard-model cosmologists chose to accommodate the CMB observations. Simply put, they executed what Popper had called a 'conventionalist stratagem': they chose to disregard the consensus that had been achieved over the previous two decades on the mean baryon density. By means of this 'immunizing tactic' (Popper's phrase), they were able to fit the CMB power spectrum, but they also created two inconsistencies which have persisted until the present day: the 'lithium problem' and the 'missing baryons problem.' Neither of these anomalies exists, or at least not necessarily, from the standpoint of a Milgromian cosmologist; both arose as a direct consequence of the decision by the standard-model cosmologists to include dark matter in their models.

§

By the early 1980s, particle physicists and astrophysicists had established a robust value for the mean density (mass per unit volume) of normal matter, or 'baryons,' in the universe. This density, ρ_b, is most often expressed via the dimensionless quantity Ω_b that is defined in Equation (6.28). The quantities Ω_b, ρ_b and H that appear in that expression are all functions of cosmological time; here the usual practice will be followed of using the symbol Ω_b to denote its current value. The current, dimensional density ρ_b is then proportional to $\Omega_b H_0^2$ with H_0 the Hubble constant, the current value of H.

The consensus value for the baryon density was based on an argument from nucleosynthesis. Nuclear species of low mass number – including ^{3}He, ^{4}He, ^{2}H (deuterium, or D) and ^{7}Li – would undergo nuclear reactions in the early universe, reaching steady-state values once the temperature had dropped below $kT \approx$ 0.01 MeV, a few seconds after the big bang. The predicted 'abundances' of the light elements – conventionally defined as number density normalized to that of hydrogen, e.g. D/H – are functions of a handful of parameters, of which only one is considered to be significantly uncertain: the number of nucleons (i.e. 'baryons') per photon:

$$\eta \equiv \frac{n_b}{n_\gamma}. \tag{6.29}$$

The quantity η is believed to remain nearly constant following the era of e^+e^- annihilation.[15] The photon density at the *current* epoch, n_γ, is well determined, and so η can be expressed in terms of the current density of *baryons* as

$$\Omega_b h^2 \approx 3.65 \times 10^{-3} \, \eta_{10}, \tag{6.30}$$

where $\eta_{10} \equiv \eta/10^{-10}$ and $h \equiv H_0/100$ km s^{-1}Mpc^{-1}; current estimates of h are $0.67 \lesssim h \lesssim 0.74$. Comparing the observed abundance of any single light element with the nucleosynthesis prediction yields a value (or confidence interval) for η, as long as the observed abundance is equal to, or can be securely related to, the primordial abundance. Repeating this exercise for each of the light species yields a set of confidence intervals; if they should overlap, then a 'concordance' value for η, hence $\Omega_b h^2$, will have been determined.[16]

The most common elements in the universe today are hydrogen and helium. The predicted helium abundance is determined essentially by the ratio of neutrons to protons just prior to the epoch of nucleosynthesis and is very insensitive to the *total* density of nucleons, hence to η. The predicted value of ^{4}He/H is about 0.24 (by mass), consistent with observed abundances. This agreement is widely interpreted as an important success of the standard model but leads to only weak constraints on η.

The tiny percentage of neutrons that do not end up in ^{4}He can be incorporated into the other, most tightly bound nuclei. Over the range $10^{-10} \le \eta \le 10^{-9}$, the predicted abundances of these nuclei just after the era of nucleosynthesis can be well approximated by the simple expressions (Sarkar, 1996)

$$\frac{D}{H} = 4.7 \times 10^{-4} \eta_{10}^{-1.6}, \tag{6.31a}$$

$$\frac{^3He}{H} = 3.3 \times 10^{-5} \eta_{10}^{-0.63}, \tag{6.31b}$$

$$\frac{^7Li}{H} = 5.5 \times 10^{-10} \left[\eta_{10}^{-2.38} + 1.02 \times 10^{-2} \eta_{10}^{2.38} \right]. \tag{6.31c}$$

The predicted abundances of D and ^{3}He are monotonic (decreasing) functions of η. The dependence of the predicted ^{7}Li abundance on η is more complicated. For high η, the abundance is determined primarily by production of ^{7}Be which electron-captures to ^{7}Li, yielding an abundance curve that first decreases, then increases, with increasing η. The result is a minimum predicted value of about 1.1×10^{-10} for ^{7}Li/H, which occurs for $\eta \approx 2.6 \times 10^{-10}$, i.e. for $\Omega_b h^2 \approx 9.4 \times 10^{-3}$. This minimum value is, of course, independent of η.

---

[15] Any entropy-producing processes after this time would imply a larger value for η during the nucleosynthesis epoch. Among other things, this would make the 'lithium problem' worse.

[16] The term 'concordance' was used with this meaning throughout the 1980s and 1990s (e.g. Olive et al., 1981; Turner, 1999). Beginning about 2002, cosmologists began associating a different meaning to the term 'concordance model,' as discussed below.

The uncertainties associated with *predicted* abundances like those of Equations (6.31) (for a given value of η) are believed to be small, and in any case, much smaller than the uncertainties associated with determinations of the *actual* primordial abundances, for at least two reasons: the many different forms (stars, cold gas, ionized gas, dust grains, etc.) in which the elements can appear ("An accounting nightmare!" in the words of Michael Turner, 2000, p. 625); and the fact that the light species are the most likely to experience changes in their abundance due to nuclear processes inside of stars and, possibly, in the interstellar medium, making the observed abundance different from the primordial value. This is particularly true for deuterium, since the weakly bound deuteron is destroyed by thermonuclear reactions more readily than any other (stable) nucleus; the strong, inverse dependence of D/H on η in Equation (6.31a) is a reflection of this fact. Indeed, deuterium has never been detected in the atmosphere of any star. Deuterium is also capable of being created, e.g. by spallation,[17] although it is usually argued (Epstein et al., 1976) that the rate of such processes is low; a typical argument is that any mechanism that produced deuterium in significant amounts would over-produce (in the sense of violating observed abundances) the other light nuclei.

The deuterium-to-hydrogen ratio is measurable from H and D Lyman-line absorption in interstellar gas observed along the lines of sight to nearby stars. The mean value is $D/H \approx 1.5 \times 10^{-5}$, although with substantial scatter; at least some of this scatter is believed to be real and to reflect true spatial variations (Linsky, 2003). However this gas is not primordial, and theoretical models of galactic chemical evolution suggest that the deuterium abundance could have been depleted by factors of two to ten due to recycling of gas through stars (Audouze and Tinsley, 1976; Scully et al., 1997).

Since deuterium is burned to ^{3}He in stars, Yang et al. (1984) suggested that the *sum* of the two abundances is insensitive to the poorly understood effects of astration.[18] Those authors, and subsequently others (e.g. Walker et al., 1991; Smith et al., 1993), found that the observed values of $(D + {}^3He)/H$ and $(^4He/H)$ were consistent with each other and with a baryon density

$$\eta \approx 3 \times 10^{-10}, \quad \Omega_b h^2 \approx 0.012. \tag{6.32}$$

This value of η is close to the value, mentioned above, at which the abundance of ^{7}Li is predicted to have its minimum. And it turns out that the measured lithium abundance is, in fact, quite consistent with this value.[19]

---

[17] 'Spallation' is the interaction of a cosmic ray with a nucleus.

[18] 'Astration' refers to a change in abundance due to thermonuclear processes within stars.

[19] It is fair to describe this agreement as a successful, novel prediction of the standard cosmological model; see e.g. Walker (1991, p. 52). But as discussed in detail below, starting around 2002, standard-model cosmologists declared the lithium measurement in error, because it conflicted with another, dark-matter-dependent prediction of their model.

The lithium abundance (as determined from a particular class of stars) has remained remarkably constant since the early 1980s, something that can not be said about the abundances of any of the other light nuclides. Furthermore, the scientific case for identifying the *observed* lithium abundance with its *primordial* value is compellingly strong for these stars. For these reasons (and some others to be presented below), the lithium abundance has played a uniquely important role in constraining the cosmic baryon density. We will therefore take a short detour and discuss how stellar astrophysicists go about the business of determining the primordial lithium abundance.

§

Lithium-7 (the common isotope of lithium) is thought to have been produced in the first few seconds after the big bang through a coupled set of reactions, the most important of which is ^{3}He (α, γ) ^{7}Be followed by electron capture of ^{7}Be to ^{7}Li. The rate constants for these reactions are well determined experimentally (e.g. Cyburt et al., 2008). Lithium, like deuterium, can be destroyed in stars, but only at higher temperatures than deuterium, above about 2.5×10^6 K (compared with 0.6×10^6 K for deuterium). And indeed lithium is detectable in the atmospheres of many stars – including the Sun – through an absorption line at 6707 Å associated with the neutral ^{7}Li atom. The fraction of the lithium that is neutral, hence detectable, drops rapidly with atmospheric temperature and so measurements of the lithium abundance are limited to dwarf and subgiant stars, i.e. stars on or near the main sequence, with effective temperatures below about 8000 K.[20]

The lithium abundance as measured in nearby stars is strongly correlated with their inferred age, consistent with the idea that the surface lithium is progressively destroyed in these stars (Duncan, 1981). Models of stellar structure suggest a compelling explanation: stars (like the Sun) on the lower main sequence contain convection zones that encompass the surface layers and extend downward to regions where the temperature is much higher, hot enough, in some cases, to remove ^{7}Li through proton capture. As a star ages on the main sequence, its surface abundance of lithium would be gradually depleted; for instance, the current surface abundance of lithium in the Sun is thought to be two orders of magnitude smaller than its primordial value. In spite of this general trend, however, wide variation is observed in lithium abundances among nearby stars of a given mass and age.

These nearby stars are so-called 'Population I' stars: stars that are relatively young and that must have formed from gas of which at least a fraction had been processed through an earlier generation of stars. A promising place to search for ^{7}Li at its *primordial* abundance is in Population II stars in the halo of the Milky

---

[20] Effective temperature, or T_{eff}, is defined as the the temperature of a black body that emits the same total radiation.

Way, stars which have metallicities[21] much lower than that of the Sun. A low metallicity indicates that the gas from which the star formed has undergone little or no pre-processing through an earlier generation of massive stars. Ideally such stars should be on or near the main sequence, hence of low mass (roughly a solar mass or less; more massive Population II stars would have evolved off the main sequence by now), but not so low that their convection zones are deep enough to deplete the surface lithium. (Convective zones in low-metallicity stars are predicted to be thinner than those in Population I stars of the same mass – another good reason for targeting these stars.) Stars should also be massive enough that they have not spent much time on the parts of the *pre*-main-sequence contraction phase when temperatures at the base of the surface convection zone can reach temperatures high enough to destroy lithium (Bodenheimer, 1965).

These various requirements restrict the useful sample of stars to the warmest metal-poor dwarf (unevolved) stars in the Galactic halo. At distances associated with the halo, such stars appear very faint, and detection of the lithium absorption feature in the spectra of such stars became technically feasible only in the early 1980s.[22] The pioneering studies, by François Spite and Monique Spite (Spite and Spite, 1982; Spite et al., 1984), measured the equivalent width of the neutral ^{7}Li feature in a sample of halo dwarf stars with temperatures in the range $5300 \, \text{K} - 6300 \, \text{K}$ and with metallicities ranging from about one-tenth solar down to one-thousandth. Spite and Spite made a remarkable discovery: for these stars, the lithium abundance was essentially *constant* as a function of temperature and metallicity – quite different from what had been observed in the nearby Population I stars. Furthermore, the observed abundance was quite close to the minimum abundance predicted by models of big bang nucleosynthesis, and, therefore, consistent with the value of η (Equation 6.32) implied by the observed abundances of the other light nuclei.

The constancy of the measured ^{7}Li abundance with respect to surface temperature is called the 'Spite plateau' (Figure 6.2). The reason that Spite and Spite chose to emphasize the constancy of the abundance with respect to temperature and not metallicity was that they *expected* to find evidence of depletion, and, if present, the degree of depletion would be expected to depend strongly on stellar mass, hence temperature.[23] In fact, for the coolest stars in their sample, below about 5400 K, the lithium abundance does begin to drop, consistent with the expectation of deep convective zones in these stars. (Stars sufficiently far down the main sequence are believed to be completely convective.) The absence of a temperature dependence

---

[21] The term 'metal' is here understood to mean any element that is not produced with significant abundance by nucleosynthesis in the early universe; in practice, 'metallicity' for stars like these is often defined as the iron abundance, Fe/H.

[22] In fact, in their initial work, Spite and Spite identified a sample of relatively *nearby* halo stars. Identification as halo stars was based on their low metallicities, and their kinematics, which indicated that they were not part of the Galactic disk population.

[23] "En effet, s'il [i.e. lithium] se détruisait partiellement, le taux de destruction dépendrait fortement de la masse de l'étoile, et on n'observe pas cet effet" (Spite, 1984, p. 376).

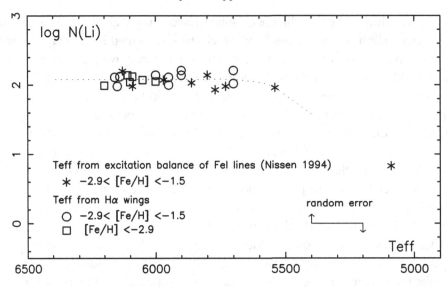

Figure 6.2 The 'Spite plateau.' Each point represents the measured abundance of ^{7}Li in a single, metal-poor dwarf star in the Galaxy's halo, plotted against that star's effective (surface) temperature. The lithium abundance is found to be constant for stars with temperatures above about 5500 K; as discussed in the text, the consensus among stellar astrophysics is that this constant value is in fact the primordial abundance. Inferring a star's current ^{7}Li abundance from the observed spectral feature requires accurate knowledge of the star's temperature; the plot shows that different techniques for inferring T_{eff} yield overlapping results. The plot also shows that the plateau location is insensitive to the stellar metallicity, expressed here in terms of the iron abundance. Reprinted with permission from Monique Spite, "Abundances of the elements in the halo stars, interaction between observation and theory," in IAU Symposium 189, p. 185–192. Copyright (1997) by Springer Nature.

among the warmer stars suggested very strongly to Spite and Spite that there had been no significant depletion in any of these stars, hence that they were measuring the primordial lithium abundance.

This conclusion was strengthened by other arguments. The scatter around the plateau was observed to be very small, consistent with measurement errors alone. It was (and remains) difficult for most stellar astrophysicists to believe that any physical mechanism could produce a significant, *uniform* depletion, independent of stellar mass and temperature, since all proposed mechanisms depend systematically on those quantities. A very different supporting argument came from the observed abundance of another lithium isotope. ^{6}Li is thought not to be produced in measurable amounts by thermonuclear reactions in the early universe, but cosmic-ray-induced spallation can produce the isotope in the cold interstellar gas from which stars form (Reeves, 1994). Detection of ^{6}Li in stellar photospheres is difficult; this isotope manifests itself only as a subtle extra depression in the red wing of the lithium resonance doublet. But there is general agreement that it is present in at least some stars on the Spite plateau. But ^{6}Li is a much more fragile isotope than ^{7}Li, and if the stellar surface material were exposed to temperatures hot enough to significantly reduce ^{7}Li, ^{6}Li should be completely destroyed (e.g. Brown and Schramm, 1988).

The sample of stars with measured lithium abundances has increased steadily in size since Spite and Spite's pioneering work, but the location of the Spite plateau

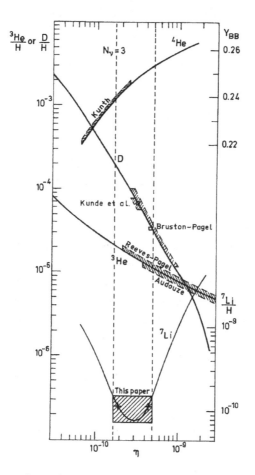

Figure 6.3 A figure from the discovery paper (Spite and Spite, 1982) of the 'Spite plateau' illustrating the concordance model as it existed in the early 1980s. Plotted are abundances of deuterium (D), ^{3}He and ^{7}Li (by number) and ^{4}He (by mass) versus the baryon-to-photon ratio η. The solid curves are from nucleosynthesis calculations by D. N. Schramm (1982). The lithium abundance measured by Spite and Spite is marked "This paper"; note that it falls close to the minimum value allowed by nucleosynthesis, at $\eta \approx 3 \times 10^{-10}$, and that this value of η is consistent with the measured abundances of the other three nuclei as well. The preferred, or 'concordance,' value for η remained remarkably close to this value throughout the 1980s and 1990s; it corresponds to a baryon density parameter $\Omega_b h^2 \approx 0.0125$. Reprinted with permission from François Spite and Monique Spite, "Abundance of lithium in unevolved halo stars and old disk stars – Interpretation and consequences," *Astronomy and Astrophysics*, 115, p. 357–366. Copyright ESO (1982).

Table 6.2 *Concordance, 1982–2000*

Reference	$\eta_{10} \equiv 10^{10} n_b / n_\gamma$	$\Omega_b h^2$
Schramm (1982)	$3 \leq \eta_{10} \leq 6$	$0.011 - 0.022$
Yang et al. (1984)	$3 \leq \eta_{10} \leq 10^a$	$0.011 - 0.037$
	$4 \leq \eta_{10} \leq 7^b$	$0.015 - 0.026$
Boesgaard and Steigman (1985)	$(3-4) \leq \eta_{10} \leq (7-10)$	$(0.011-0.015)-$
		$(0.026-0.037)$
Pagel (1986)	$3.0 \leq \eta_{10} \leq 3.6$	$0.011 - 0.013$
Walker et al. (1991)	$2.8 \leq \eta_{10} \leq 4.0$	$0.010 - 0.015$
Smith et al. (1993)	$2.86 \leq \eta_{10} \leq 3.77$	$0.011 - 0.014$
Kernan and Krauss (1994)	$2.6 \leq \eta_{10} \leq 3.0$	$0.0097 - 0.011$
Copi et al. (1995)	$2.5 \leq \eta_{10} \leq 6$	$0.0092 - 0.022$
Schramm (1998)	$2.8 \leq \eta_{10} \leq 6.8$	$0.010 - 0.025$
Ryan et al. (2000)	$1.7 \leq \eta_{10} \leq 3.9$	$0.0062 - 0.014$

[a] "Safe bet"
[b] "Best bet"

has remained remarkably unchanged, as has the consensus opinion among stellar astrophysicists that abundances on the plateau can be identified with primordial abundances. Spite and Spite noted, already in their 1982 paper, that the tight constraints on η from the measured lithium abundance were consistent with the (weaker) constraints on η from deuterium, from ^{3}He and from ^{4}He (Figure 6.3). Their preferred value was $\eta \approx 3 \times 10^{-10}$, corresponding to $\Omega_b h^2 \approx 0.012$. As Table 6.2 and Figure 6.4 document, subsequent authors reached very similar conclusions about the preferred value of η. Indeed, by about 1990, the consensus view was that Spite and Spite's original estimate was essentially correct; different authors differed mostly in terms of the confidence intervals they assigned to η, which depended primarily on the weights they assigned to the abundances of the other light elements.

Many astrophysicists who were active in the 1990s can still recall how definite were the statements, during this decade, about the baryon density that was implied by nucleosynthesis arguments. For instance, John Peacock, in his 1999 textbook *Cosmological Physics*, wrote (p. 298): "The constraint obtained from a comparison between nucleosynthesis predictions and observational data is rather tight" and quoted a range $0.010 \lesssim \Omega_b h^2 \lesssim 0.015$; he continued, "This result is one of the cornerstones of the big bang, and is often assumed elsewhere e.g. in setting the gas density at high redshifts when discussing galaxy formation." An oft-cited value (usually associated with researchers at the University of Chicago; e.g. Walker et al., 1996) was

$$\eta \approx 3 \times 10^{-10}, \quad \Omega_b h^2 \approx 0.0125, \qquad (6.33)$$

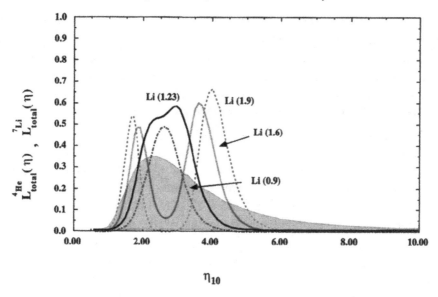

η_{10}

Figure 6.4 Illustrating the concordance model *c.* 2000. The curves are likelihood distributions for the baryon density parameter η_{10} (Equations 6.29, 6.30), derived by comparing observed values of an elemental abundance with the predictions of nucleosynthesis calculations for that element. The four curves labelled 'Li' show results assuming four values for its measured abundance; curves are labelled by the value of $10^{10} \times^7$ Li/H. The shaded region shows results for ^{4}He. The authors argue that the most likely distribution (that is, the distribution that is most consistent with observed abundances) is the one shown by the thicker curve, for which ^{7}Li/H $= 1.23 \times 10^{-10}$. This abundance is close to the minimum value allowed by nucleosynthesis for lithium; the implied confidence interval for the baryon density is $\eta = (1.7–3.9) \times 10^{-10}$ or $\Omega_b h^2 = (0.0062–0.014)$. Comparing this figure with Figure 6.3, one can see that the confidence interval assigned to η_{10} was essentially the same in 2000 as it was in 1982. Starting around 2002, standard-model cosmologists adopted a new value of η roughly twice as large; as a consequence, the observed lithium abundance fell out of agreement and was declared anomalous – the 'lithium problem.' Many Milgromian cosmologists continue to give preference to the original concordance value, and for them the 'lithium problem' does not exist. Figure reprinted with permission from Sean G. Ryan et al., "Primordial lithium and big bang nucleosynthesis," *The Astrophysical Journal Letters*, 530, p. L57–L60, 2000. Reproduced by permission of the AAS.

precisely in the middle of the range cited by Peacock, and essentially the same as Spite and Spite's original estimate from 1982.[24]

The stability over time of the 'concordance' baryon density was due in large part to the unchanging location of the Spite plateau.

§

---

[24] A colleague of the author recalls finding this value, as late as 2003, embedded as a parameter statement in a publicly-available galaxy formation code.

Such was the state of the concordance model in 1998 when McGaugh set about making his prediction. Recall McGaugh's reasoning: just as Milgrom's postulates require the dynamics to become Newtonian in the high-acceleration limit, so any relativistic variant of Milgrom's theory should reproduce the predictions of Einstein, at least approximately, in the strong-field limit. It is therefore reasonable to assume (McGaugh argued, from a Milgromian perspective) that the very early universe evolved according to Einstein's equations, at least approximately – but without any dark matter.

McGaugh used the formalism that had recently been developed by standard-model cosmologists to predict the power spectrum of temperature fluctuations in the cosmic microwave background (CMB). The CMB consists of photons that are believed to have been generated around the time of recombination. In the standard cosmological model, this would have occurred at a time of $\sim 3 \times 10^5$ yr after the big bang. The distribution of *matter* would have been nearly, but not precisely, homogeneous at this time.[25] Spatial inhomogeneities in the matter density are not directly observable but would leave imprints in the CMB, in the form of spatial inhomogeneities in the photon temperature, so-called 'primary anisotropies' ('anisotropies' rather than 'inhomogeneities' since the observable quantity is the angular distribution on the sky). 'Secondary anisotropies' would arise due to propagation of the photons through the universe to the observer.

Denoting the angular distribution on the surface of the sky of the temperature of the CMB photons as $T(\theta, \phi)$ and expanding this function in spherical harmonics,

$$T(\theta,\phi) = \sum_{\ell, |m|<\ell} a_{\ell,m} Y_{\ell,m}(\theta,\phi) = \sum_{\ell,m} a_{\ell,m} P_{\ell,m}(\cos\theta)\, e^{im\phi}. \tag{6.34}$$

In terms of the $a_{\ell,m}$, the variance in the temperature is easily shown to be

$$\langle (\Delta T)^2 \rangle \equiv \langle (T - \langle T \rangle)^2 \rangle = \frac{1}{4\pi} \sum_{\ell>1,m} |a_{\ell,m}|^2 = \frac{1}{4\pi} \sum_{\ell>1} (2\ell+1)\, C_\ell,$$

$$C_\ell \equiv \frac{1}{(2\ell+1)} \sum_m a_{\ell,m} a_{\ell,-m}. \tag{6.35}$$

The angle brackets denote averages over all observation locations, and $\langle T \rangle \equiv T_0 \approx 2.93$ K. The $a_{\ell,m}$ describe temperature variations on angular scales $\Delta\theta \approx \pi/\ell$, and the quantity $\ell(\ell+1/2) C_\ell/2\pi$ can be shown to be the contribution per interval of $\ln \ell$ to the temperature fluctuations. It turns out that a slightly more useful index is Δ_T^2 defined as

$$\Delta_T^2 = \frac{\ell(\ell+1)\, C_\ell}{2\pi}, \tag{6.36}$$

---

[25] Initially this was a postulate of the standard model, but the amplitude of the fluctuations were eventually inferred from the CMB spectrum and found to be small.

and this is the quantity in terms of which the power spectrum of observed temperature fluctuations is usually plotted.[26]

Prediction of the fluctuation power spectrum under the standard cosmological model requires a general relativistic calculation, although Einstein's equations appear only in linearized form. Gravitational infall of the primordial fluid into potential wells created by fluctuations in the matter density leads to sound waves in the baryon–photon fluid, so-called 'baryon acoustic oscillations.' Compression of the fluid in the density peaks results in higher temperatures, leading to temperature inhomogeneities; the temperature anisotropies observed from the Earth are a two-dimensional projection along the line of sight of the inhomogeneities created by these sound waves. In a universe containing no dark matter, the 'shape' of the fluctuation power spectrum – as measured, for instance, by the locations of the peaks and their amplitudes – should depend most sensitively on the baryon density, Ω_b, among other possible parameters. In particular, the amplitude of the second peak relative to the first should decrease with increasing Ω_b via the mechanism of baryonic diffusion damping (Hu et al., 1997; Hu and Dodelson, 2002).

§

At the time of McGaugh's calculation, no second peak had yet been detected in the CMB data. In the spirit of critical rationalism, McGaugh was intent on making a *testable*, i.e. refutable, prediction. And because a competing theory – the standard cosmological model – was also capable of making a prediction about the same quantity, McGaugh was in the enviable position of being able to carry out a *crucial test*: a test that can simultaneously corroborate one theory and falsify another.

Popper wrote that crucial tests

are *attempted refutations* ... they are designed – *designed in the light of some competing hypothesis* – with the aim of refuting, if possible, the theory which we wish to test. And we always try, in a crucial test, to make the background knowledge play exactly the same part – so far as this is possible – with respect to each of the hypotheses between which we try to force a decision by the crucial test (Popper, 1983, p. 188).

Popper added: "All this, clearly, cannot absolutely prevent a miscarriage of justice; it may happen that we condemn an innocent hypothesis."

McGaugh used the computer code CMBFAST, due to standard-model cosmologists Uros Seljak and Matia Zaldarriaga (1996), to compute the CMB angular power spectrum under both the Milgromian and standard-model hypotheses. The 'background knowledge' in this experiment was well defined: it consisted of the set of equations, like Einstein's, that are represented functionally in the computer code, together with the set of known parameters that serve as input to the code. One such

---

[26] The reasons are given by Weinberg (2008, p. 143).

Table 6.3 *CMB peak ratios*

Year	Experiment	Reference	$A_{1:2}$[a]
2000	*BOOMERANG*	de Bernardis et al. (2000)	_[b]
2002	*BOOMERANG*	de Bernardis et al. (2002)	2.45 ± 0.79
2003	*WMAP*	Page et al. (2003)	2.34 ± 0.09
2007	*WMAP*	Hinshaw et al. (2007)	2.26 ± 0.04

[a] Amplitude ratio: First peak to second peak
[b] Second peak not detected

parameter, of course, is the baryon density, for which a concordance value had been established by the mid 1990s, as discussed above. The predicted shape of the CMB power spectrum depends non-trivially on two other parameters. The first is the so-called curvature constant, k, in Einstein's field equations. McGaugh adopted the consensus view of standard-model cosmologists (e.g. Turner, 1999) that the universe was 'flat,' or Euclidean, i.e. $k = 0$. He noted that the choice of geometry had relatively little influence on the predicted amplitude ratio of the first two peaks; it mostly affects the peak locations (as measured by ℓ). The other important parameter, called n, determines the functional form of the power spectrum describing infinitesimal density perturbations in the matter at early times. A postulate of the standard model is that the perturbation spectrum is 'scale-invariant,' that is, that it has the form $P(k) \propto k^n$ with k the comoving wave number of the perturbation.[27] Moreover, standard-model cosmologists at the time argued that $n = 1$, a so-called 'Harrison–Zeldovich spectrum'. McGaugh therefore set $n = 1$. Thus, the only difference between the Milgromian and standard-model calculations was the choice for the 'dark matter' density parameter, $\Omega_{DM}h^2$.

Assuming $\Omega_{DM}h^2 = 0$, McGaugh (1999a) calculated the following values for the first-to-second peak amplitude ratio, $A_{1:2}$, as a function of the assumed baryon density:

$$\Omega_b h^2 = 5.63 \times 10^{-3}, A_{1:2} = 2.57,$$
$$\Omega_b h^2 = 1.13 \times 10^{-2}, A_{1:2} = 2.37,$$
$$\Omega_b h^2 = 1.69 \times 10^{-2}, A_{1:2} = 2.40. \tag{6.37}$$

For the range of $\Omega_b h^2$ values consistent with the nucleosynthesis calculations of the mid to late 1990s (Table 6.2), McGaugh (1999a) noted that $A_{1:2}$ never differed greatly from 2.4, and he proposed this value as the Milgromian prediction.

McGaugh's prediction was confirmed. Table 6.3 gives measured values of $A_{1:2}$. The first group to claim to detect and measure the amplitude of the second peak was

[27] Many standard-model textbooks present the scale invariance not as a postulate in its own right, but rather as a consequence of another postulate, the existence of an 'inflationary epoch.' See for instance Schneider (2015, p. 388). Since McGaugh's calculation in 1999, standard-model cosmologists have often resorted to treating n as a free parameter.

the '*BOOMERANG*'[28] collaboration in 2001–2002 (Lange et al., 2001; Netterfield et al., 2002; de Bernardis et al., 2002); these authors found $A_{1:2} = 2.45 \pm 0.79$. Observational determinations of $A_{1:2}$ since 2002 have trended slightly downward but have remained consistent with McGaugh's 1999 prediction.

Was McGaugh's prediction novel? It certainly was according to Popper's criterion (P) because the second peak had not yet been detected at the time of the prediction. Because the theory was not designed to give a known result – it could not possibly have been, since the result was not yet known – Zahar's criterion (Z) is also satisfied. And for the same reason, Carrier's criterion (C) does not apply.

We are left with Musgrave's criterion – the novelty criterion that standard-model cosmologists would probably find most relevant here:

M: A corroborated fact provides evidential support for a theory if the best rival theory fails to explain the fact, or if the fact is prohibited or improbable from the standpoint of that theory.

I will argue that McGaugh's prediction was, in fact, novel according to Musgrave's criterion. But here a judgment is required: because while standard-model cosmologists undeniably failed to correctly predict the peak ratio before its observational determination, they did eventually claim success in explaining the CMB spectrum, including the amplitude of the second peak.

We can set the stage for the discussion to follow by referring to Figure 6.5, which includes two computed curves: McGaugh's Milgromian prediction assuming $\Omega_{DM}h^2 = 0$, and a curve computed assuming $\Omega_{DM}h^2 = 0.11$ – approximately the consensus value of standard-model cosmologists in 1999 (and still today). The latter curve illustrates what a standard-model cosmologist *would have predicted* for the CMB spectrum prior to discovery of the second peak. (Apparently, no standard-model cosmologist published the results of such a calculation prior to McGaugh.) The addition of dark matter has the effect of increasing the amplitude of the second peak compared with the Milgromian (no dark matter) prediction, for reasons that will be discussed below. For the range of baryon and dark matter densities allowed at the time, the standard model predicted a first-to-second peak ratio in the range ~1.5 to ~1.9. In order to explain the observed peak ratio under the standard model, some adjustment would seem to be required. And as we will see, starting around 2002, standard-model cosmologists chose to make such an adjustment: they *doubled* the assumed value of $\Omega_b h^2$, disregarding all nucleosynthesis constraints on the baryon density that had been published prior to 1998. In so doing, they managed to fit the CMB fluctuation spectrum, but they simultaneously created two

---

[28] The *BOOMERANG* experiment (*Balloon Observations Of Millimetric Extragalactic Radiation And Geophysics*) consisted of three high-altitude (~40000 m) balloon flights in 1997, 1998 and 2003. The balloons were launched from McMurdo Station in the Antarctic.

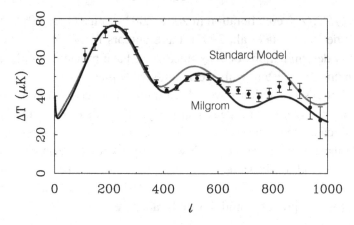

Figure 6.5 McGaugh's (1999a) successful, novel prediction of the height of the second peak in the cosmic microwave background (CMB) fluctuation spectrum. Data points are from observations made by the *BOOMERANG* balloon-borne observatory soon after McGaugh's prediction had been published (Netterfield et al., 2002). The *BOOMERANG* data were the first that clearly showed the second peak, at $\ell \approx 540$. The lower curve is McGaugh's prediction assuming no dark matter; McGaugh predicted a first-to-second peak amplitude ratio of 2.4. The measured value was 2.45 ± 0.79 (Table 6.3). The upper curve, also computed by McGaugh, shows the *a priori* standard-model prediction, assuming the same baryon density but including dark matter with $\Omega_{DM}h^2 = 0.11$, the value preferred by standard-model cosmologists both then and now. Neither model accurately predicts the form of the spectrum beyond the second peak; only the Milgromian model correctly predicts both the first and second peaks. Standard-model cosmologists eventually managed to fit the spectrum including the third peak, but only by arbitrarily increasing $\Omega_b h^2$ to ~ 0.022, twice the concordance nucleosynthesis value (Table 6.2), and by allowing the index n of the primordial fluctuation spectrum to be different from one. By doubling the assumed baryon density, standard-model cosmologists created two inconsistencies that have persisted until the present day: the 'lithium problem' and the 'missing-baryons problem.' Neither anomaly exists from the standpoint of a Milgromian researcher. (Figure based on Figure 45 from Famaey and McGaugh, 2012, with data points from Table 3 of Netterfield et al., 2002.)

inconsistencies in their model that have persisted until the present day: the 'lithium problem' and the 'missing-baryons problem.'

§

By the mid 1990s, standard-model cosmologists had decided that the largest contribution to the mass density of the universe takes the form of dark matter with a mean density several times greater than that of normal matter (or 'baryons'). The preferred value for the dark matter density parameter – determination of which was based initially on the observed dynamics of large, gravitationally bound systems such as galaxy clusters (e.g. White et al., 1993) – was $\Omega_{DM}h^2 \approx 0.1$.

The existence of dark matter was (and remains) conjectural. As discussed in Chapter 1, a statement such as 'dark matter exists,' in isolation, is metaphysical; it has no empirical consequences. To be scientific, statements about dark matter must take the form of postulates that (singly or in combination with other postulates) can be used to relate the existence of dark matter to the behavior of normal matter. One such postulate, which cosmologists invoke when making predictions about the CMB or the large-scale distribution of galaxies, was called DM-2 in Chapter 1:

DM-2: Beginning at some early time, the universe contained a (nearly) uniform dark component, which subsequently evolved, as a collisionless fluid, in response to gravity.

Standard-model cosmologists usually make three additional assumptions about dark matter:

DM-3: The dark matter consists of elementary particles.

DM-4: The dark matter is 'cold,' that is, the velocity dispersion of the dark particles, at a time before the formation of gravitationally bound structures such as galaxies, was small enough that it did not impede the formation of those structures.

DM-5: The spectrum of density fluctuations in the dark matter at early times was 'scale invariant,' that is, perturbations of all sizes had the same average density contrast.[29]

In his *Milgromian* prediction of the CMB spectrum, McGaugh adopted a postulate similar to DM-5, except that in his calculation, the scale invariance referred to fluctuations in the 'baryon' density.

By virtue of DM-4, standard-model cosmologists are free to assign a very small value to the initial random velocity of the dark particles when computing the evolution of density perturbations; they can assume that all dark particles at any given point moved initially with the mean velocity field of the cosmic 'fluid' at that point, before clumping together under the influence of their mutual gravitational force.

If dark matter dominates the total mass density of the universe, it must have a substantial influence on the form of the CMB power spectrum. Without dark matter, baryonic diffusion damping dominates, and one predicts that each peak in the spectrum should be lower in amplitude than the preceding one. Dark matter, if present with the density assumed by standard-model cosmologists, will dominate perturbations in the gravitational potential and drive the oscillations in the photon–baryon fluid, with photon pressure providing most of the restoring force. The net

---

[29] In the textbooks, postulates DM-4 and DM-5 are generally presented as consequences of two other postulates: the 'coldness' of the dark matter particles is said to be a consequence of their large mass; the scale invariance of the fluctuation spectrum is said to be a consequence of an earlier epoch of 'inflation.' Of course, one could continue even further down this road, crafting postulates from which a large particle mass and an era of inflation necessarily follow, etc. As far as I know, no one has ever attempted to write down a complete list of standard-model postulates, much less a well-formulated set that are independently testable.

result (e.g. Hu et al., 1997) is to increase the predicted amplitude of the second peak relative to the first.

The first publication of results from the *BOOMERANG* and *MAXIMA-1* collaborations[30] reported measurements of the first peak in the CMB power spectrum, at $\ell \approx 200$, but no clear evidence of a second peak. De Bernardis et al. (2000, p. 955) wrote:

The power spectrum is dominated by a peak at $\ell_{peak} \approx 200$, as predicted by inflationary cold dark matter models. These models additionally predict the presence of secondary peaks. The data at high ℓ limit the amplitude, but do not exclude the presence, of a secondary peak. The errors in the angular power spectrum are dominated at low multipoles ($\ell \underset{\sim}{<} 350$) by the cosmic/sampling variance, and at higher multipoles by detector noise.

There followed a rash of papers by standard-model cosmologists interpreting these early results. A key concern of these papers was the apparent lack of a second peak. In order to compensate for the predicted, amplifying effect of dark matter on the second peak, standard-model cosmologists found that they needed to increase the assumed value of the *baryon* density well above the value implied by nucleosynthesis arguments. For instance, Tegmark and Zaldarriaga (2000, p. 2240) wrote:

The lack of a significant second acoustic peak in the new *BOOMERANG* and *MAXIMA* data favors models with more baryons than big bang nucleosynthesis predicts, almost independently of what prior information is included

and Bridle et al. (2001, p. 339) concluded:

In the region of the power spectrum where a second acoustic peak is predicted, we note that our best-fitting models are not a good fit to the data, producing more power than observed by both *BOOMERANG* and *MAXIMA-1*. The easiest way to reconcile this is to increase $\Omega_b h^2$ to a value approximately double that found from nucleosynthesis.

One might think that the 'nucleosynthesis value' for $\Omega_b h^2$ referred to by these authors was the concordance value, $\Omega_b h^2 \approx 0.0125$. In fact it was not. Starting with this set of papers from 2000–2001, and continuing until the present day, standard-model cosmologists have routinely *disregarded* all nucleosynthesis constraints on $\Omega_b h^2$ published prior to about 1998. Figure 6.6 suggests a possible reason why. Two papers that were published in that year (Burles and Tytler, 1998a,b) claimed 'nucleosynthesis' values of the baryon density, derived from measurements of the deuterium abundance, of $\Omega_b h^2 \approx 0.019$ – substantially larger than any value that had been claimed up until that time, and nearer, therefore, to the value of $\Omega_b h^2 \approx 0.035$ that was thought to be needed to explain the non-detection of the second CMB peak. When Bridle et al. cite a need to "increase $\Omega_b h^2$ to a value approximately double that found from nucleosynthesis," they are referring to the Burles and Tytler

---

[30] *MAXIMA (Millimeter Anisotropy eXperiment IMaging Array)* was a balloon-borne detector. There were two flights, known as *MAXIMA-I* and *MAXIMA-II*, launched in August 1998 and June 1999, respectively.

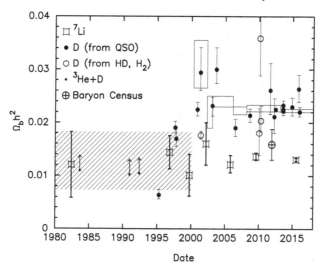

Figure 6.6 Estimates of the baryon density parameter, $\Omega_b h^2$, derived using three independent methods: (*i*) nucleosynthesis calculations combined with measured abundances of the light nuclides ^{7}Li, ^{2}H($=$ D), ^{3}He, and ^{4}He; (*ii*) direct inventory of matter in the local universe ('Baryon Census'); and (after about 2000) (*iii*) analyses of the CMB temperature fluctuation spectrum and other data sets relating to large-scale structure. The first two methods are not affected by the postulates relating to dark matter in the standard cosmological model; while in the third method, the inferred baryon density depends strongly on the assumed dark matter density. Horizontal axis is the date on which the corresponding research article was listed as 'Received' by the publishing journal. The cross-hatched region shows the 'concordance model' as it existed in the two decades prior to 2000 (Table 6.2, Figure 6.4). The nuclides that are most useful for constraining $\Omega_b h^2$ are ^{7}Li and ^{2}H; the plotted symbols are based on measurements of these two elements. In the case of ^{7}Li, each point represents a large number of stars on the 'Spite plateau' (Figure 6.2), while in the case of ^{2}H, each point represents a single measurement. The rectangular regions show the evolving CMB-based (i.e. standard-model) estimates. Estimates of $\Omega_b h^2$ based on ^{7}Li remained constant throughout the time interval shown here, while estimates derived from measurements of deuterium in distant gas clouds trended upward over time, in a manner that seems to track the CMB-based estimates. In any case, after about 2000, standard-model cosmologists chose to disregard all abundance-based constraints on $\Omega_b h^2$ that were published prior to 1998, and results from deuterium published after 2000 were accepted only to the extent that they were deemed consistent with the CMB analysis. A Milgromian researcher is likely to prefer the pre-2000 concordance value of $\Omega_b h^2$ since it is independent of the dark matter postulates in the standard cosmological model.

numbers. Had they compared their results with the *concordance* value of $\Omega_b h^2$, the discrepancy would have been a factor of three rather than a factor of two.

These new deuterium measurements by Burles and Tytler were based on a technique that was first suggested by Thomas F. Adams (1976). Adams's idea was to observe atomic absorption features along the lines of sight to quasars; the absorbing (intergalactic) systems ('Lyman-alpha clouds') are believed to consist of unprocessed gas and hence to reflect the primordial abundance more closely than gas clouds within the Milky Way. In the next few years, four other determinations of D/H using this technique were published, and the inferred values of $\Omega_b h^2$ were even larger – in two papers (Pettini and Bowen, 2001; Crighton et al., 2004), the authors claimed very low values of D/H, implying $\Omega_b h^2 \approx 0.025$–$0.03$, large enough to be fully consistent with the lack of a second CMB peak (see Figure 6.6).

The situation changed dramatically in spring of 2001 when new CMB anisotropy measurements from three experiments (*BOOMERANG, MAXIMA* and *DASI*) were released.[31] For the first two experiments, the observational data were unchanged from the previous year, but a new analysis had been applied using an improved instrumental calibration. The power spectrum now revealed the second peak, and a third peak was detected as well, although not with high significance (see Figure 6.5); de Bernardis et al. (2002) estimated that the third peak was detected at a 2.2σ level of certainty. The revised analysis from the *BOOMERANG* group, entitled "Multiple peaks in the angular power spectrum of the cosmic microwave background" (de Bernardis et al., 2002), found a first-to-second peak amplitude ratio of 2.45 ± 0.79, confirming McGaugh's Milgromian prediction (Table 6.3). Standard-model cosmologists were also able to explain (in a post hoc sense) this amplitude ratio, by adjusting $\Omega_b h^2$ to the value needed to counteract the amplifying effect of dark matter on the second peak. The value they chose was $\Omega_b h^2 \approx 0.022$ (e.g. Netterfield et al., 2002), a value that has not changed substantially since.

Bernardis et al. (2002, p. 560) described the situation as follows:

the new *BOOMERANG* spectrum gives a value for the baryon fraction $\Omega_b h^2$ that is in excellent agreement with independent constraints from standard big bang nucleosynthesis (BBN), eliminating any hint of a conflict between the BBN and the CMB-derived values for the baryon density.

This is extremely misleading. Prior to 2000, *no* researcher had claimed, based on nucleosynthesis arguments, a best-fit value of $\Omega_b h^2$ as large as 0.02, and almost all such determinations had been below 0.015 (Figure 6.6). Cyburt et al. (2008, p. 14), writing a few years later, after publication of results by the satellite observatory *WMAP*, described the situation more accurately (italics added):

What has emerged is rather excellent agreement between the predicted abundance of D/H as compared with the determined abundance from quasar absorption systems.... Indeed *what is often termed the success in cosmology between BBN and the CMB is in reality only the concordance between theory and observation for D/H at the WMAP value of η.* Currently, there is no discrepancy between theory and observation for ^{4}He. But this success is tempered by the fact that ^{4}He is a poor baryometer, it varies only logarithmically with η, and the observational uncertainty is rather large.

§

---

[31] *DASI (Degree Angular Scale Interferometer)* is an interferometric telescope built in 1999–2000 at the Amundsen–Scott South Pole Station in Antarctica.

In his 2010 textbook *Fundamentals of Cosmology*, standard-model cosmologist James Rich summarizes the likely sources of systematic error in determinations of $\eta \equiv n_b/n_\gamma$ using deuterium abundances:

> As is often the case in astrophysics, the cited error [in η] is purely formal because the real uncertainty comes from the hypotheses necessary to interpret the data. In this case, it is necessary to suppose that the two [absorption] lines ... near 555.8 nm are correctly identified and to suppose that the measured abundances are primordial. If either hypothesis is false, the measurement must be reinterpreted. For instance, the "deuterium" line could be a hydrogen line of a second cloud of a slightly different redshift. This would cause the observers to overestimate the deuterium and therefore underestimate Ω_b. On the other hand, if the measured deuterium is not primordial, the primordial deuterium is underestimated since stellar processes generally destroy deuterium. This would cause an overestimation of Ω_b (Rich, 2010, p. 224).

So far, so good. Rich continues (italics mine): *"Debates over which clouds were the right ones to use were closed with the WMAP measurement. . . . The agreement* between the observed primordial ^{2}H abundances and the prediction of nucleosynthesis theory using the *WMAP* value of η is a great triumph for cosmology." In other words: experimental determinations of the deuterium abundance that disagree with the predictions of the current, concordance cosmological model are *ipso facto* incorrect.

The *assumption* that the CMB-based determination of Ω_b supersedes all others is very commonly made by standard-model cosmologists. For instance, Fields (2011, p. 48) writes: *"WMAP* measurements of the cosmic microwave background (CMB) radiation have precisely determined the cosmological baryon and total matter contents. . . It is difficult to overstate the cosmological impact of the stunningly precise CMB measurements by *WMAP* and other experiments." Coc et al. (2015, p. 2) write: "The observations of the anisotropies of the cosmic microwave background by *WMAP*, and more recently the *Planck* space missions, enabled the extraction of cosmological parameters and, in particular, the baryonic density of the Universe. It was the last free parameter in BBN calculations, now measured with an uncertainty of less than 1%." And Olive (2004, p. 190) writes: "With the value of the baryon-to-photon ratio determined to relatively high precision by *WMAP*, standard BBN no longer has any free parameters."

The aim here is not to decide which value of $\Omega_b h^2$ – the pre-2000 'concordance' value, or the post-2002 'concordance' value – is (more nearly) correct. Rather, we wish to establish on what epistemic or evidential ground standard-model cosmologists decided in favor of the latter, since it was this decision that allowed them to explain the same peak ratio that McGaugh had predicted on the basis of Milgromian theory in advance of its measurement.

Peter Schneider, in his 2015 textbook *Extragalactic Astronomy and Cosmology* (p. 448–449), presents a common justification for the primacy of the CMB-based determination:

Cosmological parameter estimates from the CMB are considered to be more robust than those from most other methods, because predicting the CMB properties requires only well-understood physical processes which, due to the small density fluctuations in the early universe, need to be considered only in the linear regime. The underlying physics of most other methods is far more complicated.

This is true as far as it goes, but it skirts over the epistemically more important point: that parameters derived from the CMB under the standard cosmological model are meaningful only to the extent that the hypotheses in that model – which include the correctness of Einsteinian gravity, and the dark matter postulates – are valid. Determining Ω_b from nucleosynthesis arguments, or from a direct inventory of baryons, are two techniques that remain physically meaningful even if those postulates of the standard cosmological model are incorrect.[32]

Another common justification is the one given by Rich (2010) and Cyburt et al. (2008) in the passages quoted above: consistency of the CMB-based value of $\Omega_b h^2$ with the value derived from deuterium abundances. But as Figure 6.6 illustrates, only some of the (evolving) deuterium measurements have been consistent with the (evolving) CMB value of $\Omega_b h^2$, and there is consensus even now that the scatter in measured values of D/H is too large to be explained by known sources of error (Kirkman et al., 2003; Ivanchik et al., 2010; Pettini and Cooke, 2012; Riemer-Sørensen et al., 2015; Balashev et al., 2016). Furthermore (a critical rationalist would likely say), there is little in Figure 6.6 to suggest that extragalactic observers were on track to find the 'correct' deuterium abundance before 2002.[33]

It is instructive to examine the papers associated with the 'discrepant' points in Figure 6.6 – the deuterium-based determinations of $\Omega_b h^2$ that are most in conflict with the CMB results. Crighton et al. (2004, p. 1051) write that their measurement of the deuterium abundance

disagrees at the 99.4 per cent level with the predicted D/H value using the Ω_b value calculated from the *WMAP* cosmic background radiation measurements ... Some early mechanism for D astration may be the cause for the scatter in D/H values (Fields et al. 2001), although in this case we would expect to see a correlation between D/H and [Si/H], which is not observed.

---

[32] Similar remarks apply to some of the other concordance parameters; for instance, the Hubble constant can be derived from the local redshift–magnitude diagram, and there is a long-standing discrepancy between the value of H_0 determined in this way and via the CMB (e.g. Riess et al. (2016)).

[33] That statement may strike some as gratuitously harsh, but observational astrophysics is rife with instances of confirmation bias, the most notorious being the determination of the Hubble constant; see Trimble (1996). The reader is also invited to examine Figure 10 of Steigman (2007) which shows that observational determinations of the ^{4}He mass fraction have trended steadily upward, from $\lesssim 0.23$ c. 1990 to ~ 0.25 by 2005: as in the case of deuterium, achieving only an ex post facto consistency with the cosmologists' $\Omega_b h^2$. In Steigman's words (p. 480): "When it comes to using published estimates of Y_p [the helium mass fraction], *caveat emptor.*"

Srianand et al. (2010, p. 1899) argue in a similar way that their discrepant value must be due to astration:

> We measure the deuterium abundance in this DLA [damped Lyman-α cloud] to be $\log(D/H) = \log(D_I/H_I) = -4.93 \pm 0.15$. This is a factor of 2 lower than the primordial value, $[D/H]_p = -4.59 \pm 0.02$, derived from 5-yr data of the *WMAP* (Komatsu et al. 2009). We therefore conclude that astration factors can vary significantly even at low metallicity.

These authors *assume* – like Rich – that a measurement of D/H that conflicts with the CMB value of Ω_b is necessarily in error. There is little basis here for claiming that measurements of D/H support the concordance value of $\Omega_b h^2$. At least in the minds of standard-model cosmologists, the evidentiary arrow appears to point in the opposite direction.

In order to minimize the possibility of systematic errors in measurements of D/H, Cooke et al. (2014) established a set of criteria for the selection of absorbing clouds; in their words (p. 2): "Our goal is to identify the small handful of systems currently known where the most accurate and precise measures of [primordial] D/H can potentially be obtained." Balashev et al. (2016) subsequently measured D/H in an absorbing system chosen according to the Cooke et al. criteria. They found a deuterium abundance that differed significantly from the expected value (D/H $= 1.97 \times 10^{-5}$ vs. 2.58×10^{-5}). Rather than draw any physical conclusion from this result, Balashev et al. noted that after *averaging* their value with the fourteen other values in the existing Cooke et al. sample, there was no statistically significant change in the *mean* value of D/H. They concluded on this basis that their measurement of a discrepant value for D/H, hence $\Omega_b h^2$, did not conflict with the CMB results. Rather than treat their measured deuterium abundance as a potential refutation of the standard-model prediction, Balashev et al. chose to present their data in a manner that minimized its disagreement with that prediction.

§

A cynic might wonder whether the verificationist approach to hypothesis-testing displayed in these examples – giving credence only to data that are consistent with the hypothesis being tested – is simply the norm for scientists. But the methodology of the extragalactic observers who measure the deuterium abundance contrasts in a striking way with the methodology of the *stellar* astrophysicists who measure the *lithium* abundance. Even during the pre-*WMAP* era, stellar astrophysicists – recognizing the importance of lithium for testing models of big-bang nucleosynthesis and Galactic chemical evolution – were engaged in a concerted, collective effort to *test* the hypothesis that the abundances they measured along the Spite plateau were a valid guide to $\Omega_b h^2$. That is: they behaved like critical rationalists. Their skeptical stance with regard to their own results was, if anything, intensified after 2002, when standard-model cosmologists began telling them that their abundance determinations were not consistent with the revised value of $\Omega_b h^2$.[34]

Writing early in the post-*WMAP* era, stellar astrophysicist David Lambert summarized the difficulties created by the new 'concordance' value of $\Omega_b h^2$:

Although one may find a dissident or two among observers, the collective view is that there is a gap of about 0.5 dex [i.e. a factor of 3] between the lithium abundance of the Spite plateau and the *WMAP*-based prediction for standard primordial nucleosynthesis. ... An observer cannot but fail to be impressed by the variety and depth of study of processes affecting the surface lithium abundances of stars on the Spite plateau ... Yet, the challenge is not only to find a process – more likely, a combination of processes – that reduces the surface lithium abundance by the required 0.5 dex but does so with great uniformity for the observed stars on the plateau (Lambert, 2004, p. 218).

Lambert identified three possible explanations for the apparent discrepancy between the measured lithium abundance and the CMB-based value of $\Omega_b h^2$: (1) systematic errors in the interpretation of lithium absorption features in stellar atmospheres; (2) reduction in the surface abundance of lithium due to processes inside of stars; (3) errors in the equations of big-bang nucleosynthesis (BBN).[35] Each of these avenues has been, and continues to be, explored, as summarized in the remainder of this section. The current consensus among stellar and particle astrophysicists, however, is that no one has yet produced a convincing refutation of the hypothesis that the lithium abundance on the Spite plateau is primordial, or that the value of $\Omega_b h^2$ implied by the Spite plateau abundance is accurate, in spite of its disagreement with the cosmologists' preferred value.

(1) As mentioned above, ^{7}Li is detected in the atmospheres of stars via an absorption feature at 6707 Å, a so-called 'resonance doublet.' Atomic data for the doublet have been well established in the laboratory. The absorption feature is due to *neutral* lithium ('LiI' in astronomical parlance), but in the atmospheres of the observed stars, most of the lithium is singly ionized ('LiII'). The ratio of neutral to singly-ionized lithium varies exponentially with temperature, and so an independent, and accurate, measurement of a star's atmospheric temperature is required to convert the measured absorption line width into an estimate of the total lithium content, summed over all ionization states. Here, 'accurate' means having an error less than about 100 K; in order for the measured line strengths to be consistent with

---

[34] At least one, major international conference has been devoted entirely to 'the lithium problem.' "Lithium in the Cosmos" was held at the Institut d'Astrophysique de Paris [IAP], 27–29 February 2012. The IAP web page announcing the meeting stated: "By gathering together observers, stellar theorists and cosmologists we aim to discuss the latest findings in this long-lasting puzzle, hopefully shedding new light on this historic and yet fascinating problem of physics in the cosmos." I am not aware of any conference having been devoted to 'the deuterium problem.'

[35] To his credit, Lambert also mentions the possibility that the CMB-based value of $\Omega_b h^2$ might be wrong, but writes (p. 221): "Discarding the WMAP estimate of $\Omega_b h^2$ introduces a different set of problems." Indeed.

the cosmologists' prediction, the stellar effective temperatures would need to be raised by something like 500 K (Fields et al., 2005).

Fortunately, a number of independent techniques are available for measuring T_{eff}:

(a) The 'wings' of the Balmer absorption lines, Hα and Hβ, depend sensitively on photospheric temperature; at the same time they are *in*sensitive to the other two parameters that are commonly used to characterize stellar atmospheres: the 'gravity' parameter, log g, and the metallicity. This technique uses model atmospheres to compute the predicted shapes of the Balmer lines as a function of temperature, which are then compared with high-dispersion spectra to determine the best-fitting T_{eff} (e.g. Gehren, 1981). Errors are believed to be mostly systematic, due for instance to uncertainties in the theory of spectral line broadening, and to have a magnitude of about 100 K in T_{eff} (Asplund et al., 2006).

(b) A second method compares the equivalent widths of different absorption lines for a single species, e.g. FeI, with the predictions of the Boltzmann equation. The latter contains an exponential term in $\epsilon_i/(k_B T)$ where ϵ_i is the excitation energy of the ith energy level. There is generally a single temperature that yields the same, inferred abundance for each ϵ_i in a single star (Peterson and Carney, 1979). Systematic errors are believed to be dominated by departures from local thermodynamic equilibrium (LTE) and to average less than about 100 K (Hosford et al., 2009, 2010).

(c) Photometric, or 'color,' methods compare the integrated flux emitted in two different wavelength intervals. A well-calibrated variant, the 'infrared flux method' (IRFM; Blackwell et al., 1980, 1990), compares the flux in a narrow infrared band with the total flux integrated over all wavelengths; the ratio is compared with predictions of model atmospheres to determine T_{eff}. This method is particularly well suited to stars in the spectral classes that are represented on the Spite plateau; for such stars, random errors in T_{eff} are believed to be smaller than those of the other two techniques, perhaps as small as 60 K (Alonso et al., 1999).

The IRFM technique is widely considered to be the least model-dependent of the three, so much so that it is sometimes used to calibrate the other two. And so it was very exciting when Meléndez and Ramírez (2004), after applying the IRFM to data from a sample of about 40 stars on the Spite plateau, concluded that previous estimates of T_{eff} had in fact been systematically in error. The revised temperatures were on average about 200 K higher – roughly half of what would be needed to bring the inferred lithium abundance in line with the value preferred by standard-model cosmologists.

However the Melendez and Ramirez result was immediately called into question by other groups (Hosford et al., 2010; González Hernández and Bonifacio, 2009). The latter authors, who also applied the IRFM technique, suggested that the discrepancy was due to an absolute-flux calibration error – a suggestion that was soon confirmed (Casagrande et al., 2010). There is currently a consensus that the three methods for determining T_{eff} yield results that are consistent, within the expected errors, and that the inferred abundance of ^{7}Li for stars on the Spite plateau is essentially the same as the value first inferred by Spite and Spite.

Interestingly, this work yielded another result which is potentially important but which is not yet generally accepted. Recall that one argument in favor of the Spite-plateau abundances being primordial was the detection, in some stars, of an absorption feature attributed to ^{6}Li, the less-common isotope of lithium. ^{6}Li is destroyed by thermonuclear reactions at lower temperatures than ^{7}Li, and so its presence in stellar atmospheres, at measurable levels, strongly militates against the lithium abundance having been decreased by transport of gas to deeper layers. However Lind et al. (2013) claimed that when stellar atmospheres are modeled with a code that accounts for departures from local thermodynamic equilibrium and relaxes the assumption of a plane-stratified atmosphere, "no star in our sample has a significant (2σ) detection of the lighter isotope." On the other hand, Mott et al. (2017), who used a very similar computer algorithm to model spectral line formation, claimed to definitely detect ^{6}Li in at least one, metal-poor star.

(2) The second possibility considered by stellar astrophysicists for reconciling the observed lithium abundances with a higher primordial value is nuclear 'burning' of the surface lithium at greater depths. The mechanism that comes naturally to mind is convective transport, but models of main-sequence, Population II stars in the T_{eff} range of the Spite plateau do not have convective zones that reach downward to regions where temperatures are high enough to burn lithium (Deliyannis et al., 1990). But already in the 1980s (e.g. Michaud et al., 1984), stellar astrophysicists had begun to consider models in which vertical transport of heavy elements might occur *below* the outer convective zone due to the mechanism of microscopic diffusion: the slow 'settling' of ions more massive than hydrogen in response to the combined influence of gravitational, radiative and electrostatic forces. This process must certainly occur, but it is often ignored in stellar models, on the ground that even small amounts of turbulence – generated, for instance, by a star's differential rotation – would render such diffusion irrelevant. In addition, stellar models that include atomic diffusion are difficult to treat numerically due to the extremely steep, and probably unphysical, concentration gradients that form. The approach that stellar modelers have taken is to include *both* atomic diffusion (which is well understood) as well as an extra, ad hoc diffusion term which is intended to mimic the effects of some sort of turbulence that acts in regions outside of the convectively unstable zones. That turbulence could have one of any number of physical origins, including the differential rotation just mentioned, internal gravity waves etc. No one has ever claimed to have a good, first-principles understanding of such turbulence and it is typically represented in the stellar evolution codes by a diffusion coefficient, D_T, that is related in some simple way to the (known) atomic diffusion

coefficients. For instance, Richard et al. (2005), whose stellar models have been widely used in the interpretation of stellar surface abundances, parameterize the turbulent diffusion coefficient as

$$D_T = C \times D_{\text{He}}(T_0) \left[\frac{\rho}{\rho(T_0)} \right]^{-3}, \tag{6.38}$$

with D_{He} the (known) atomic diffusion coefficient for helium and ρ the total gas density. The strong, inverse dependence of D_T on ρ ensures that the turbulent mixing is confined to a thin layer below the convective zone. To their credit, stellar astrophysicists have always clearly acknowledged the ad hocness of this approach. For instance, Richard et al. (2005, p. 539) write: "Our aim in calculating models with turbulence is to determine the level of turbulence that the ^7Li and ^6Li observations require. This may be used later to determine the physical mechanism causing this turbulence." Or even more bluntly, in the words of Korn et al. (2007, p. 409): "As no parameter-free physical description of turbulent mixing is available, it is introduced into the stellar evolution models in an ad hoc manner." These models are fine-tuned: the predicted surface abundance of lithium for stars of a given surface temperature, after some 10^{10} years of main-sequence evolution, can be either increased or decreased compared with models lacking atomic diffusion.[36]

Using the Richard et al. stellar models, a possible signature of atomic diffusion was claimed by Andreas Korn, Karin Lind and collaborators in two Galactic globular clusters, NGC 6397 and NGC 6752 (Korn et al., 2006; Gruyters et al., 2013). Korn et al. used data from samples of stars – presumably all formed at similar times, from gas with similar chemical compositions – to evaluate trends of surface abundance with evolutionary stage: from stars near the main-sequence turnoff (where most stars on the Spite plateau sit) to 'subgiant' stars that have evolved some way up the red giant branch. The data were compared with the Richard et al. evolutionary models to see if some choice of D_T, together with assumed values for the primordial abundances, could be found that reproduced the data. Only weak ("1σ") trends of abundance with evolutionary stage were observed in NGC 6752; in the case of NGC 6397, the trends were stronger: the 'turnoff' stars were found to exhibit slightly lower (~ 0.1 dex) abundances of iron, calcium and other heavy elements than in the more evolved stars. These trends were found to be crudely reproducible using the Richard et al. models, given certain choices for the parameters defining D_T, and assuming primordial abundances (in the case of lithium) that were as great as twice the value on the Spite plateau: not quite large enough to convincingly reconcile the data with the cosmologists' $\Omega_b h^2$.

There are a number of other reasons why the models of lithium destruction in stars are considered unsuccessful: (*i*) In fitting the abundance data, different

---

[36] Certain parameter choices imply that lithium will be 'stored' just below the convective zone, before being brought again to the surface in the first 'dredge-up' that occurs in the middle of the subgiant branch; see Nordlander et al. (2012).

values for the parameters defining the turbulent diffusion are required in different clusters (Korn et al., 2007; Mucciarelli et al., 2011; Nordlander et al., 2012). (*ii*) Some extra turbulence is required to limit diffusion in the hotter stars and maintain uniform lithium abundance along the Spite plateau (Richard et al., 2005). (*iii*) Given the sensitive dependence of the predicted abundances on the choice of D_T, a star that happened to have (say) a slightly higher amplitude of turbulence should have retained a lithium abundance well above the Spite plateau. Such stars are not seen (Spite et al., 2012). (*iv*) The models are not able to explain the peculiarly high abundance of ^{6}Li. For these reasons, there is a general consensus among stellar astrophysicists (Fields, 2011; Spite et al., 2012; Frebel and Norris, 2013) that a compelling case can not be made that physical processes inside stars are able to reconcile the Spite plateau with the higher primordial abundance required by cosmologists.

The models fail in yet another way. Around 2009, a *second* regime was discovered in which the Spite plateau 'breaks down.' Recall that in their original study, Spite and Spite had found that the plateau extends to stars with surface temperatures as low as ~ 5000 K; at lower temperatures, surface lithium abundances began to scatter below the constant value on the plateau (Figure 6.2). This behavior was (and is) considered natural due to the existence of deeper convective zones in cooler stars. The new regime was one of ultra-low 'metallicity': stars with iron abundances less than about 2×10^{-3} of the solar value sometimes also fall below the plateau (and they never fall above) (Aoki et al., 2009; Sbordone et al., 2010). None of the stellar models predict a lower, surface lithium abundance for such stars.

(3) Even if the primordial lithium abundance is assumed known, inferring the cosmic baryon density from it requires an additional assumption: that the equations describing nucleosynthesis in the early universe (BBN, or 'big-bang nucleosynthesis') are correct and complete. Those equations are part of the standard model of particle physics, and there would seem to be little justification for questioning them. In the words of particle physicists David Schramm and Michael Turner:

The basic calculation, a nuclear reaction network in an expanding box, has changed very little... The predictions of BBN are robust because essentially all input microphysics is well determined: The relevant energies, 0.1 to 1 MeV, are explored in nuclear-physics laboratories and the experimental uncertainties are minimal (Schramm and Turner, 1998, p. 303).

Nevertheless, particle physicists (including Schramm himself; see e.g. Schramm, 1991) had begun already in the 1990s to question the correctness of the BBN calculations. And, like the stellar astrophysicists who worked on the lithium abundance, their efforts in this direction only intensified after about 2002, when standard-model cosmologists began claiming an inconsistency with their concordance model.

The following modifications to the standard model of BBN have been considered in the context of the 'lithium problem': (a) revising the rates of the dozen or so reactions in the standard BBN network; (b) 'boosting' reactions that are not usually considered important by means of overlooked resonances; (c) postulating the existence of new particles or new interactions that can contribute to the net destruction of $A = 7$ nuclei. It is fair to say that none of these suggestions has gained strong support, and only the briefest of summaries of each will be given in what follows.

(a) In the standard model of BBN, ^{7}Li is created and destroyed by the reactions:

$$^4\text{He} + {}^3\text{H} \to {}^7\text{Li} + \gamma \tag{6.39a}$$

$$^7\text{Li} + p \to {}^4\text{He} + {}^4\text{He} \tag{6.39b}$$

$$^4\text{He} + {}^3\text{He} \to {}^7\text{Be} + \gamma \tag{6.39c}$$

The first two, competing reactions dominate for lower baryon densities, $\eta_{10} \lesssim 3$, and the last for higher baryon densities, $\eta_{10} \gtrsim 3$. If the pre-2000 concordance value for the baryon density ($\eta_{10} \approx 3$) is correct, both channels are equally important, and (as noted above) the predicted ^{7}Li abundance is close to its minimum allowed value. Whereas if the post-2002 concordance value ($\eta_{10} \approx 6$) is correct, only the second channel is important.[37] In this case, most ^{7}Li would be produced in the form of ^{7}Be; only much later, when the universe has cooled sufficiently for nuclei and electrons to combine into atoms, does ^{7}Be decay to ^{7}Li through electron capture.

Of course, the equilibrium abundance of ^{7}Li depends as well on any other reactions that create or destroy the *reactants* in Equations (6.39): that is, ^{2}H, ^{3}He, ^{4}He and ^{7}Be. Altogether about a dozen such reactions are included in the standard model of BBN (Kolb and Turner, 1994). The cross sections for most of these are accessible to laboratory experiment at the energies (a few 100 keV) relevant to BBN and have in fact been measured (Descouvemont et al., 2004; Coc et al., 2012; Pizzone et al., 2014), leaving little 'wiggle room' for adjusting the reaction rates.

There are two exceptions to that statement, and interestingly the two reactions in question are just the ones on which the predicted ^{7}Li abundance turns out to depend most sensitively: $p + n \to {}^2\text{H} + \gamma$ and $^3\text{He} + {}^4\text{He} \to {}^7\text{Be} + \gamma$ (Coc and Vangioni, 2010). The first reaction, radiative capture of a neutron on a proton, affects the neutron abundance at the time of formation of ^{7}Be and hence the rate of its conversion to ^{7}Li by neutron capture via $^7\text{Be} + n \to {}^7\text{Li} + p$. The reaction rate has been measured in the laboratory but the experimental data are sparse. However the cross section can also be calculated from theory

---

[37] Many authors give approximate analytic expressions for the predicted abundances as functions of the baryon-to-photon ratio η; for instance, Sarkar (1996), whose expressions are reproduced here as Equations (6.31). Sarkar's relations are said to be accurate for $1 \leq \eta_{10} \leq 10$ and so can be applied under both the old and new concordance values for η. But the reader is warned that, since about 2003, a common practice has been to give expressions that are valid only for a narrow range of η-values centered on the new concordance value and so can not be applied to the case $\eta_{10} \approx 3$; for instance, Serpico et al. (2004) assume $5.5 \leq \eta_{10} \leq 7.1$.

(Ando et al., 2006) and the results agree well with the experimental data at the energies where data are available (Coc, 2013).

The second reaction, $^3\text{He} + ^4\text{He} \rightarrow ^7\text{Be} + \gamma$, is just Equation (6.39c): it is the main path for formation of ^7Li at high η. Its cross section was for a long time problematic because two experimental techniques gave systematically different results. Recently the discrepancy appears to have been resolved (Di Leva et al., 2016), although experimental data remain sparse.

There is a third possibility within standard BBN for a rate revision that could affect the ^7Li abundance. *Prior* to the onset of nucleosynthesis, weak interactions determine the neutron/proton ratio; this is believed to occur roughly one second after the big bang when the temperature is about 1 MeV, much higher than during nucleosynthesis. The rates of these weak reactions are computed using the standard model of particle physics, but the theory requires as an input parameter the neutron lifetime τ_n. Laboratory experiments have long implied $\tau_n \approx 880$ s, but in the last few years a systematic difference of about 4 s, many times the expected uncertainty, has emerged when comparing results of two different sorts of experiment (Greene and Geltenbort, 2016). As of this writing the discrepancy remains unexplained. The main consequence of changing τ_n by this amount is to change the predicted abundance of ^4He, but the change is smaller than the uncertainty in the measured helium abundance.[38] The consequences for the other light elements, including lithium, are believed to be even smaller (Mathews et al., 2005).

Based on this and other work, there is intersubjective agreement among particle physicists (e.g. Steigman, 2007; Cyburt et al., 2008; Pizzone et al., 2014) that "the so-called lithium problem cannot be due to nuclear reaction rate uncertainties" (Bertulani and Kajino, 2016, p. 64).

(b) Assuming the post-2002 concordance value for the baryon density ($\eta_{10} \approx 6$), the main channel for production of ^7Li would be reaction (6.39c) followed (much later) by electron capture of ^7Be to ^7Li. If there were an additional reaction that efficiently destroyed ^7Be before electron capture could occur, the net production of ^7Li would be reduced, tending to alleviate the lithium problem. Coc et al. (2004) pointed out the possibility that the rate of the compound reaction

$$^7\text{Be} + ^2\text{H} \rightarrow ^9\text{B}^* \rightarrow ^8\text{Be}^* + p \rightarrow ^4\text{He} + ^4\text{He} + p, \tag{6.40}$$

which is not included among the standard set of BBN reactions, might be higher than normally assumed if there were previously undetected resonances that increased the cross section for the first step of the reaction. (Two resonances were known to exist based on experimental data, but those experiments did not extend to the low energies relevant to BBN.) Coc et al. estimated that an

---

[38] But see footnote 33.

increase of $O(10^2)$ in the reaction cross section would be required to alleviate the lithium problem.

In a splendid display of critical rationalist methodology, the Coc et al. hypothesis was soon put to the test (Angulo et al., 2005; Scholl et al., 2011; O'Malley et al., 2011) and found to be incorrect (Kirsebom and Davids, 2011). But the suggestion that overlooked resonances in nuclear reactions involving ^{7}Be might contribute to the destruction of $A = 7$ isotopes continues to be pursued. So far, the most promising candidate reactions have been ^{7}Be + ^{3}He → ^{10}C (Chakraborty et al., 2011) and ^{7}Be + ^{4}He → ^{11}C (Broggini et al., 2012; Civitarese and Mosquera, 2013). One set of experiments (Hammache et al., 2013) found that upper limits on the presence of new energy levels in ^{10}C and ^{11}C were low enough to exclude the existence of resonances strong enough to have a significant impact on the lithium abundance.

(c) A third way to reduce the ^{7}Li abundance is to postulate the existence of particles and/or interactions for which there is currently no basis in the standard model of particle physics. A wide variety of such possibilities have been considered in the context of the lithium problem. Here we consider just one class of proposals, which are designed to increase the density of neutrons around the time of ^{7}Be formation, when the temperature is $\sim$50 keV. It was recognized already by Reno and Seckel (1988) that such an 'injection' of neutrons could have the effect of reducing the abundance of ^{7}Be at the time of freeze-out. The mechanism works by enhancing the conversion of beryllium to lithium, ^{7}Be + n → p + ^{7}Li, immediately after ^{7}Be is created, followed by more efficient 'proton burning' of ^{7}Li, ^{7}Li + p → 2^4He. Among the non-standard-model postulates that have been discussed for producing extra neutrons are: (*i*) The existence of a hypothetical particle, call it X, that can decay in a way that produces neutrons, i.e. X → n + $\cdots$; (*ii*) a hypothetical particle (like the supersymmetric particles that were once considered prime candidates for the dark matter) that constitutes its own antiparticle and that self-annihilates via a reaction like $X + X$ → n + $\cdots$; (*iii*) the existence of a 'mirror world' from which mirror-neutrons can oscillate into our world. It is assumed that the microphysics are identical in both worlds but the temperatures and baryonic densities are different.

Fortunately, there is at least one way in which these rather vague hypotheses can be made susceptible to refutation. A decrease in the ^{7}Li abundance due to neutron injection is generally accompanied by an *increase* in the deuterium abundance, via the reaction ^{1}H + n → γ + ^{2}H (Jedamzik, 2004; Albornoz Vásquez et al., 2012; Coc et al., 2014). One study (Kusakabe et al., 2014) finds

$$\frac{D}{H} > \left[2.56 + 0.227 \left(5.24 - \frac{^7\text{Li}}{H} \times 10^{10} \right) \right] \times 10^{-10} \quad (\eta_{10} = 6.2). \quad (6.41)$$

The extent to which a relation like (6.41) can be used to falsify the neutron-injection hypothesis depends critically, of course, on how well one believes that the primordial deuterium abundance has been determined observationally. All of the cited papers adopt the view of standard-model cosmologists that the deuterium abundance is in fact well determined and that its value is just what one would predict, under *standard* BBN, given the concordance value of $\Omega_b h^2$. The neutron-injection hypothesis is then tested by asking whether the parameters defining the injection (which include quantities such as the annihilation cross section, particle decay time etc.) can be varied so as to reduce the equilibrium ^{7}Li abundance by a factor of three or four without changing the predicted deuterium abundance by more than a few percent. The current consensus is that "no model [of neutron injection] can be in agreement with both lithium-7 and deuterium" (Coc and Vangioni, 2017, p. 1741015).

§

Recall the definition of a critical rationalist: someone who believes that hypotheses should be judged on the basis of how well they stand up to attempts to refute them. The hypothesis that the value $\Omega_b h^2 \approx 0.0125$ has been correctly inferred from the abundance of lithium measured along the Spite plateau has been subjected to almost four decades of attempted refutation, and so far, no one has managed to find a convincing basis on which to reject it. On that ground (a critical rationalist would say), the hypothesis remains viable. That the hypothesis has held its own even in the face of a prolonged and concerted critical onslaught does not prove that it is correct, of course. But in deciding whether we should favor the lithium-based estimate of $\Omega_b h^2$ over a competing one (a critical rationalist would say), we are justified in asking whether the competing estimate has been subjected to a comparable degree of critical scrutiny.

§§

The issue at hand is how standard-model cosmologists explained the first-to-second peak ratio, the quantity that McGaugh successfully *predicted* in advance of its measurement. As we have seen, standard-model cosmologists disregarded (and have continued to disregard) nucleosynthesis constraints on $\Omega_b h^2$ when accommodating their theory to the CMB data. How, exactly, did they accommodate those data?

Return for a moment to 2001, when the first robust detection of the second peak was announced (de Bernardis et al., 2002; Netterfield et al., 2002); the same papers also included a detection, at lower significance, of the third peak.[39] In fitting models

---

[39] The height of the *third* peak is sometimes cited as a successful, novel prediction of the standard cosmological model. I do not find support for this view in the published literature. As Figure 6.5 shows, the standard-model

to those data, cosmologists found that there were significant degeneracies between the parameter values. For instance, there is a nearly exact degeneracy between Ω_m and Ω_Λ, the parameters that define the energy density in matter (normal plus dark) and 'dark energy': any combination that gives the same angular diameter distance to the last scattering surface results in nearly identical predictions for the CMB fluctuation spectrum. There is also a strong degeneracy between $\Omega_b h^2$ and n, the exponent that defines the spectrum of primordial density fluctuations. (Recall that McGaugh set $n = 1$, a so-called Harrison–Zeldovich spectrum.) Decreasing n from 1 to 0.8 allows $\Omega_b h^2$ to be reduced from 0.022 to $\sim$0.015, consistent with the pre-2000 concordance value, without a significant degradation of the overall fit to the first and second peaks (de Bernardis et al., 2002). Yet another degeneracy exists between the amplitude of density fluctuations at the epoch of recombination and the scattering optical depth after reionization, as discussed above.

Degeneracies like these persist to the present day, and standard-model cosmologists deal with them in a number of ways. One strategy is simply to fix some of the parameters; for instance, the condition $\Omega_{tot} = \Omega_m + \Omega_\Lambda = 1$ is sometimes imposed, whether or not such a value is required by the data.[40] A second strategy is to adopt 'priors,' in the Bayesian sense, on the values of the cosmological parameters or on combinations of them; for instance, the Planck Collaboration XVI (2014) adopt a 'flat' prior for the value of n, the exponent of the primordial perturbation spectrum, within an allowed range of $[0.9, 1.1]$. A third approach, which is now standard, is to combine the CMB data with other data sets that relate in some way to the large-scale distribution of matter. These include the galaxy correlation function, the redshift–magnitude ('Hubble') diagram, and data related to gravitational lensing. Figure 6.7, from Kowalski et al. (2008), shows how three such data sets can be used to constrain the values of the two cosmological parameters $(\Omega_m, \Omega_\Lambda)$. In current studies, the number of independent data sets that are used to constrain the cosmological parameters is typically about the same as the number of parameters that are determined from them. Stated differently: few if any of the fitted parameters can be independently derived from different data sets; all of the data sets are implicated in the determination of any single parameter.

---

prediction, based on accepted values of the baryon and dark matter densities *c.* 1999, was far from the value measured in 2001–2002; the Milgromian prediction of the third peak came closer. And right from the start (in the two cited papers from 2002), standard-model cosmologists fit *both* the observed second and third peaks. They did not claim to be (nor were they) testing predictions of either peak's amplitude. Thus, the statements made here with regard to the second peak apply equally to the third peak: both were accommodated, *after* measurement, by adjustment of parameters; and those adjustments resulted in a value of $\Omega_b h^2$ that was inconsistent with existing determinations based on other sorts of data.

[40] The reader is warned that tabulations of 'measured' cosmological parameters in textbooks and review articles often present values that have been *adjusted* so that $\Omega_{tot} = 1$. Indeed, so strong is the preference on the part of standard-model cosmologists for $\Omega_{tot} = 1$ models that some authors do not bother to point out that the tabulated numbers have been adjusted. (An example is Table 8.1 of Schneider (2015). I thank Peter Schneider for pointing this out to me.) This practice can give a false impression that fitted parameters like Ω_b have remained nearly constant since 2000, when in fact they have changed substantially from experiment to experiment.

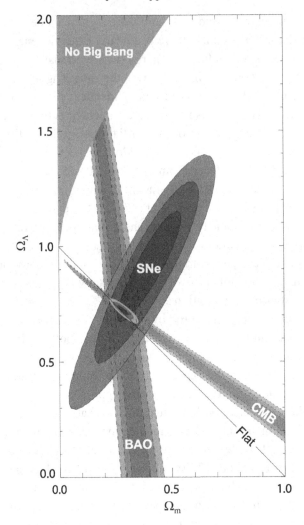

Figure 6.7 Determination of the cosmological parameters Ω_Λ and Ω_m from three data sets; shown are the inferred 68.3%, 95.4% and 99.7% confidence contours. "CMB" refers to measurements of the power spectrum of temperature anisotropies in the cosmic microwave background; "SNe" refers to measurements of redshift as a function of distance using a sample of supernovae; and "BAO" ('baryonic acoustic oscillations') refers to measurements of the galaxy correlation function. This figure illustrates two of the three methods described in the text for reducing degeneracies in the inferred values of the cosmological parameters: combining the results from independent data sets, and fixing the values of certain parameters (in this case, the value of the 'equation of state' parameter for dark energy, w, which was set to $w = -1$, corresponding to the assumption that dark energy takes the form of a 'cosmological constant.'). The line labelled "Flat" plots $\Omega_{tot} = \Omega_\Lambda + \Omega_m = 1$. Figure reprinted with permission from Marek Kowalski et al., "Improved cosmological constraints from new, old, and combined supernova data sets," *The Astrophysical Journal*, 686, p. 749–778, 2008. Reproduced by permission of the AAS.

The term 'concordance model' refers, nowadays, to the best-fitting set of cosmological parameters as determined from studies like these. It differs from the 'concordance model' of 1982–2000 which (as discussed earlier) was defined by the value of $\Omega_b h^2$ for which the abundances of four light nuclei, as predicted from nucleosynthesis calculations, agreed with measured abundances. Constraints on the cosmological parameters based on nucleosynthesis arguments are explicitly *excluded* from the data that define the current concordance model.

How do epistemologists view theory adjustments of this sort? Elie Zahar elaborated on the case of a theory containing parameters that are determined from data:

Consider the following situation. We are given a set of facts and a theory T $[\lambda_1, \ldots, \lambda_m]$ which contains an appropriate number of parameters. Very often the parameters can be adjusted so as to yield a theory T* which 'explains' the given facts ... In such a case we should certainly say that the facts provide little or no evidential support for the theory, since *the theory was specifically designed to deal with the facts* (Zahar, 1973, p. 102–103).

When writing these words, Zahar could not have had in mind the post-2000 'concordance' model since that model did not exist in 1973. But the hypothetical situation described by Zahar applies perfectly to the use of cosmological data sets, like the CMB spectrum, to determine the values of the parameters, like $\Omega_b h^2$, that define the concordance model. According to Zahar's argument, such data are of little or no relevance for establishing the *correctness* of that model.

John Worrall argues in a similar way that the development of theory T can not be said to be driven by the research program's heuristic if "Some constant appears in T in a place where a free parameter had occurred in T's predecessor, and the value of this constant was not dictated by the heuristic of the programme but rather had to be 'read off' from the facts" (Worrall, 1978b, p. 325). Elsewhere, Worrall restates this condition as "one can't use the same fact twice: once in the construction of a theory and then again in its support" (Worrall, 1978a, p. 48).[41]

Neither Worrall nor Zahar is objecting to the use of experimental data to set the parameters of a theory. Their point is rather that such data become part of the 'background knowledge' that the theory was designed to reproduce. Determination of parameters from data *completes* a theory; those same data can not then be said to *corroborate* the theory.

What *would* constitute corroboration of a theory after its parameters have been determined in this way? Recall Popper's words:

apart from explaining all the *explicanda* which the new theory was designed to explain, it must have new and testable consequences (preferably consequences of a *new kind*); it must lead to the prediction of phenomena which have not so far been observed (Popper, 1963, p. 241).

---

[41] It is clear from context that Zahar and Worrall are using 'fact' in the sense defined by Martin Carrier: as quoted in Chapter 2, Carrier defined a 'fact' as "a law-like relation between two variables that are observationally or experimentally detectable" – for instance, the relation between ΔT and ℓ that is plotted in Figure 6.5.

The "new theory" here is, of course, the theory with its parameters newly determined from data. Popper's prescription directs the theorist to generate predictions from that theory *with its parameters now fixed* – and preferably, these should be predictions "of a new kind," that is, predictions of phenomena that are as different as possible from the the phenomena that were used in setting the theory's parameters. If those predictions are confirmed (a critical rationalist like Popper would say), the theory has been corroborated, and one is justified in accepting the adjusted parameters, at least provisionally. Whereas if the predictions are refuted by data, one should view the newly adjusted theory with suspicion.

This is precisely the reasoning that led, in Chapter 2, to our decision about how to judge novelty in cases where a confirmed Milgromian prediction is accommodated, eventually, under the standard cosmological model. The relevant criterion here is that of Musgrave:

M: A corroborated fact provides evidential support for a theory if the best rival theory fails to explain the fact, or if the fact is prohibited or improbable from the standpoint of that theory.

It was ruled, in Chapter 2, that Musgrave's criterion is satisfied if the modifications of the rival theory weaken the explanation of some other known fact or result in a theory that is inconsistent. That is: if the modified theory makes an incorrect prediction. And this was, in fact, the case for the standard cosmological model, after it had been adjusted to accommodate the first three peaks of the CMB spectrum.

§

Standard-model cosmologists were able to fit the CMB spectrum by doubling (approximately) the assumed value of $\Omega_b h^2$, from its pre-2000 concordance value of ~ 0.0125 to ~ 0.022.[42] But this adjustment in the baryon density created two inconsistencies in the model that remain unresolved today: the 'lithium problem' and the 'missing baryons problem.'

Consider first the lithium problem. Inserting the current concordance value for $\Omega_b h^2$ into the nuclear reaction rate equations yields a prediction for the primordial lithium abundance ^{7}Li/H: $(5.12^{+0.71}_{-0.62}) \times 10^{-10}$ (Cyburt et al., 2008) when using the *WMAP* results, or $(4.89^{+0.41}_{-0.39}) \times 10^{-10}$ (Coc et al., 2013) with the *Planck* data. As discussed above, the lithium abundance as measured on the Spite plateau is a factor of three or four lower: $(1.23^{+0.34}_{-0.16}) \times 10^{-10}$ (Ryan et al., 2000) or $(1.58 \pm 0.31) \times 10^{-10}$ (Sbordone et al., 2010).

Standard-model cosmologists (sometimes) acknowledge the discrepancy but they routinely disregard its implications. For instance, the Planck Collaboration,

---

[42] As Figure 6.6 illustrates, there has been a general trend, downward, in the 'concordance' value of $\Omega_b h^2$ since about 2000: from $\Omega_b h^2 \approx 0.031$ *c.* 2000, to ~ 0.024 *c.* 2003, to ~ 0.022 after about 2013.

when comparing their results for $\Omega_b h^2$ with the nucleosynthesis constraints, write (italics mine):

We consider only the ^{2}He and D abundances in this paper. We do not discuss measurements of ^{3}He abundances since these provide only an upper bound on the true primordial ^{3}He fraction. Likewise, *we do not discuss lithium*. There has been a long standing discrepancy between the low lithium abundances measured in metal-poor stars in our Galaxy and the predictions of BBN [i.e. big-bang nucleosynthesis]. At present it is not clear whether this discrepancy is caused by systematic errors in the abundance measurements, or has an "astrophysical" solution (e.g., destruction of primordial lithium) or is caused by new physics (Planck Collaboration XVI, 2014, p. A16).

This passage (similar versions of which appear in a number of other papers) is remarkable for a number of reasons. The sentence ending "the predictions of BBN" is a non sequitur: the equations of big-bang nucleosynthesis make no "prediction" for the lithium abundance absent an input value of $\Omega_b h^2$. The authors are here *assuming*, apparently, that their value of $\Omega_b h^2$ is above question. They are also remiss in failing to note that the "long standing discrepancy" only came into existence after about 2002 when standard-model cosmologists reset the baryon density – and this, following a longer period (roughly two decades) during which the measured lithium abundance was fully *consistent* with $\Omega_b h^2$. And in their list of possible resolutions to the 'lithium problem,' the possibility that the discrepancy is due to their value of Ω_b being incorrect is excluded without any justification.

Other authors – e.g. de Bernardis et al. (2002) and Spergel et al. (2003) – when comparing their CMB-based results for Ω_b with nucleosynthesis constraints, simply omit to mention lithium at all.[43] The disregard on the part of standard-model cosmologists of the constraints on Ω_b from measured lithium abundances is a good example of what Popper called a "conventionalist stratagem" (Chapter 2). It is the methodological equivalent of ignoring the existence of a black swan when evaluating the hypothesis 'All swans are white.'

By resetting the cosmic baryon density, standard-model cosmologists also created a second inconsistency in their model. The value of $\Omega_b h^2$ can be determined observationally – independent of any assumptions about dark matter – by carrying out an 'inventory' of matter in some large volume of the universe. This is a difficult exercise, primarily because 'baryons' can exist in so many different forms, all

---

[43] One also finds statements in the standard-model literature to the effect that the lithium problem is ignorable since solutions are known to exist. For instance, Steven Weinberg, in his 2008 text *Cosmology* (p. 172–173), referring to the mechanism of atomic diffusion (which he calls, incorrectly, "convection") writes: "it is plausible that . . . the Li^7 abundance has been depleted by convection in stellar atmospheres. Observation of Li^7 abundances in stars of varying temperature in the globular cluster NGC 6397 gave results in agreement with a theory of convective depletion of Li^7, with an assumed initial Li^7/H ratio equal to the expected [i.e. *WMAP*] value" and he cites the Korn et al. (2006) study. Korn himself, in discussing his own work, is careful to stress the ad hoc nature of the 'solution': "The extra-mixing efficiency is a free parameter of our present-day modeling capabilities. The pessimistic view is that it removes all predictive power from the atomic-diffusion calculations. . . . we should use it, but use it with care; and try to replace it with a physical description as soon as we can" (Korn, 2012, p. 69).

requiring different sorts of observations for their detection. Attempts at a complete census of baryons have been made a number of times since the 1980s (Fabian, 1991; Gnedin and Ostriker, 1992; Bristow and Phillipps, 1994; Fukugita et al., 1998; Fukugita and Peebles, 2004). All studies agree that only a small fraction – less than about 10% – of normal matter is located inside of galaxies. The majority exists in the form of gas (mostly hydrogen) around galaxies, between galaxies, or in intergalactic space. When sufficiently cold, hydrogen gas can be detected in absorption (the same basic technique that is used when determining the deuterium abundance along the lines of sight to quasars). When sufficiently hot, as in clusters of galaxies, the gas can be detected from its free-free emission. The most difficult problem is posed by the so-called 'warm-hot intergalactic medium,' gas at temperatures in the range 10^5–10^7 K; such gas is believed to have been heated by energetic events like stellar outflows or explosions. Due to its relatively low density, such gas is difficult to observe in emission; a more successful approach is to look for absorption lines originating with high-ionization states of 'heavy' (compared with hydrogen) elements such as carbon, nitrogen and oxygen. Such measurements are (like the deuterium abundance measurements described above) model dependent; they require assumptions about, for instance, how to relate the observed absorption in the heavy species to the density of hydrogen, the dominant element.

Since about 1998, almost all such studies have found a baryon density that is substantially less than the current concordance value. Table 6.4 reproduces a tabulation from the study of Shull et al. (2012), considered by many astrophysicists to be the most careful and complete of its kind. Shull et al. conclude $\Omega_b h^2 = 0.0159 \pm 0.0029$, which is a fraction 58–85% of the current concordance value (0.0221). At the same time, the Shull et al. value is quite *consistent* with the baryon density required, by way of nucleosynthesis, to reproduce the measured abundances

Table 6.4 *Baryon census c. 2012*

Component	$\Omega_b h^2$
Lyα forest	$(6.33 \pm 2.5) \times 10^{-3}$
Warm-hot intergalactic medium	$(5.65 \pm 1.8) \times 10^{-3}$
Galaxies	$(1.58 \pm 0.45) \times 10^{-3}$
Circumgalactic medium	$(1.13 \pm 0.68) \times 10^{-3}$
Intercluster medium	$(9.04 \pm 3.4) \times 10^{-4}$
Cold gas	$(3.84 \pm 0.90) \times 10^{-4}$
Total	0.0159 ± 0.0029

Data from Shull et al. (2012)

of $^7\mathrm{Li}$, $^3\mathrm{He}$ and $^4\mathrm{He}$ – and (like the baryon census) these predicted abundances are independent of any assumptions about dark matter.[44]

Neither the 'lithium problem' nor the 'missing-baryons problem' existed prior to about 2002, when standard-model cosmologists chose to revise the concordance value of $\Omega_b h^2$ in response to observations of the CMB. From the standpoint of a Milgromian researcher, there is no compelling reason to interpret the current concordance parameter Ω_b as a measure of the baryon density, since its value is a function of the value assigned to the dark matter density parameter, Ω_{DM}, and that parameter has no obvious physical interpretation if dark matter does not exist. It is reasonable for a Milgromian researcher (like McGaugh, in 1999) to continue using the pre-2002 'concordance' value of Ω_b, which was derived via methods that are independent of the 'dark matter density,' and which remains consistent with the baryon density inferred from a direct census.

In any case: I conclude that the ability of standard-model cosmologists to fit the temperature fluctuation spectrum of the CMB, including the ratio of the first-to-second peak heights, is based on ad hoc or unreasonable modifications to the theory, and therefore that McGaugh's successful, Milgromian prediction of the peak ratio constitutes a successful novel prediction according to Musgrave's criterion. The conclusions about the novelty of McGaugh's prediction are summarized in the final line of Table 6.1.

§§

We are now in a position to judge the progressivity of the Milgromian research program including theory variant T_2. The requirement of heuristic progress ("successive modifications of the protective belt must be in the spirit of the heuristic") is *not* satisfied. As we have seen, the assumed form of the relativistic Lagrangian has been modified a number of times in response to unexpected observational results; in other words, the heuristic that guides theory development is apparently not 'powerful' enough (to use Lakatos's expression) to anticipate all of the observations that the theory would be expected to explain.

The lack of a unique, or even preferred, relativistic generalization complicates the generation of testable predictions, but two predictions were identified that are likely to be only weakly dependent on the form of the relativistic Lagrangian: an early epoch of cosmic reionization, and a particular value for the amplitude ratio of the first and second peaks in the power spectrum of cosmic microwave

---

[44] One might think that estimates like that of Shull et al. are only lower limits, but that is not necessarily the case. Shull et al. argue persuasively that some previous 'baryon' inventories had *over*-estimated $\Omega_b h^2$ by double-counting the gas detected via Lyman-α absorption, on the one hand, and via absorption lines due to highly ionized 'metals' such as OVI on the other. See also Nicastro et al. (2017).

background (CMB) temperature fluctuations. On this basis, the research program including T_2 can be said to be 'theoretically progressive.' The first prediction was judged 'possibly corroborated' at best. However, the second prediction has been beautifully confirmed by measurements of the CMB spectrum. On this basis, the research program including T_2 may be said to be 'empirically progressive' as well.

Unfortunately, theory T_2 does not solve the problem with galaxy cluster dynamics discussed in Chapter 5. Furthermore, the theory does not correctly predict features in the CMB spectrum beyond (that is, at higher values of ℓ than) the second peak.

§§

We would be justified in concluding our discussion of theory variant T_2 here. But before moving on, in the next chapter, to a description of theory variant T_3, I want to pause for a moment and discuss a 'transitional' model that was influential in motivating T_3.

As Table 6.1 suggests, theory variant T_2 is modestly progressive according to Lakatos's criteria. But even most Milgromian researchers would be quick to acknowledge that the theory is disappointing. In spite of its successes, theory variant T_2 fails to meet the goals that were set out near the start of this chapter by Bekenstein and Milgrom: "to provide tools for investigating cosmology in light of MOND." Neither of the two, novel predictions that are discussed in this chapter makes explicit use of any particular relativistic generalization of Milgrom's theory. One could reasonably take the point of view that these two predictions are not tests of theory variant T_2 so much as they are tests of the need for dark matter on cosmic scales – that is: of the need for a postulate like DM-2. That both predictions were successful is certainly deserving of our attention: those successes tend to strengthen the case (at least in the minds of Milgromian researchers) that dark matter does not exist, or at least that its existence is not compelled by the data.

But one would like more. McGaugh's successful *prediction* of the height of the second CMB peak was impressive, but standard-model cosmologists have managed to *fit* all of the observed peaks. As discussed in this chapter, this success of the standard model was not (like McGaugh's) a *predictive* success: standard-model cosmologists achieved their fit to the CMB spectrum via a set of auxiliary hypotheses and parameter adjustments, and they disregarded the fact that those adjustments brought their model *out* of agreement with some other, well-established facts.

To their credit, Milgromian cosmologists have largely avoided such 'conventionalist stratagems.' But it is natural to wonder how easily a Milgromian cosmologist can accommodate data like the CMB spectrum – if she is allowed, like the standard-model cosmologists, to invoke auxiliary hypotheses.

Garry Angus showed, in 2009, how this could be done. Angus's starting point was the fact that the height of the *third* peak in the CMB spectrum is larger than predicted by (most of) the relativistic generalizations of Milgrom's theory that have been considered so far. As discussed earlier in this chapter, the large amplitude of that third peak is interpreted, by standard-model cosmologists, as being due to 'forcing' of the acoustic oscillations by dark matter. Angus asked: What form of dark matter could achieve this, *without* contributing significantly to the rotation curves of individual galaxies? – thus allowing Milgrom's three postulates to remain in effect and conserving the successes of theory variants T_0 and T_1.[45]

Remarkably, there is a natural candidate for an elementary particle that meets these requirements. It is the *sterile neutrino* – a hypothetical particle that has (of course) not been detected in the laboratory, but which has long been considered by physicists to be a natural addition to the list of known particles in the standard model of particle physics (Drewes, 2013). The argument (in simplified form) goes as follows: With the exception of neutrinos, all known fermions come in pairs of left and right chirality: roughly speaking, this means that a particle's spin can either be aligned (right-handed) or antialigned (left-handed) with its momentum. But neutrinos are always observed to be left-handed, and antineutrinos are always observed to be right-handed. One possible interpretation is that we have failed to observe (for instance) right-handed neutrinos because their interaction with other kinds of matter is very weak – hence the name 'sterile.' (The three known species of neutrino are 'active.')

The mass of the putative sterile neutrino is a free parameter, although anomalous results from accelerator experiments have been interpreted as being due to a sterile neutrino with mass in the range 4–18 eV (Giunti and Laveder, 2008). Standard-model cosmologists have considered the possibility that a sterile neutrino with a much larger mass – in the $\sim$keV range – could constitute the dark matter, or some part of it (Dodelson and Widrow, 1994; Shaposhnikov, 2010). Such a particle mass implies that sterile neutrinos would be 'warm dark matter,' i.e. their random velocities would be large enough to limit how effectively they can cluster under gravity into systems with the masses of galaxies and galaxy clusters. Ordinary neutrinos are known to have masses $\lesssim$ 2eV; at least under the assumptions of the standard cosmological model, they would constitute 'hot' dark matter (no significant clustering on galaxy scales) and would contribute only slightly to Ω_{DM}.

Angus noticed that a sterile neutrino with a mass of order 10 eV would be 'hot' enough to avoid clustering on galactic scales, while at the same time, its presence in the early universe could enhance the third CMB peak, in much the same way as the cold dark matter favored by standard-model cosmologists. Angus repeated McGaugh's 1999 calculation of the CMB fluctuation spectrum, this time allowing for the presence of a sterile neutrino component. He was able to achieve a good fit to the observed spectrum – including all five of the peaks that had been observed

---

[45] Note that Angus is here exploiting the independence of the dark matter postulates DM-1 and DM-2; see the discussion in Chapter 1.

at that time – if he assumed a particle mass of 11 eV and set the cosmological density of sterile neutrinos to $\Omega_{\text{sterile}}h^2 = 0.117$, quite similar to the value assumed by standard-model cosmologists for their dark matter.

Recall from Chapter 5 that Milgromian theory variant T_1, while correctly predicting galaxy rotation-curve data, *failed* to correctly predict the central gas temperatures in galaxy clusters: the mass required by the Milgromian equations of equilibrium was roughly a factor of two greater than the mass inferred from the observed stars and gas. Angus argued (following Sanders, 2003, 2007) that the sterile neutrino 'dark matter' in his model could account for this discrepancy, since particles of mass ~10 eV can cluster effectively on the scale of galaxy groups.

We should not allow ourselves be too impressed by the ability of a model like Angus's to reproduce the CMB spectrum, since it was "cleverly engineered" (Zahar's phrase) to do so (as was, of course, the concordance cosmological model). As has been noted again and again in this book, there will always be an infinite number of theories that can correctly reproduce any finite set of data. But there is nothing to stop us from applying Lakatos's criteria of progressivity to Angus's modified version of theory variant T_2 – call it T_2'. The modified theory makes one, clearly novel, prediction: the existence of a new 'sterile' (i.e. weakly interacting) lepton with a mass of about 10 keV, and with a cosmological density $\Omega_{\text{sterile}}h^2 \approx 0.12$. That prediction suffices to make theory variant T_2' 'theoretically progressive' according to Lakatos, although the fact that the postulated particle has not yet been detected in the laboratory means that the theory is not yet 'empirically progressive.'

But theory variant T_2', like theory variant T_2, fails to satisfy Lakatos's condition that "successive modifications of the protective belt [of auxiliary hypotheses] must be in the spirit of the heuristic." According to the Milgromian heuristic, data like the CMB fluctuation spectrum ought to be explained by some relativistic generalization of theory variant T_1. Angus's sterile neutrino postulate may turn out to be correct, of course, but it falls outside the range of hypotheses that are allowed by the Milgromian heuristic: it is 'ad hoc$_3$' according to Lakatos's classification scheme (Chapter 2).

This is more than just empty word-play. The ability of Milgrom's research program to successfully predict so many novel facts, in advance of their experimental confirmation, has always been considered worthy of notice by Milgromian researchers; Lakatos's criterion of "heuristic progress" is simply an acknowledgement that we should, in fact, be impressed by successes of this sort. And so an auxiliary hypothesis such as Angus's should give us pause. What it suggests is that the Milgromian heuristic lacks the power to guide theory development past variant T_1 – that is: that some revision is called for in the constitution of the Milgromian research program. As we will see in the next chapter, this is just the interpretation adopted by the theorists who have gone on to develop theory variant T_3.

7

Theory Variant T_3: A Modified Hard Core

In the late nineteenth century, Peter Clark (1976) argues, the atomic-kinetic research program was in a degenerating state. Predictions had been made that could not be confirmed (e.g. fluctuations due to finite particle numbers), and some known experimental results (e.g. the ratio of specific heats) could only be explained with the help of ad hoc hypotheses. As a consequence, Ludwig Boltzmann was led to abandon his belief in the existence of atoms and molecules in favor of a purely thermodynamic description:

That Boltzmann gave up his realism in regard to atoms and their interactions is a corroboration of the prediction made on the basis of the methodology of research programmes that scientists *believe* the hard cores of their programmes just to the extent that those programmes are objectively *progressive*. In this case Boltzmann was forced into abandoning his realistic position on atoms and molecules, because his own hypothesis as to the constitution of molecules and their interactions was incompatible with the laws of mechanics (Clark, 1976, p. 90).

Clark's view that the hard-core postulates can and should be re-evaluated, as Boltzmann did, under conditions of research program degeneration was endorsed by William Herbert Newton-Smith:

In point of fact what should be recognized is that the scientist's faith is a faith that there is something important in the basic theoretical assumptions and not that those assumptions are exactly right as they stand. ... the constraint that operates is the following weaker one: while progress is being made only those variants on the basic assumptions which preserve the observational successes of the programme should be explored (Newton-Smith, 1981, p. 84).

Indeed even Lakatos was not, apparently, completely wedded to the concept of an inviolate hard core, as when he stated, "The actual hard core of a programme does not actually emerge fully armed ... It develops slowly" (Lakatos, 1970, p. 48, n. 4) and "if and when the programme ceases to anticipate novel facts, its hard core might have to be abandoned" (Lakatos, 1970, p. 49). Lakatos argued, in a similar way, that the *heuristic* may also be subject to revision: "it occasionally happens that when a research programme gets into a degenerating phase, a little revolution or a *creative shift* in its positive heuristic may push it forward again" (Lakatos, 1970, p. 51). Of course, modification of a research program's hard-core assumptions would almost certainly require a re-evaluation of its heuristic as well.

§§

In the opening decade of the twenty-first century, the Milgromian research program, in spite of its early successes, appeared – like the Boltzmann research program in the late nineteenth century, or the Bohr research program in the early twentieth century – to be in a state of stagnation. Theory variant T_2 had failed to improve on the discrepant predictions of theory variant T_1 for the dynamics of galaxy clusters, and some of its predictions about the universe on cosmological scales, e.g. the angular power spectrum of the cosmic microwave background (CMB), were either ill-defined or in conflict with observations. As a number of authors noted around this time (Milgrom, 2011a; Dodelson, 2011; Kroupa, 2012; Famaey and McGaugh, 2012), these failures of prediction seemed not to be random, but instead seemed to respect a sort of spatial hierarchy: novel predictions about systems the size of galaxy groups or smaller were mostly corroborated; while predictions (novel or otherwise) about larger systems were often found to be incorrect, and furthermore, the magnitude of the discrepancy seemed to correlate with system size.

Theory variant T_3 was a response to this state of affairs. In 2015, Lasha Berezhiani and Justin Khoury proposed a modification to the hard core of the Milgromian research program. Berezhiani and Khoury acknowledged the progressivity of Milgrom's program in its early stages, and they sought (with limited success, as we will see) to preserve its successes in the modified theory. But instead of viewing Milgrom's postulates as asymptotic expressions of a modified theory of gravity or motion, they assumed that Einstein's theory (and Newton's, in the appropriate regime) was correct. They postulated instead the existence of an additional dark component that couples to normal matter, in such a way as to generate Milgrom's asymptotic relations in the outskirts of galaxies. The same component would behave on cosmological scales like cold, collisionless dark matter.[1]

Berezhiani and Khoury proposed that both these properties could arise naturally from a single dark component if it was able to be in a *superfluid* state:

The superfluid nature of DM [dark matter] dramatically changes its macroscopic behavior in galaxies. Instead of behaving as individual collisionless particles, the DM is more

---

[1] Given that Berezhiani and Khoury's theory adopts elements from the standard cosmological model, one could argue that it 'belongs' to that research program rather than to Milgrom's. I would not find such a view unacceptable. However, I know of no standard-model cosmologist who views the situation in this way nor are there any textbooks or review articles that take such a position. Whereas Berezhiani and Khoury are quite explicit, in their published papers, that the *raison d'être* of their theory is the record of successful predictions of Milgrom's theory.

aptly described as collective excitations, in particular phonons. The DM phonons couple to ordinary matter, thereby mediating a long-range force (beyond Newtonian gravity) between baryons. In contrast with theories that propose to fundamentally modify Newtonian gravity, in this case the new long-range force mediated by phonons is an *emergent* property of the DM superfluid medium (Berezhiani et al., 2017, p. 3).

In the approach of Berezhiani and Khoury, the long-range force between particles of normal matter is mediated, in regions where the dark matter is condensed, by a collective excitation of the dark condensate, similar to the phonon–electron (Fröhlich) interaction in a laboratory superfluid. The conditions for formation of a condensate are assumed to be satisfied in the central parts of galaxies; the dark 'fluid' in this regime would be more aptly described as a collective excitation, with wavelength comparable to the size of the galaxy. In the same way that phonons mediate a long-range force between particles (electrons) in a laboratory superfluid, particles of normal matter orbiting in a galaxy would be subject to phonon-mediated forces generated by the dark component, in addition to the Newtonian gravitational force from the normal (as well as the dark) matter. By making appropriate choices for the condensate equation of state, and for the coupling between dark and normal matter, Berezhiani and Khoury showed that Milgrom's asymptotic laws could be reproduced in this regime.

At sufficiently low densities or high temperatures (that is, velocity dispersions), the same dark matter would behave like a self-gravitating 'fluid' of collisionless particles. If the primordial velocities of those particles were sufficiently low – roughly speaking, non-relativistic – their behavior could mimic that of the dark matter postulated by standard-model cosmologists to permeate the universe. Thus, Berezhiani and Khoury included among their set of hard-core postulates two of the same hypotheses that are part of the standard cosmological model:

DM-2: Beginning at some early time, the universe contained a (nearly) uniform dark component, which subsequently evolved, as a collisionless fluid, in response to gravity.

DM-4: The dark matter is 'cold,' that is, the velocity dispersion of the dark particles, at a time before the formation of gravitationally bound structures like galaxies, was small enough that it did not impede the formation of those structures.

The demonstrated ability of the standard model to reproduce the statistical properties of the large-scale galaxy distribution, and the spectrum of temperature fluctuations of the CMB (Chapter 6), could therefore be reproduced without additional assumptions.

The particles constituting the dark fluid are left unspecified in Berezhiani and Khoury's theory (as they are, of course, in the standard cosmological model), but certain of their properties appear in relations that describe the fluid's behavior. A necessary condition for the dark matter to form a condensate is that the de Broglie wavelength $\lambda_{dB} = h/(mv)$ of the particles exceed the interparticle separation

$l = (m/\rho)^{1/3}$; here v is a typical particle velocity and ρ is the dark matter density. This implies an upper limit to the particle mass of $m \lesssim (\rho/v^3)^{1/4}$, or

$$mc^2 \lesssim 50 \left(\frac{\rho}{0.015 \, M_\odot \, \text{pc}^{-3}}\right)^{1/4} \left(\frac{v}{200 \, \text{km s}^{-1}}\right)^{-3/4} \text{eV}. \tag{7.1}$$

(The normalizing values chosen for ρ and v in Equation (7.1) are appropriate for dark matter near the Sun.) The particles must therefore be light. For a given m, the condition (7.1) is best satisfied in regions of high density and/or low velocity dispersion. Compared with galaxies, galaxy *clusters* have lower mean ρ and higher v, hence the dark matter in these systems could fail to satisfy Equation (7.1) even if that condition is well satisfied in individual galaxies (Berezhiani and Khoury, 2015).

As is the case for the particles that are postulated to make up the dark matter in the standard cosmological model, no known particles satisfy these conditions. Berezhiani and Khoury suggested that *axions*, the elementary particles postulated by Peccei–Quinn theory to resolve the strong CP problem in quantum chromodynamics, might have the necessary properties. Axions, if they exist, could be produced through a phase transition in the early universe and would remain thermally decoupled from normal matter; furthermore the axions would have an extremely low random velocity at the time of their formation. The axions would therefore constitute 'cold dark matter' in the language of a standard-model cosmologist.[2]

Berezhiani and Khoury based their dynamical model of the dark component on the existing formalism for describing laboratory superfluids. In the low-temperature, mean-field limit, the wave function of a superfluid condensate evolves as

$$\psi(x,t) = \Psi_0(x)e^{-i\theta} = \Psi_0(x)e^{-i\mu t/\hbar} \tag{7.2}$$

with μ the chemical potential. The time dependence in Equation (7.2) is basically Josephson's relation; μ can be derived (for instance) as the eigenvalue of the time-independent Gross–Pitaevskii equation. The dynamics of the condensate subject to an external force are conveniently encapsulated in an effective Lagrangian. Greiter et al. (1989) and Son and Wingate (2006) use symmetry and invariance arguments to show that the most general form for the Lagrange density describing a superfluid condensate near the ground state is

$$\mathcal{L} = P(X), \tag{7.3}$$

$$X = \hbar\dot{\theta} - \frac{\hbar^2}{2m}(\nabla\theta)^2 - m\Phi(x) = -\hbar\dot{\phi} + \mu - \frac{\hbar^2}{2m}(\nabla\phi)^2 - m\Phi(x)$$

---

[2] In the standard cosmological model, the particles postulated to make up the dark matter are assumed to be in thermal equilibrium with the photons and the normal matter at early times, and they achieve their 'coldness' by virtue of being massive.

with $\Phi(x)$ the external potential; the second expression for X has replaced θ by $\mu t/\hbar - \phi$ where $\phi(x,t)$ represents small fluctuations, i.e. phonon excitations. The functional form of P is arbitrary and determines the equation of state: the pressure is given, in the hydrostatic description, by $P(X = \mu)$ and the number density of condensed particles by $n = dP/d\mu$.

Berezhiani and Khoury's postulate is that the dark matter forms a superfluid inside galaxies, and that its interaction with normal matter – via Newtonian gravity, but also via coupling of the superfluid phonons to normal matter – results in a test-body acceleration that approximates Milgrom's relation under the appropriate limiting conditions. This is achieved via two additional ansatze:

(*i*) The superfluid Lagrangian is assumed to have the form

$$P(X) = \frac{2}{3} \frac{\Lambda c^2}{\hbar^3} |2m X|^{3/2} \tag{7.4}$$

with Λ a parameter of dimension mass. Setting $X = \mu$ yields the equation-of-state relations

$$P(\mu) = \frac{2}{3} \frac{\Lambda c^2}{\hbar^3} (2m\mu)^{3/2}, \quad n(\mu) = \frac{\Lambda c^2}{\hbar^3} (2m)^{3/2} \mu^{1/2}, \tag{7.5}$$

i.e.

$$P(\rho) = \frac{\hbar^3}{12c^4 \Lambda^2 m^6} \rho^3 \tag{7.6}$$

($\rho \equiv mn$), a polytropic equation of state of index $1/2$. Thus, a steady-state, isolated, self-gravitating, low-temperature condensate will have a density that depends on radius in the same way as that of a classical $n = 1/2$ polytrope. The density falls to zero beyond $r = R_0$, and is nearly homogenous (constant density) within a distance $\sim 0.2 R_0$ from the center, where

$$R_0 \approx 2.75 \sqrt{\frac{\hbar^6 \rho_0}{32\pi G c^4 \Lambda^2 m^6}} \tag{7.7a}$$

$$\approx 45 \left(\frac{M_{\rm DM}}{10^{12} M_\odot}\right)^{1/5} \left(\frac{mc^2}{\rm eV}\right)^{-6/5} \left(\frac{\Lambda c^2}{\rm meV}\right)^{-2/5} {\rm kpc} ; \tag{7.7b}$$

here ρ_0 is the central mass density of the condensate and $M_{\rm DM}$ is the total dark mass.

(*ii*) As analog of the Fröhlich phonon–electron interaction in a laboratory superfluid, Berezhiani and Khoury postulate a phonon–matter interaction described by an additional piece of the Lagrangian,

$$-\alpha \frac{\Lambda}{M_{\rm Pl}} \theta \rho_{\rm normal} c^2 \tag{7.8}$$

with α a second (dimensionless) parameter; ρ_{normal} is the mass density of normal matter and $M_{\text{Pl}} = \sqrt{\hbar c/(8\pi G)}$ is the Planck mass.

The functional forms of Equations (7.4) and (7.8) were chosen in order to reproduce Milgrom's effective force law in the low-acceleration limit, as follows: The phonon equation of motion is obtained as the extremum of the combined Lagrangian; after imposing spherical symmetry and setting $\theta = \mu t/\hbar - \phi$, one finds

$$\nabla \cdot \left(\sqrt{2m\,|X|} \nabla \phi \right) = \frac{\alpha \hbar}{2M_{\text{Pl}}} \rho_{\text{normal}}, \quad X \equiv \mu - m\Phi(r) - \frac{\hbar^2}{2m} \left(\frac{d\phi}{dr} \right)^2. \quad (7.9)$$

The first integral is

$$\sqrt{2m\,|X|} \frac{d\phi}{dr} = \frac{\alpha \hbar M_{\text{normal}}(<r)}{8\pi M_{\text{Pl}} r^2} \equiv \kappa(r). \quad (7.10)$$

Assuming for the moment $X < 0$, this yields

$$\hbar \frac{d\phi}{dr} = \sqrt{m} \left(\hat{\mu} + \sqrt{\hat{\mu}^2 + \hbar^2 \kappa^2/m^2} \right)^{1/2} \quad (7.11)$$

with $\hat{u} = \mu - m\Phi$. The low-acceleration limit corresponds to $\kappa/m \gg \hat{u}$. In this limit,

$$\hbar \frac{d\phi}{dr} \approx \sqrt{\hbar \kappa(r)}, \quad (7.12)$$

and the acceleration of a particle of normal matter due to the Fröhlich force is

$$a(r) = \alpha c^2 \frac{\Lambda}{M_{\text{Pl}}} \frac{d\phi}{dr} \approx \frac{\alpha c^2}{\sqrt{\hbar}} \frac{\Lambda}{M_{\text{Pl}}} \sqrt{\kappa(r)} \approx \sqrt{\frac{\alpha^3 c^3 \Lambda^2}{\hbar M_{\text{Pl}}} \frac{G M_{\text{normal}}(<r)}{r^2}}. \quad (7.13)$$

This matches Milgrom's asymptotic form if the first term under the square root is identified with a_0, which requires

$$\alpha \approx 0.86 \left(\frac{a_0}{10^{-8} \text{ cm s}^{-2}} \right)^{1/3} \left(\frac{\Lambda}{\text{meV}/c^2} \right)^{-2/3} \quad (7.14)$$

($\text{meV}/c^2 \approx 1.78 \times 10^{-36}$ grams, about 2×10^{-9} times the mass of the electron).

As Berezhiani and Khoury (2015, 2016) point out, there is a problem with this derivation. Solutions with $X < 0$ are dynamically unstable, in the sense that the density of a self-gravitating condensate described by the $X < 0$ Lagrangian would increase monotonically over time. It is argued that this deficiency is an artifact of the assumption of zero temperature and can be removed by making appropriate assumptions about the form of the Lagrangian describing finite-temperature condensates. Following Nicolis (2011), Berezhiani and Khoury (2015) propose a simple, ad hoc modification to Equation (7.4):

$$P(X) = \frac{2}{3} \frac{\Lambda c^2}{\hbar^3} (2m)^{3/2} X \sqrt{|X - \beta Y|}, \quad Y \equiv \mu - m\Phi + \hbar \dot{\phi} + \hbar v \cdot \nabla \phi, \quad (7.15)$$

where v is the superfluid velocity and β is an additional dimensionless parameter. Clusters of galaxies have systematically higher internal velocity dispersions than single galaxies ($\sim 10^3$ km s^{-1} compared with $\sim 10^2$ km s^{-1}), and given an appropriate choice of the parameter β, the dark matter in galaxy clusters could be sufficiently 'hot' to be in a mixed or entirely normal phase. To be successful, Berezhiani and Khoury's hypothesis would need to imply that the same assumptions about the dark condensate that yield Milogrom's laws on galaxy scales can also correctly reproduce the not-quite-Milgromian dynamics of galaxy clusters, by assigning correct fractions of a cluster's dark mass to condensed and non-condensed components.

§§

In judging the progressivity of the Milgromian research program including theory variant T_3, some issues arise that were not relevant, at least not centrally, when judging the progressivity of the three earlier variants of the theory. These issues result from the modification of the hard core assumptions, and in particular, from the adoption of some postulates from the standard cosmological model.

(*i*) To be judged progressive, theory variant T_3 should account for the successes of the previous theories in its own research program (in addition to making new, novel predictions). One of the most remarkable of those successes was the prediction of rotation curves for individual galaxies, given only the observed distribution of the *normal* matter. Now, under T_3, the non-Newtonian behavior of galaxy rotation is ascribed (directly or indirectly) to the postulated *dark* matter. It follows that any prediction (successful or otherwise) of a particular galaxy's rotation curve, under T_3, will require an assumption about (for instance) the total mass of the galaxy's dark component.[3] The same is true with regard to some of the other successful, novel predictions discussed earlier in this book. For instance, the two predictions of theory variant T_1 involving central surface densities of disk galaxies (Chapter 5) are expressed purely in terms of the distribution of normal matter. Under T_3, these relations are *not* (necessarily) predicted, and explaining them would require some additional set of postulates regarding the behavior of the (collisionless) dark matter. Another example, discussed later in this chapter, is the internal dynamics of tidal dwarf galaxies, which seem to be correctly predicted under Milgrom's theory but not under the standard model. In this respect, at least, theory variant T_3 is epistemically weaker than the theories that preceded it in the Milgromian research program: there is a loss of explanatory content.[4]

---

[3] Alternatively, one could ask: What mass of dark halo yields the best prediction, under T_3, of this galaxy's rotation curve? This is tantamount to using the rotation-curve data to fix a parameter of the theory (the galaxy's dark mass), resulting in the situation discussed under (*iv*).

[4] The expressions 'Kuhn loss' and 'explanatory overlap' are sometimes used to describe situations like this. Feyerabend (1970) has emphasized that these are expected concomitants of a transition from one research program to another. Future philosophers of science, with the benefit of hindsight, may judge that Berezhiani

(*ii*) Theory variant T_3 loses some of the testability of theory variants T_0–T_2, as just discussed, but it also *gains* a great deal of new empirical content, which of course is a favorable circumstance according to the criteria of Popper and Lakatos. Now, Berezhiani and Khoury designed T_3 explicitly with the goal of explaining certain facts that the standard cosmological model also explains ("We propose a novel theory of dark matter (DM) superfluidity that matches the successes of the ΛCDM model on cosmological scales"; Berezhiani and Khoury, 2015, p. 103510). One feels intuitively that T_3 does not deserve 'credit' for those explanatory successes, and that intuition is consistent with the criteria adopted in Chapter 2 for deciding whether a theory's predictions are 'novel,' i.e. capable of providing evidential support for the theory. For instance, the ability of T_3 to explain the CMB fluctuation spectrum (which it does, in precisely the same way that the standard cosmological model does) fails to satisfy novelty criterion P (since the observations preceded the theory) or criteria Z or M (since T_3 was designed to explain those observations).

But there is another issue here, and one that is not so easily dealt with. Among the excess content claimed for their theory, Berezhiani and Khoury include some components of the standard cosmological model that do *not* follow, necessarily, from the set of postulates on which T_3 is based. For instance, in their papers, Berezhiani and Khoury assume that gravitationally bound systems like galaxies and galaxy clusters are always associated with extensive 'haloes' of collisionless (i.e. non-superfluid) dark matter, having the same structure (e.g. dependence of density on radius) and the same statistical properties (e.g. mean relation between normal and dark mass) claimed by standard-model cosmologists on the basis of their simulations. But those simulations require (or at least, adopt) a large number of assumptions in addition to postulates DM-2 through DM-4. One example: Berezhiani and Khoury routinely invoke 'abundance matching,' a standard-model postulate (not based on any physical principle) that allows them to associate a set of dark matter haloes in a simulation uniquely with an observed set of galaxies.

Berezhiani and Khoury never clearly state their position with regard to standard-model methodology. But one infers that they accept all of the explanatory successes claimed by standard-model cosmologists – with the exception, of course, of explanations that might be affected by the presence of superfluid dark matter.[5]

The issue at hand is how to judge the progressivity of Milgromian theory variant T_3. I propose to take a conservative, that is a narrow, view. I will not infer progressivity on the basis of any predictions of T_3 that are simply a consequence of its having incorporated assumptions that standard-model cosmologists also adopt. 'Excess content' will only be considered evidentiarily relevant for T_3 if it differs

---

and Khoury's theory is properly seen as the beginning of a new research program rather than as a variant of Milgrom's. See also note 1.

[5] For instance, Berezhiani and Khoury appear to accept the post-2000 revision of the baryon density parameter Ω_b, since it was only with the help of this revision that standard-model cosmologists were able to fit the CMB spectrum; see Chapter 6.

from the content of the standard cosmological model – that is: if it results from the incorporation of superfluid dark matter.

To cite a concrete example: By associating extended dark matter haloes with observed galaxies, Berezhiani and Khoury would predict a statistical rate of mergers between galaxies that is the same, or essentially the same, as in the standard cosmological model, and much *higher* than predicted by Milgromian theory variants T_0–T_2 (since those theories contain no dark matter). But I will not consider the latter fact evidentiarily relevant.

(*iii*) We are thus interested in predictions of T_3 that arise as a consequence of the presence of superfluid dark matter. But in comparing T_3 with earlier theory variants in the Milgromian research program, we must be careful to construct those predictions so that they are meaningful under both T_3 and (say) T_2. For instance: one could imagine predictions, under T_3, of the form 'The superfluid dark matter has such-and-such a property.' Such a prediction may or may not be testable under T_3, but it can not possibly be testable under T_0–T_2, because dark matter (much less *superfluid* dark matter) does not exist in those theories; a prediction about the properties of superfluid dark matter would be akin to the prediction 'All unicorns have cloven hoofs.' To be useful in assessing progressivity, such predictions must be recast. For instance, if T_3 predicts that systems of a certain recognizable type (e.g. dwarf galaxies) are dominated by superfluid dark matter, one could use T_3 to predict something observable about the behavior of *normal* matter in such systems. Such a prediction could remain meaningful under T_2 and could be used to assess progressivity.

(*iv*) Theory variant T_3 adds a number of undetermined parameters to the single parameter a_0 that appears in Milgrom's T_0. These include particle mass m; the constants Λ and α that appear in the zero-temperature Lagrangian (7.8); and the constant β that appears in the finite-temperature Lagrangian (7.15). Complete specification of the theory requires a specification of those parameters. Since a fundamental underlying description of the dark component is lacking, the route so far followed has been to determine or constrain those parameters using observational data. Of course, this is the same situation that standard-model cosmologists found themselves in during the first decade of the twenty-first century, when they freely adjusted the half-dozen or so parameters of *their* model in order to accommodate new observations of the CMB and the large-scale distribution of galaxies.

In the discussion of the 'concordance model' in Chapter 6, it was argued (following John Worrall and Elie Zahar) that experimental or observational data that are used in setting the unknown parameters of a theory constitute part of the 'background knowledge' that went into theory construction; the ability of the theory to reproduce those data does not count as evidential support for the theory. The same argument will be considered valid here. Thus: If the parameter values of T_3 are adjusted in response to data, then those data can no longer be considered novel according to criteria such as those of Popper, Zahar or Musgrave, since criterion P precludes facts that were known prior to theory construction and criteria Z and M preclude facts that a theory was designed to explain.

Suppose on the other hand that one observation, or set of observations, is used to set the parameters of a theory, and that this theory – with its parameters now treated as known, fixed quantities – successfully predicts another, independent fact. In this case, corroboration of the *second* fact *can* be said to provide evidential support for the theory, as long as the second fact satisfies one of the other criteria for novelty defined in Chapter 2. We encountered instances of this kind earlier in the book; for instance, Milgrom's second postulate was designed to yield asymptotically flat rotation curves, but the theory so constructed (in that case, T_0) then successfully predicted (among other things) the baryonic Tully–Fisher relation.

Now, in the case of T_3, Berezhiani, Khoury and collaborators argue that the most useful sorts of data from which to determine the additional free parameters of their theory are the rotation curves of (as it turns out) just two disk galaxies (Hodson et al., 2017). I accept this suggestion, and accordingly I will judge the progressivity of T_3 based on predictions made using the parameter values that have been established from rotation-curve fitting.

§§

Calculating rotation curves of individual galaxies in T_3 is mathematically complicated; by the end of 2017, data for only two galaxies had been analyzed in this way, and the range of parameter values explored was modest. Berezhiani et al. (2017) attempted to fit data for two disk galaxies, one a 'high surface brightness galaxy' and the other a 'low surface brightness galaxy.' As noted above, T_3 says nothing about the total amount of dark matter to be associated with a given galaxy, and so Berezhiani et al. tested two very different assumptions: that the dark mass is ten times the mass in stars and gas, and that it is roughly fifty times this value. The parameter β that appears in the finite-temperature Lagrangian (7.15) was fixed at $\beta = 2$. The equation of state of the zero-temperature condensate, Equation (7.6), is determined by the product $m^3\Lambda$ and various values were tried for this quantity. Berezhiani et al. argued that – even at zero temperature – the existence of a superfluid state requires another condition to be satisfied in addition to wave function overlap: the relaxation time determined by collisions between dark matter particles must be shorter than the local 'crossing time' $\sim (G\rho)^{-1/2}$. They showed that the ratio between these two times is determined by $m\sigma^{-1/4}$, with σ an additional parameter that sets the hard-sphere interaction cross section (assumed velocity independent). This quantity enters into the modeling by determining the size of the 'thermalized' region, R_T. Roughly speaking, R_T should be large enough to encompass the region containing normal matter so that Milgrom's relation is reproduced via the phonon force; Berizhiani et al. achieved this by fixing $m = 1$ eV and

$\sigma/m = 0.01 \text{ cm}^2 g^{-1}$. Beyond R_T, the density of the dark fluid was 'matched' to the same family of fitting functions that standard-model cosmologists conventionally use to model their dark matter 'haloes.' Finally, Equation (7.14) was used to replace the free parameter α by a_0, which was assigned a value falling in a range centered on the value established in T_0–T_2 ($\sim 1.2 \times 10^{-8} \text{ cm s}^{-2}$).

Berezhiani et al. found that the rotation-curve data for both galaxies could be well reproduced, under T_3, with the parameter values

$$\Lambda = 0.05 \text{ meV}/c^2 \quad (\text{i.e. } \Lambda m^3 = 5 \times 10^{-5} \text{ eV}^4/c^8),$$
$$a_0 = 0.87 \times 10^{-8} \text{ cm s}^{-2} \quad (\text{i.e. } \alpha \approx 6), \tag{7.16}$$

together with the fiducial values already given for the other parameters. They further demonstrated that values of a_0 in the range

$$0.6 \times 10^{-8} \text{ cm s}^{-2} \lesssim a_0 \lesssim 1.2 \times 10^{-8} \text{ cm s}^{-2} \tag{7.17}$$

with $\Lambda = 0.05 \text{ meV}/c^2$, i.e. $5.3 < \alpha < 6.7$, were reasonably consistent with data for both galaxies. Finally, Λ itself was varied over the range 0.02–$0.1 \text{ meV}/c^2$, at fixed m, but the predicted rotation curve was found to be insensitive to this variation and so the allowed range of Λ was not constrained. Interestingly, results were found to be almost independent of the *total* mass assumed for the dark matter.

§§

We are now in a position to begin assessing the progressivity of the Milgromian research program including theory variant T_3. Taking as given the set of parameters (or parameter ranges) established as described above, the theory makes the following predictions:

1. A predicted dependence of the temperature of intracluster gas on distance from the center in clusters of galaxies, given an assumed value for the cluster's total dark mass;
2. In any system that, under T_3, contains dark matter only in a superfluid state, the dynamical friction acting on an object moving through the system on a bound orbit should be due entirely to the normal matter;
3. Absence of the external field effect (EFE) in the case of satellite systems that do not orbit inside the superfluid region of the parent system.

§§

1. *A predicted dependence of the temperature of intracluster gas on distance from the center in clusters of galaxies, given an assumed value for the cluster's total dark mass*: The gravitational acceleration as a function of radius in a (spherical) cluster

of galaxies can be related to the observable quantities $n_X(r)$ and $T_X(r)$, the number density and temperature of the X-ray emitting intracluster gas, via the equation of hydrostatic equilibrium. In a number of galaxy clusters, such data have been shown to be inconsistent with the predictions of Milgrom's T_0–T_2, in the sense that the gravitational acceleration computed (via Milgrom's relations) from the observed mass is insufficient, particularly near the cluster center, to produce the observed gas temperatures (Gerbal et al., 1992; Sanders, 1999, 2003, 2007; Aguirre et al., 2001). As discussed above, this failure of prediction was one motivation for theory variant T_3.

Under T_3, the gravitational acceleration has three components:

$$a = a_{\text{normal}} + a_{\text{dark}} + a_{\text{phonon}}, \tag{7.18}$$

due respectively to the (Newtonian) gravitational force from the normal and dark matter and the phonon (Fröhlich) force from the dark matter. The third term is appreciable in regions where the dark matter is in a superfluid state; if in addition $a_{\text{phonon}} > \max[a_{\text{normal}}, a_{\text{dark}}]$, the gravitational acceleration in this region will reproduce Milgrom's asymptotic laws. These two conditions are often simultaneously satisfied in the outer parts of individual galaxies. In galaxy clusters, one finds under T_3 that the conditions for dark matter superfluidity are sometimes satisfied near the cluster center, but that the second condition typically is not; another way to state this is to note that a is typically of order a_0 or greater in galaxy clusters. The force that appears in the hydrostatic equilibrium equation is therefore predicted to be due mostly to Newtonian gravity from the normal and dark components, with little contribution from the phonon force. Now, it is an observed fact that the internal velocity dispersion, or equivalently the mean gas temperature, in galaxy clusters is an increasing function, on average, of cluster size or richness (number of galaxies). Hence one expects under T_3 that the fraction of superfluid dark matter will tend to decrease with increasing cluster richness, and that the richest clusters will be entirely in the collisionless regime.

The test consists of: (*i*) measuring $n_X(r)$ and $T_X(r)$ in a galaxy cluster, procedures for which are well established; (*ii*) measuring or estimating the (normal) mass associated with individual galaxies in the cluster; this can be especially important in clusters that contain a very massive, central galaxy; (*iii*) postulating a value for the total dark mass of the cluster, and/or for the dependence on radius of the dark matter density outside the superfluid region; (*iv*) using T_3 and Poisson's equation to compute the dark matter distribution in the region where the conditions for superfluidity obtain; (*v*) using the hydrostatic equilibrium equation to predict the temperature profile implied by the observed gas density profile $n_X(r)$ given the gravitational force as computed in (*iv*) from the dark and normal matter; (*vi*) comparing predicted and observed temperature profiles.

This program is straightforward in principle but so far has only been carried out in a very approximate way, using data from four galaxy clusters (Hodson et al.,

2017). Hodson et al. assume spherical symmetry and ignore the phonon force; the latter simplification means that their models are, at least formally, inconsistent. In lieu of a (numerically) exact calculation of the dark matter distribution, they adopt a simplified model consisting of a zero-temperature, superfluid 'core' inside a radius R_c, with an abrupt transition to a collisionless dark matter 'halo' at $r > R_c$; their model includes no intermediate region with a mixture of superfluid and normal fluid. The equation of state of the superfluid is taken to be Equation (7.6), which depends on the theory parameters only via Λm^3. In computing R_c, the authors do not appear to impose a requirement that the conditions for superfluidity be satisfied at $r < R_c$. Instead they treat R_c simply as a 'matching' radius and require that the dark matter density and pressure at this radius match the corresponding values in the collisionless halo. Two different assumptions are tried for the latter, both based on analytic descriptions of dark matter haloes that are commonly used by standard-model cosmologists; this procedure indirectly determines the total dark mass. Their computation of a_{normal} includes the gravitational force from the gas itself as well as an estimate of the force from the central galaxy (if present); the gravity produced by all the other galaxies in the cluster is ignored, a reasonable approximation since the galaxies probably contribute only a small percentage of the total (gas plus galaxies) mass. Finally, rather than compare measured and predicted temperature profiles, as described above, Hodson et al. choose to compare mass distributions, $M(< r)$, determined once from the measured $T_X(r)$ using the hydrostatic assumption, and again from their computed dark matter density.

Hodson et al. were only partially successful. All four clusters were predicted to have superfluid cores, with radii 10^1–10^2 kpc, implying that only a small fraction of the total (dark) cluster mass is in the form of a condensate. This is qualitatively in agreement with expectations for T_3, since it implies that cluster dynamics should differ only modestly from that of clusters containing only collisionless dark matter. However, the parameter values found necessary to reproduce the cluster data were somewhat different than the values derived from the rotation-curve analysis of Berezhiani et al. (2017). Hodson et al inferred

$$\Lambda m^3 \approx 1\text{--}3 \times 10^{-4} \text{ eV}^4/c^8, \tag{7.19}$$

substantially greater than the value 5×10^{-5} eV$^4/c^8$ of Equation (7.16). Values for this parameter more in keeping with the rotation-curve data were found to yield a superfluid core of too low mass to fit the temperature data.

Given the approximate nature of the Hodson et al. modeling, and the limited quality of the constraints from the rotation-curve data, it is not necessarily the case that these results contradict the predictions of T_3 for galaxy clusters. But it is certainly correct to conclude that, as of this date, corroboration of the prediction is partial at best.

The prediction is not novel according to criterion P (galaxy cluster temperature data existed prior to 2015), but criterion Z *is* satisfied, since T_3 was not specifically

Table 7.1 *Corroborated excess content, theory variants $T_0 - T_3$*

Theory variant	Prediction	Status	Novelty			
			P	Z	M	C
T_0	$V(R) \to V_\infty$	confirmed	no	no	–	yes
	$V_\infty = \left(a_0 G M_{\mathrm{gal}}\right)^{1/4}$	confirmed	yes	yes	yes	–
	$a = f\left(g_{\mathrm{N}}/a_0\right) g_{\mathrm{N}}$	confirmed	yes	yes	yes	–
	Renzo's rule	corroborated	yes	yes	yes	–
T_1	$\Sigma_{\mathrm{ph}}(0) \lesssim a_0/(2\pi G)$	corroborated	yes	yes	yes	–
	Central surface density relation	confirmed	yes	yes	yes	–
	Vertical kinematics in Milky Way	corroborated	yes	yes	yes	–
	$V_{\mathrm{rms}} \approx (a_0 G M)^{1/4}$	partially corroborated	yes	yes	yes	–
	External field effect	possibly corroborated	yes	yes	yes	–
	Low merger rate	neither confirmed nor refuted	no	–	yes	–
T_2	$z_{\mathrm{reion}} \gtrsim 15$	possibly corroborated	yes	yes	yes	–
	$A_{1:2} \approx 2.4$	confirmed	yes	yes	yes	–
T_3	Galaxy rotation curves	corroborated	no	no	–	yes
	Galaxy cluster $T(r)$	partially corroborated	no	yes	no	–
	$\langle \Delta v_\parallel \rangle_{\mathrm{condensate}} = 0$	possibly corroborated	no	yes	yes	–
	No EFE outside of superfluid	neither confirmed nor refuted	yes	–	yes	–

designed to reproduce those data. One could plausibly make the case that the prediction is novel with respect to Musgrave's criterion, given that the rival theory (i.e. the standard cosmological model) is not completely successful at explaining the galaxy cluster data (e.g. Figure 4 of McGaugh, 2015, and the discussion in Chapter 5). But I choose not to do so. These conclusions are summarized in Table 7.1.

§§

2. *In any system that, under T_3, contains dark matter only in a superfluid state, the dynamical friction acting on an object moving through the system on a bound orbit should be due entirely to the normal matter*: Lev Davidovich Landau argued (1941, 1947) that the speed of an object moving through a superfluid in its ground state must exceed the sound speed, c_s, if the object is to transfer its kinetic energy to sound waves. The absence of friction originates from the fact that sound waves are the only low-energy excitations present in the superfluid.

The sound speed is given by $c_s^2 = dP/d\rho$. Combining the equation of state (7.6) with the equation of hydrostatic equilibrium for a spherical system, $dP/dr = -\rho(r)d\Phi/dr$, it is easy to show that

$$c_s^2(r) = \frac{dP}{d\rho} = -2\Phi(r), \tag{7.20}$$

where r is the distance from the center and the zero of the potential has been set at the outer edge of the condensate. Note that $\sqrt{-2\Phi(r)}$ is the escape velocity at radius r; thus Landau's condition can be stated as: In order for an object moving through a condensate to experience friction, it must be moving faster than the escape velocity determined by the condensate's gravitational potential.

According to T_3, much of the dark matter surrounding the luminous parts of galaxies would be in a superfluid form; beyond some radius, the dark matter is assumed to be in a non-condensed state and to behave dynamically like ordinary matter. An object – for instance, a dwarf galaxy – passing through a larger galaxy would experience a local acceleration due to dynamical friction, $\langle\Delta v_\|\rangle,$[6] with contributions from each of these components:

$$\langle\Delta v_\|\rangle = \langle\Delta v_\|\rangle_{\text{normal}} + \langle\Delta v_\|\rangle_{\text{dark}}. \tag{7.21}$$

If the dark matter is in a non-condensed state, its frictional force obeys the same functional relations as the frictional force from normal matter. Berezhiani et al. (2017) assumed that the superfluid 'core' of the dark matter should encompass, at least, the region traced by the observed rotation curve and enforced this condition by their choice of parameters. In the case of the low surface brightness galaxy modeled by them, the density of the superfluid 'core' exceeded that of the normal matter everywhere; while in the case of the high surface brightness galaxy, the density near the center was due mostly to normal matter, with the superfluid dominating beyond about 20 kpc.

The standard approximation for the frictional force due to motion through normal matter (e.g. stars) or collisionless dark matter is derived assuming Newtonian gravity, and so it remains valid under T_3. The result (Merritt, 2013, p. 219–220) is

$$\langle\Delta v_\|\rangle = -16\pi^2 G^2 M\rho \int_0^\infty \left(\frac{v_f}{v}\right)^2 f(v_f)H\left(v, v_f\right)dv_f, \tag{7.22a}$$

$$H = \begin{cases} \log\Lambda, & v > v_f, \\ 0, & v < v_f. \end{cases} \tag{7.22b}$$

In this expression, ρ is the mass density of objects (e.g. stars) responsible for the frictional force and $f(v_f)$ is their velocity distribution, both quantities evaluated

---

[6] The strange notation used in Equation (7.21) is standard; see e.g. Chapter 5 of Merritt (2013). For our purposes, it is enough to note that $\langle\Delta v_\|\rangle$ denotes the magnitude of the acceleration due to the frictional force and that that acceleration is opposite in direction to the object's motion through the decelerating medium. The total acceleration would be given by the sum of this (vector) quantity and $-\nabla\Phi$.

at the instantaneous position of the object that experiences the frictional force; the latter has mass M and velocity v. The quantity $\log \Lambda$, the 'Coulomb logarithm,' is ill-defined but is often set equal to $\log(p_{max} v^2_{rms}/(GM))$ where v_{rms} is the rms velocity of stars in the larger system and p_{max}, the maximum 'impact parameter' for stars that contribute to the frictional force, is set to the linear size of the larger system.

As applied to the case of a dwarf galaxy moving through a larger galaxy, Equation (7.22) implies that all of the frictional force is due to stars that are moving more slowly than the dwarf galaxy, i.e. stars with $v_f < v$.[7] In a realistic model galaxy, stars would be moving with a range of velocities at every point, from $v = 0$ to $|v| \approx v_{esc} = \sqrt{-2\Phi(r)}$, where $v_{esc}(r)$ is the escape velocity at r, and so the frictional force would typically be nonzero everywhere. If one assumes that the velocity distribution of the stars is everywhere locally Maxwellian with the same rms velocity, $v_{rms} \equiv \sqrt{3}\sigma$ (this is the 'isothermal sphere' defined in Chapter 5), the integral in Equation (7.22) can be evaluated. Figure 7.1 plots $\langle \Delta v_\parallel \rangle$ as a function of v/σ. The frictional force increases from zero at $v = 0$ to a maximum when $v \approx \sqrt{2}\sigma$, then falls off as $\sim v^{-1}$.

Any steady-state system is expected to contain no matter that is moving faster than the escape velocity at every point. An object bound to the galaxy and moving through the condensate would necessarily also be moving with less than the local escape velocity, hence by Landau's criterion it would feel no frictional force from the dark matter (Berezhiani et al., 2017).

A nonzero frictional force from the condensate is, however, predicted in some other cases. A dwarf galaxy might 'fall through' a larger galaxy with a speed higher than the escape velocity from the larger galaxy if the two galaxies are initially unbound; for instance, the dwarf galaxy could have acquired a large initial velocity through a 'slingshot' interaction with a third galaxy. In such cases, Landau's criterion implies that dissipation of kinetic energy could occur, due to both the outer (collisionless) and inner (superfluid) dark matter systems.

It was argued above that to be testable, under both T_3 and earlier theory variants, predictions should say something about the behavior of *normal* matter in systems of a *recognizable type*. Based on the Hodson et al. modeling, low surface brightness galaxies might be a good target for predictions about dynamical friction, since under T_3 they would be dominated throughout by superfluid dark matter. The dynamical friction acting on a body passing through such a system would therefore be predicted to be due entirely to the normal matter, with no contribution from either superfluid or collisionless dark matter. Furthermore, since the density of normal matter in such systems is low, the *total* frictional force should also be small.

---

[7] A more exact calculation (Merritt, 2001) reveals that a small fraction of the frictional force comes from stars with $v_f > v$ for all v.

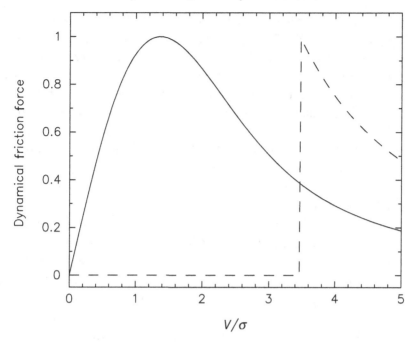

Figure 7.1 Dependence of the dynamical friction force on the velocity, v, of the object that experiences the force (Equation 7.22). The solid curve is the prediction in the case of an object that moves through a homogeneous background of normal matter (e.g. stars) or collisionless (non-superfluid) dark matter. The velocity distribution of the matter in the background system is assumed to be Maxwellian with rms velocity $v_{rms} \equiv \langle v^2 \rangle^{1/2} = \sqrt{3}\sigma$. The curve is normalized to a maximum value of one. The predicted form of this curve is the same under T_3 and under the standard cosmological model, both of which assume Newton's theory of gravity and motion in the regime assumed in deriving Equation (7.22). If v_{rms} is independent of position (the 'isothermal sphere'), then the escape velocity from the system has an average value of $v_{esc} = 2v_{rms} = 2\sqrt{3}\sigma$. The dashed line is a schematic representation of the frictional force due superfluid dark matter, assuming this value for v_{esc}.

Corroboration of this prediction has been claimed (in the absence, so far, of detailed modeling) based on one system that seems to have avoided the consequences of dynamical friction (Berezhiani and Khoury, 2015). The so-called Fornax dwarf galaxy is a satellite system of the Milky Way galaxy that is located at a distance of about 140 kpc from the Sun and that has a (stellar) luminosity of about $10^7 L_\odot$. The Fornax dwarf is unique among known satellite galaxies in having a retinue of globular clusters: the latter are compact, nearly spherical systems consisting of mostly old stars and having masses of $\sim 10^5 M_\odot$. About a half-dozen globular clusters have been identified with projected distances from the Fornax galaxy of less than one kiloparsec, and having low enough velocities relative to the galaxy that they are probably bound to it (e.g. Mackey and Gilmore, 2003).

Standard-model cosmologists would associate the Fornax galaxy with a dark matter halo having a mass of order $10^8 M_\odot$, much greater than the mass observed in stars. To them, the system represents a puzzle, in the sense that dynamical friction due to that dark matter should have brought the globular clusters to the center of the galaxy in a time much less than the age (~ 10 Gyr) of the clusters. Standard-model cosmologists have proposed a number of possible resolutions to the puzzle, all rather implausible: for instance, that the dark matter consists of black holes with masses comparable to globular cluster masses so that dynamical friction is countered by gravitational scattering (Oh et al., 2000) or that the standard approximation for the dynamical friction force (Equation 7.22) is substantially in error (Read et al., 2006).

Survival of the globular clusters in their current orbits would be a natural consequence of T_3. Given the lack of detailed modeling of this system under T_3, however, I consider the existence of the Fornax system to constitute 'possible corroboration,' at best, for the prediction of a vanishing dynamical friction force. The prediction is not novel according to Popper's criterion (P) since the Fornax system and its properties were well known prior to 2015, but it is novel according to Zahar's criterion (Z) since the theory was not designed in order to yield systems with zero dynamical friction.[8]

Table 7.1 summarizes the evidentiary status of this prediction.

§§

3. *Absence of the external field effect (EFE) in the case of satellite systems that do not orbit inside the superfluid region of the parent system*: As discussed in Chapters 4 and 5, a prediction of theory variants T_0 and T_1 is the 'external field effect' (EFE): under MOND, the *internal* dynamics of a self-gravitating system in free fall are affected by the *external* gravitational field that is responsible for the body's motion. This is true even if the external field is completely uniform.

Under T_3, the modified dynamics is a consequence of phonon-mediated forces generated by superfluid dark matter, and it manifests itself only in regions that contain the superfluid. One consequence is that the EFE only applies within the superfluid 'core' of a galaxy or galactic system. A satellite that is moving freely *outside* of the core is not subject to the EFE. Of course, if the satellite is embedded in its own superfluid halo, its internal motions can still be Milgromian.

---

[8] Of course, as Musgrave (1974) and Carrier (1988) point out, the historical record is not always clear with regard to whether data played a role in a theory's development. In this case, I note that in their first published discussion of the Fornax system (Berezhiani and Khoury, 2015), the authors thank Lam Hui for having brought the system to their attention.

Berezhiani et al. (2017, p. 4) point out some potentially testable consequences (italics added):

... globular clusters, such as Pal 5 (Thomas et al. 2017), and satellite galaxies *residing within the superfluid core of the Milky Way* should follow MOND predictions with the EFE. On the other hand, globular clusters at large distances from the Milky Way, such as NGC 2419 (Ibata et al. 2011) or Pal 14 (Jordi et al. 2009), are expected to be Newtonian *as long as they do not harbor their own DM halo*. The same applies to tidal dwarf galaxies resulting from the interaction of massive spiral galaxies. Those *are not expected to harbor a significant DM halo*, and thus should be Newtonian (Lelli et al. 2015a,b) as long as they are located outside the superfluid core of their host.

Note the qualifications. Under earlier variants of Milgromian theory, there was no dark matter, and predictions could be formulated entirely in terms of the properties of the *observed* matter. Under T_3, such predictions are often contingent on additional assumptions about dark matter.

Recall that in Chapter 5 a prediction was formulated, under T_1, of an observable effect due to the EFE based on the properties of dwarf spheroidal galaxies. Berezhiani et al. (2017, p. 15) discuss how such a prediction could be formulated under T_3:

With a core of ~60 kpc, most dwarf spheroidals would sit outside of the core, and should thus not be affected by the phonons of the superfluid core. In this case, if the dwarf spheroidals are primordial (hence not tidal dwarfs), the expected DM masses from abundance matching being very large, they would mostly be expected to display higher velocities than predicted in MOND, even without the EFE (Berezhiani et al., 2017, p. 15).

This passage is important and it is worth taking time to unpack it. (*i*) "Core" refers to the superfluid core of the Milky Way. No one has yet modeled the Milky Way, under T_3, in sufficient detail to know whether the size predicted for its superfluid region is greater or less than 60 kpc. (*ii*) The distinction between 'primordial' and 'tidal' dwarf galaxies is due to Pavel Kroupa (2012). Kroupa notes that standard-model cosmologists routinely associate the retinue of *observed* dwarf galaxies around a giant galaxy like the Milky Way with the *simulated* 'subhaloes' that form in their computer models of dark matter clustering. These objects (assuming that the postulated association is correct) would be the 'primordial dwarfs.' But Kroupa points out that there exists a category of dwarf galaxies that are structurally indistinguishable from the dwarf galaxies in the Local Group, but which almost certainly are of much more recent origin, consisting of matter pulled out of tidally interacting disk galaxies.[9] The latter would not be expected, by standard-model cosmologists, to contain significant amounts of dark matter. In the quoted passage,

---

[9] The existence of tidal dwarf galaxies appears to be universally acknowledged; I am not aware of any objection on the part of standard-model cosmologists to the existence of dwarf galaxies that form in this way.

Berezhiani et al. are assuming that the dwarf spheroidal galaxies in the Local Group are of the first type, hence that they contain massive dark matter haloes. (*iii*) In reference to 'primordial dwarfs,' Berezhiani et al. state, "the expected DM masses from abundance matching being very large, they would mostly be expected to display higher velocities than predicted in MOND." 'Abundance matching' has been mentioned a number of times previously in this book, and we now consider it in a little more detail.

§

   Computer simulations by standard-model theorists of the growth of structure in the universe always include dark matter, which (according to their theory) is the dominant, gravitating component. Simulating the behavior of the *normal* matter is much more difficult, for the reasons discussed in Chapters 2, 4 and 5, and a common practice is to 'put in' the normal matter at the very end, by adopting some prescription that associates (for instance) the mass of a dark matter halo with the mass of the galaxy that is presumed to form inside of it. One such prescription is abundance matching; in its simplest form, it consists of ranking the dark haloes in a simulation by mass, ranking a set of observed (or imagined) galaxies by luminosity, then going down the two lists, associating the pairs in sequence, from greatest to smallest. In the case of populations of giant galaxies in the field, the technique works tolerably well; that is, one finds that the dark masses that are assigned to galaxies of a given luminosity are comparable to what one might have expected based on empirical relations such as the Tully–Fisher or Faber–Jackson laws (e.g. Desmond and Wechsler, 2015).

   Abundance matching faces a stringent test when applied to simulations of a *single* giant galaxy like the Milky Way – or rather, to simulations of the galaxy's dark matter halo. In this case, the technique is used to associate the dark matter 'subhaloes' in the simulation with the observed population of dwarf satellite galaxies. The test is stringent, because (in the case of the Milky Way, for instance) we have a good inventory of the dwarf satellites, at least the brighter ones, and the measured values of their internal velocity dispersions provide (under standard-model assumptions) a direct route to estimating the required masses of their dark haloes. And one finds that, in this application of the method, abundance matching fails dramatically. The first problem to be noticed (Moore et al., 1999; Klypin et al., 1999) was the enormous mismatch between the thousands of subhaloes that accompany galaxies like the Milky Way in the simulations, and the few dozen satellites that are actually observed. (This is the 'missing-satellites problem' to a standard-model cosmologist; the 'dwarf over-prediction problem' to a Milgromian researcher. Of course, no such problem exists in theory variants T_0–T_2.) The problem has persisted; e.g. Silk and Mamon (2012, p. 939) write: "The excessive predicted numbers of dwarf galaxies are [sic] one of the most cited problems with ΛCDM. The discrepancy amounts to two orders of magnitude," and Del Popolo and Le Delliou (2017, p. 21) write: "every cosmological simulation predicts Milky Way-like galaxies surrounded with at least one order of magnitude more small subhalos (dwarf galaxies) than observed."

Standard-model theorists have dealt with this problem in the way that they often do: by invoking auxiliary hypotheses (hopefully plausible) that can reconcile theory with data. The favored explanation exploits the ease of heating or removing gas from systems, such as dwarf galaxies, that have shallow potential wells. One such mechanism is energy input from the earliest generation of stars; another is photoionization of star-forming gas during the epoch of cosmic reionization. To be successful, such models must predict that the suppression of star formation is most efficient in low-mass subhaloes, and that the great majority of subhaloes fail completely to form stars so that they would remain dark and unobserved.

Standard-model theorists (e.g. Bull et al., 2016) argue that some combination of these mechanisms is probably able to account for the 'missing satellites.'[10] But even they acknowledge a stubborn discrepancy. Plausible mechanisms for removing or heating the gas would be least effective in the most massive subhaloes. Thus, the most luminous dwarf satellites around a galaxy like the Milky Way ought to exhibit internal velocities that are consistent with their having been drawn, directly, from the high-mass end of the subhalo population of a computer simulation. Stated differently: Abundance matching should 'work' at the high-mass end of the satellite/subhalo mass spectrum. But this is not the case. The simulations are found to contain of order ten subhaloes that are too massive and dense, by a factor of about five, to host even the brightest satellites of the Milky Way (Kroupa et al., 2010; Boylan-Kolchin et al., 2011; Garrison-Kimmel et al., 2014). This is called, by standard-model cosmologists, the 'too big to fail' (TBTF) problem.[11] As defined by Papastergis et al. (2015, p. A113): "In a lambda cold dark matter (ΛCDM) cosmology, dwarf galaxies are expected to be hosted by halos that are significantly

---

[10] The dwarf-galaxy literature provides some of the starkest insights into standard-model methodology. For instance, Bullock (2010, p.12), after listing the various mechanisms that have been proposed for suppressing star formation in the subhaloes, remarks: "each imposes a different mass scale of relevance . . . If, for example, we found evidence for very low-mass dwarf galaxies $V_{max} \sim 5$ km s^{-1} then these [galaxies] would be excellent candidates for primordial H_2 cooling 'fossils' of reionization in the halo." As Karl Popper (1983, p. 168) remarked in his critique of Freud's theory: "*every conceivable case will become a verifying instance*" (italics his).

[11] The expression 'too big to fail' can be difficult to parse, particularly for someone who has not internalized standard-model methodology. The intended meaning is: "Star formation needs *to fail* in the largest of the simulated subhaloes, since we do not observe dwarf galaxies of comparable mass, but these simulated subhaloes are *too big* for the mechanisms that are postulated to suppress star formation in smaller subhaloes to work."

more massive than indicated by the measured galactic velocity" – that is: abundance matching assigns dark masses that are greater than what would be inferred based on the galaxies' internal kinematics and postulate DM-1.[12]

§

That last sentence explains the statement of Berezhiani et al. (2017) that motivated our digression, i.e. "...the expected DM masses from abundance matching being very large, [dwarf galaxies] would mostly be expected to display higher velocities than predicted in MOND".[13] Berezhiani et al. are here noting, correctly, that the standard cosmological model is internally inconsistent: dark matter postulate DM-1, which states that the dark matter in any galaxy is distributed in such a way as to reproduce the galaxy's internal kinematics, conflicts with the simulations. And so Berezhiani et al. are forced to make a choice when assigning the dark mass to an observed dwarf galaxy: they can do so based on the galaxy's measured kinematics, or they can accept the results of standard-model simulations that predict higher masses. In the cited paragraph, and throughout their work, Berezhiani, Khoury and collaborators choose the second alternative as their default.

This situation can be seen as just another example of the 'loss of content' of T_3: that is, of the fact that T_3 loses some of the explanatory power of theory variants T_0–T_2. But this loss of explanatory power is particularly severe in the case of predictions about the EFE. Recall from Chapter 5 that the predicted change in the internal rms velocity of a dwarf galaxy due to the EFE is small, and the only reason the prediction is testable is that, under T_1, the internal velocities are dependent only on the observed distribution of stars and gas. But, under T_3, the change in V_{rms} predicted by the EFE is likely to be much smaller than the changes that would result from different assumptions about the dark mass of the galaxy. One could 'explain' any small discrepancy in V_{rms} in the manner favored by standard-model cosmologists, that is, by postulating a slightly different mass for the dark matter subhalo.

---

[12] The multitude of problems associated with dwarf galaxies under the standard cosmological model invites the question: How should one view these galaxies under MOND? Pavel Kroupa (2012) made a bold suggestion. 'Primordial' dwarf galaxies, he proposed, *do not exist*; all dwarf galaxies are tidal dwarfs. That is, they consist of clumps of stars and gas that were 'pulled out' of disk galaxies during close interactions with other galaxies. Kroupa notes that the dwarf satellites of the Local Group are dynamically and morphologically indistinguishable from known tidal dwarfs; for instance, both populations sit on the same Tully–Fisher and Faber–Jackson relations. Even more striking is the fact that almost all of the Local Group satellites are observed to be distributed in two thin planes, one of which circulates about the Milky Way galaxy and the other around the Andromeda galaxy. No standard-model theorist has come up with any, even remotely, plausible explanation for this extraordinary degree of correlation; in their simulations, the dwarf satellites are predicted to be distributed isotropically in position and velocity space around their host galaxies (Pawlowski et al., 2014). This anomaly is called, by standard-model cosmologists, the 'problem of the satellite planes.' But Kroupa (2014) argues that *tidal* dwarf galaxies would *necessarily* be distributed, at least initially, in rotating planar structures. Kroupa's brilliant synthesis works only under theory variants T_0–T_2; it is lost under T_3.

[13] By "MOND," Berezhiani et al. mean, apparently, theory variants T_0 or T_1.

One could argue that in the case considered in Chapter 5 – two dwarf galaxies that were chosen to be structurally similar – the predicted dark masses would be the same. But under standard-model methodology, that assumption is not necessarily correct. Abundance matching is only assumed to be *statistically* valid; two dwarf galaxies with the same observed structure would be expected to have different dark masses, if only slightly. And standard-model simulations predict that the masses of dark subhaloes at the current epoch are different than their 'primordial' values due to environmental effects, e.g. tidal stripping, and these effects could be very different in two satellites located at very different distances from their host – as indeed is the case for the two satellites of the Andromeda Galaxy considered in Chapter 5.

I conclude that all predictions of observable consequences of the EFE, under T_3, are currently uncorroborated, due to the freedom allowed in assigning dark matter to the galaxies used in the test.

Recall from Chapter 2 that unconfirmed predictions can contribute, at most, to the 'theoretical progressivity' of a research program. Following the guidelines set out there for judging the novelty of unconfirmed predictions, I conclude that this prediction is novel according to criterion P, since the prediction differs from the prediction under T_2; and it is novel according to criterion M, since the standard cosmological model fails to make this prediction. These conclusions are summarized in Table 7.1.

§§

In summary: Theory variant T_3 makes at best a modest contribution to the progressivity of the Milgromian research program. Three predictions, which test the influence of superfluid dark matter on the behavior of normal matter in galaxies or galaxy clusters, contribute to the 'theoretical progressivity' of the research program. Of these, it was argued that two are possibly or partially corroborated and the third is uncorroborated, and so it remains unclear whether the addition of T_3 is an *empirically* progressive step. In addition, by adopting standard-model postulates about (collisionless) dark matter, theory variant T_3 loses much of the explanatory power of theory variants T_0–T_2: it fails to satisfy Lakatos's requirement of 'incorporation.'

8

Convergence

Writing around 1912, a time when the atomistic model was still being challenged, the physicist Jean Baptiste Perrin made an interesting argument for the existence of atoms and molecules (Perrin, 1916). Perrin pointed out that kinetic theory implies the existence of a *new constant of nature* that relates the macroscopic to the microscopic. Such a constant had already been discussed by Amedeo Avogadro, who proposed in 1811 that the volume of a gas at a given pressure and temperature is proportional to the number of atoms or molecules, regardless of the nature of the gas. One hundred years later, Perrin christened this constant Avogadro's number, or N, which he defined as the number of particles in one gram-molecule (exactly 32 grams) of oxygen. Perrin pointed out that there were a number of independent experiments that could be used to determine N. For instance, Perrin's own research on Brownian motion gave estimates of the mean molecular mass, implying $N \approx 6.8 \times 10^{23}$ molecules per gram molecular weight. Perrin obtained a similar value for N via experiments on diffusion and viscosity. Yet another example was black-body radiation: Planck's law for the black-body spectrum contains N in the argument of the exponential term, as $Nh\nu/RT$. Perrin summarized these and other experimental results in a table, which is reproduced here as Table 8.1.

Perrin then wrote:

Our wonder is aroused at the very remarkable agreement found between values [of N] derived from the consideration of such widely different phenomena. Seeing that not only is the same magnitude obtained by each method when the conditions under which it is applied are varied as much as possible, but that the numbers thus established also agree among themselves, without discrepancy, for all the methods employed, the real existence of the molecule is given a probability bordering on certainty (Perrin, 1916, p. 206–207).

In his 1920 Nobel Prize in Physics lecture, Max Planck argued, in a similar way, that the convergence of experimental determinations of another new constant of nature – what we now call Planck's constant, or h – was compelling evidence for energy quantization. Planck's constant appears in equations that describe the

Table 8.1 *Jean Perrin's (1916) tabulation of experimental determinations of Avogadro's constant*

Phenomena observed		$N/10^{22}$
Viscosity of gases (van der Waal's equation)		62
Brownian movement	Distribution of grains	68.3
	Displacements	68.8
	Rotations	65
	Diffusion	69
Irregular molecular distribution	Critical opalescence	75
	The blue of the sky	60 (?)
Black-body spectrum		64
Charged spheres (in a gas)		68
Radioactivity	Charges produced	62.5
	Helium engendered	64
	Radium lost	71
	Energy radiated	60

photoelectric effect, ionization, the specific heat of solids, and of course black-body radiation:

After all these results, towards whose complete establishment still many reputable names ought essentially to have been mentioned here, there is no other decision left for a critic who does not intend to resist the facts, than to award to the quantum of action, which by each different process in the colourful show of processes, has ever-again yielded the same result, namely, 6.52×10^{-27} erg sec, for its magnitude, full citizenship in the system of universal physical constants (Planck, 1922, p. 496–500).

To which Planck added: "To be sure, the introduction of the quantum of action has not yet produced a genuine quantum theory."

Note that neither Perrin nor Planck was arguing for the correctness of any particular theory. Perrin was arguing for "the real existence of the molecule"; Planck was arguing for the reality of "the quantum of action." As John Losee (2005, p. 166) has emphasized, the convergence of different experimental determinations of N or h was seen by these scientists as "a warrant for a *type* of theory": that is, as sufficient justification for (in Lakatos's language) a *problemshift*, or a migration to a new research program:

The convergence of various determinations of the value of Avogadro's number on 6.02×10^{23} molecules/gram molecular weight warrants as progressive the transition from theories of the macroscopic domain to the atomic-molecular theory of its microstructure. And the convergence of various determinations of the value of Planck's constant on 6.6×10^{-27} erg-sec warrants the transition from classical electromagnetic theory to the theory of the quantization of energy (Losee, 2004, p. 156–157).

Losee added: "I know of no plausible countercase in which convergence of this kind is achieved in the case of a transition judged not to be progressive on other grounds. ... Unfortunately, opportunities to apply the convergence condition are rare within the history of science."[1]

Losee wrote those words in 2004. He was apparently not aware, at that time, of a third "opportunity to apply the convergence condition": namely, to Milgrom's constant, a_0.

Milgrom's constant appears in almost every testable statement that is derivable from his theory. But there are three predictions of the theory that are especially useful when it comes to determining the value of a_0 from data. These are:

1. the baryonic Tully–Fisher relation (BTFR),
2. the central surface density relation (CSDR),
3. the radial acceleration relation (RAR).

Observational corroboration of these three predicted relations was discussed in Chapters 4 and 5. In what follows, I discuss the relations again, this time with emphasis on the determination of Milgrom's constant. Results are summarized in Table 8.2.

§§

1. *The baryonic Tully–Fisher relation*: As discussed in Chapter 4, a novel prediction of theory variant T_0 is the so-called 'baryonic Tully–Fisher relation' (BTFR), which relates the asymptotic value of the circular rotation speed to the total mass of a disk galaxy:

$$V_\infty^4 = a_0 G M_{\text{gal}}. \tag{8.1}$$

The BTFR provides the potentially most direct route to determination of Milgrom's constant – or, more precisely, to determination of the quantity $a_0 G$ – since the relation is defined in terms of asymptotic quantities only: the rotation velocity at large distances from the galaxy center, where the motion is fully in the low-acceleration regime; and the total galaxy mass. But Equation (8.1) is 'asymptotic' in one additional sense: it was derived on the assumption the star or gas cloud whose velocity is measured is orbiting so far from the center of the galaxy that the galaxy's (Newtonian) gravitational potential can be well approximated as that of a point mass. That condition may not be satisfied, even if the motion is comfortably in the Milgromian regime. This is important, because a given mass, distributed in a disk,

---

[1] Note that Losee is making a very strong claim here. He argues that *every* existing proposal for a criterion of scientific progress, including Lakatos's *Methodology*, is "open to countercases," with a sole exception: "the convergence of diverse experimental determinations [of a new constant of nature] upon a single value" (2004, p. 156). One can remain skeptical about this claim, while still being impressed by examples of convergence when they occur.

Table 8.2 *Determinations of Milgrom's constant*

Prediction	Reference	N^a	$a_0{}^b$
Baryonic Tully–Fisher relation[c]	Begeman et al. (1991)	10	1.21 ± 0.24
$(a_0 G)$	Stark et al. (2009)[e]	28	1.18^d
	+ Trachternach et al. (2009)[e]	34	1.30^d
	McGaugh (2011)[e]	47	1.24 ± 0.14
	Lelli et al. (2016c)	118	1.29 ± 0.06^f
Central surface density relation	Donato et al. (2009)	$\sim 10^3$	1.3^g
(a_0/G)	Lelli et al. (2016b)[h]	135	1.27 ± 0.05^i
			1.27 ± 0.05^j
Radial acceleration relation	Wu and Kroupa (2015)	74	0.94 ± 0.03^i
$(a_0/G \to a_0 G)$			1.21 ± 0.03^k
	McGaugh et al. (2016)	153	$1.20 \pm 0.02^l \pm 0.24^m$
	Lelli et al. (2017)		

[a] Number of galaxies in sample
[b] Units: 10^{-10} m s^{-2}
[c] Values of a_0 are based on fits to data *assuming* a log-log slope of exactly four for the BTFR.
[d] Taken from McGaugh (2012), who performed the analysis using first the Stark et al. sample alone, then combined these data with the data of Trachternach et al.
[e] Gas-rich galaxies only
[f] Adjusted from the value given in the paper to account for the departure of a galaxy's potential from a point mass, as discussed in the text. I thank Federico Lelli for computing this corrected value.
[g] Based on Donato et al.'s characteristic value of the projected 'dark matter density' for high surface brightness galaxies
[h] Best-fitting a_0 values computed by the author via least-squares minimization using the Lelli et al. (2016b) data plotted in Figure 5.2.
[i] $v(y)$ given by Equation (8.4) with $n = 1$
[j] $v(y) = \left(1 - e^{-\sqrt{y}}\right)^{-1}$
[k] $v(y)$ given by Equation (8.4) with $n = 2$
[l] random error
[m] systematic error

generates a larger gravitational force in the plane of the disk than if the same mass were concentrated at the origin, and the dependence of that force on radial distance is not precisely inverse-square. For an infinitely thin disk, the force is augmented by roughly 25% compared with the point mass case, at radial distances that correspond to typical measurements of V.

McGaugh and de Blok (1998b) deal with this issue by writing

$$M_{\text{gal}}(\le R) = \chi(R)\frac{V^4(R)}{a_0 G}, \tag{8.2}$$

a relation which assumes that the motion is in the low-acceleration regime but not necessarily in the point-mass-potential regime. The 'correction factor' $\chi(R)$ can be estimated in various ways; for instance, using a parameterized model for the disk surface density. Those authors show that a disk with an exponential surface density profile has $\chi \approx 0.76$ at a distance of four disk scale lengths from the center. All of the a_0 estimates in Table 8.2 have taken into account the necessary correction.[2]

A second set of complications arise in regard to M_{gal}. In many disk galaxies, the largest contribution to the total mass comes from stars. The stellar mass of a galaxy is determined from its measured (optical) luminosity, L, via an expression such as $M_\star = \Upsilon L$ where Υ is an estimate of the 'mass-to-light ratio' of the stars: the factor that converts a measured luminosity (in whatever passband is observed) to the mass in stars, *including* stellar remnants (neutron stars, black holes) which may produce no light. An estimate of Υ is usually taken from 'stellar population models' together with optical spectra. The population models embody a number of assumptions. Most critical is probably the assumed 'initial mass function,' the distribution function describing the masses of stars at the start of their energy-producing lifetimes. Other uncertainties include the dependence of the star formation rate on time, and the relation between the initial and final masses of the (mostly massive) stars that end their lives as dark remnants.

The other major contributor to M_{gal} is gas. Neutral atomic hydrogen, or 'HI,' dominates the gas component of disk galaxies and its total mass follows directly from the intensity of the 21 cm line emission due to the spin-flip transition of hydrogen. But gas exists in other forms as well, some of which are not so easily observed. A typical parameterization is $M_{\mathrm{gas}} = \eta M_{\mathrm{HI}}$, where M_{HI} is the observed atomic gas mass and the factor η accounts for other forms of gas (molecular hydrogen or other elements). The mass in H_2 can be determined in some galaxies via microwave observations of rotational transitions,[3] and empirical scaling relations have been derived that relate M_{H_2} to M_{HI}. Typical estimates of η are ~ 1.3.

Most astrophysicists would agree that the observational determination of gas masses in galaxies is subject to less uncertainty than the determination of stellar masses. Ideally, therefore, one would try to extract estimates of a_0 from a sample of disk galaxies in which almost all of the mass is in the form of gas. Happily, such galaxies exist: they are the late-type (i.e. Hubble types Sc,d or Irr), low surface brightness galaxies, which frequently have gas masses in excess of their stellar mass (Schombert et al., 2001). Many such galaxies have measured 'line widths' in

---

[2] Standard-model cosmologists often discuss the BTFR, but almost never do they acknowledge the Milgromian prediction or cast their analysis as a test of that prediction. One consequence is that they are free to define the BTFR in whatever way they find most convenient; for instance, the velocity that appears in that relation can be defined as the maximum measured value of $V(R)$, the peak circular velocity in a model of a dark matter halo, etc. This is an excellent example of how the existence of a *testable* prediction – in this case, the prediction due to Milgrom – can guide the scientist to look for, and find, relations that otherwise might never have been discovered. It also explains why none of the BTFR studies by standard-model cosmologists contain results that can usefully be included in Table 8.2.

[3] Direct detection of emission from molecular hydrogen is difficult and the mass in H_2 is typically inferred from observations of the $J = 1 \rightarrow 0$ transition of carbon monoxide gas, together with an assumed value for the M_{H_2}/M_{CO} ratio.

HI, yielding approximate estimates of the rotation velocity, and some studies of the BTFR (e.g. Verheijen, 2001; Gurovich et al., 2010; Catinella et al., 2012) have used line widths as proxies for V_∞. But spatially resolved rotation curves are far more precise for this purpose (Trachternach et al., 2009). Three of the entries in Table 8.2 (from Stark et al., 2009; Trachternach et al., 2009; and McGaugh, 2011) are based on samples of galaxies for which most of the mass is in the form of gas.

The other entries in that table are from data sets that contain both gas- and star-dominated galaxies. Begeman et al. (1991) used detailed rotation curves of star-dominated galaxies; those authors treated the stellar mass-to-light ratio as a free parameter for each galaxy. The entry due to Lelli et al. (2016c) is based on a sample of 118 galaxies for which two sorts of data are available: detailed rotation curves based on spatially resolved neutral hydrogen or ionized hydrogen rotation curves; and disk surface photometry at 3.6 μm (infrared) wavelengths (Lelli et al., 2016a). The latter data are especially useful for estimating stellar masses, since a number of studies suggest that the mass-to-light ratio at near-infrared wavelengths depends only weakly on galaxy type or mass.

All of the BTFR-derived a_0 values in Table 8.2 lie in the interval $1.2 \lesssim a_0 \lesssim 1.3$, in units of 10^{-10} m s^{-2}.

§

2. *The central surface density relation*: Recall from Chapter 5 that this prediction comes in two versions, both derived from theory variant T_1. The first, more approximate version predicts a characteristic, or maximum, central surface density of $\sim 1.5\Sigma_0$ for the 'dark matter' around high surface brightness galaxies, where $\Sigma_0 \equiv a_0/(2\pi G)$ was expressed in physical units in Equation (5.18) as

$$\Sigma_0 \approx 138 \left(\frac{a_0}{1.2 \times 10^{-10} \text{ m s}^{-2}} \right) M_\odot \text{ pc}^{-2}.$$

The factor 1.5 comes from the lower panel of Figure 5.1, which shows that $\sim 1.5\Sigma_0$ is the approximate maximum central surface density of 'phantom dark matter' for galaxies of high surface brightness. Table 8.2 includes one estimate of a_0 based on this prediction. It was derived from the Donato et al. (2009) result that 'dark matter haloes' of high surface brightness galaxies have a characteristic central surface density of $\Sigma(0) \approx 220 M_\odot \text{ pc}^{-2}$, with little dependence on galaxy luminosity or type. (This is the value that is plotted as the filled square in Figure 5.1.) Setting

$$\Sigma(0) \approx 220 M_\odot \text{ pc}^{-2} = 1.5\Sigma_0 \approx 207 \left(\frac{a_0}{1.2 \times 10^{-10} \text{ m s}^{-2}} \right) M_\odot \text{ pc}^{-2}$$

yields $a_0 \approx 1.3 \times 10^{-10}$ m s^{-2}. This value should be considered very approximate, for the reasons outlined in Chapter 5; for instance, Donato et al. do not use Milgrom's theory to compute the structure of phantom dark matter haloes for the galaxies in their sample,[4] choosing instead to fit a parameterized model of a 'dark matter halo' extracted from standard-model simulations.

The second version of the prediction states that, for precisely axisymmetric and thin disks, there is a functional relation between $\Sigma_{ph}(0)$, the central surface density of phantom dark matter, and $\Sigma(0)$, the central density of matter. This prediction, due to Milgrom (2016c), was given in Equation (5.22) as

$$\Sigma_{ph}(0) + \Sigma(0) = \Sigma_0 S\left[\Sigma(0)/\Sigma_0\right], \quad S(y) \equiv \int_0^y \nu(y)dy,$$

and was called the 'central surface density relation' (CSDR). Milgrom's constant appears in this relation via $\Sigma_0 \equiv a_0/(2\pi G)$. A determination of Σ_0, hence a_0, from the CSDR is possible given measurements of $\Sigma(0)$ for a sample of disk galaxies, together with calculated values of $\Sigma_{ph}(0)$ for each galaxy. Ideally the latter would be computed from the observed surface densities using T_1. As discussed in Chapter 5, the nearest thing to such an analysis that is currently available comes from the study of Lelli et al. (2016b), who used a more approximate, Newtonian method that derives estimates of the central 'dark matter' density from an observed rotation curve. The data set used in that study (and also in two of the studies of the radial acceleration relation described in the next subsection) was the so-called SPARC database, for "Spitzer Photometry and Accurate Rotation Curves" (Lelli et al., 2016a). "Spitzer" refers to the *Spitzer Space Telescope*, an infrared telescope launched in 2003. Lelli et al. used surface photometry at 3.6 μm to construct mass models for 175 disk galaxies; as discussed in Chapter 4, observations at near-infrared wavelengths provide the most direct route to estimates of the stellar mass density. For the same set of galaxies, extended HI (neutral hydrogen) rotation curves were available from the literature. In addition, for about a third of the SPARC galaxies, rotation curves were also available from integral-field (i.e. 2D) spectroscopic observations of the emission of HII (ionized hydrogen), data which probe the inner regions at higher spatial resolution than the HI data. Lelli et al. note that the SPARC galaxy sample spans a wide range in luminosity, surface brightness, rotation velocity, and Hubble type.

Milgrom (2016c) argued that a multiplicative factor of $\sim 4/\pi$ should be applied to the Newtonian surface densities computed by Lelli et al., and this adjustment has been made to the data as plotted in Figure 5.2.

While the estimates of $\Sigma_{ph}(0)$ by Lelli et al. (2016b) were Newtonian, a determination of a_0 via Equation (5.22) still requires a specification of the transition function $\nu(y)$ that appears in T_1. In principle, one could *determine* the function $\nu(y)$ given rotation-curve data of sufficient quality. In practice, as discussed in Chapter 4, one finds that a number of functional forms for $\nu(y)$ fit the available

---

[4] In fact, Donato et al. do not seem to be aware that they are corroborating a prediction of the Milgromian research program.

data equally well. And here an important point should be made. The definition of a_0 differs, slightly, between the different theory variants in the Milgromian research program. In theory variant T_0, a_0 appears, unambiguously, in predictions like the baryonic Tully–Fisher relation, which applies only to the low-acceleration regime in galaxies. This prediction remains unchanged under theory variant T_1, but T_1 also includes a transition function as auxiliary hypothesis, and a_0 appears as a parameter in that function. Experimental determinations of a_0 based on predictions from T_1 will yield slightly different results depending on the choice of transition function. Nor is there any guarantee that this ambiguity will be removed in later theory variants, since there may not be a single 'transition function' that turns out to be correct for all applications of the theory.[5] For these reasons, we should not necessarily expect exact convergence of the measured values of a_0, even ignoring the usual problems associated with errors and finiteness of data sets.

Two estimates of a_0 based on the Lelli et al. (2016b) published data set are included in Table 8.2, derived using two different expressions for $v(y)$. (The solid curve in Figure 5.2 is based on the first of these.) The value of a_0 that minimizes the χ^2 deviation between data and model is $a_0 \approx 1.27 \times 10^{-10}$ m s^{-2} for both choices of transition function.

§

3. *The radial acceleration relation*: Recall from Chapters 3 and 4 that the radial acceleration relation (RAR) is a consequence of T_0, i.e. of Milgrom's three postulates, which imply that, in the plane of a disk galaxy, the acceleration of a star in a circular orbit, a, is uniquely related to the Newtonian gravitational acceleration, g_N, as

$$a \equiv \frac{V^2}{R} = v\left(\frac{g_N}{a_0}\right) g_N, \tag{8.3}$$

where $V(R)$ is the circular orbital velocity at distance R from the disk center. While the prediction (8.3) follows from T_0, the form of the function $v(g_N/a_0) \equiv v(y)$ is not specified in T_0 except in the asymptotic, low-acceleration limit: $v \to y^{-1/2}$. At first blush, one might think that determination of a_0 by fitting data to a relation such as (8.3) would be dependent on the functional form assumed for $v(y)$. Indeed this was true in the case of the central surface density relation, as just discussed (although the dependence appeared not to be strong). On the other hand, if a particular data set extends to regimes of sufficiently low a, the prediction (8.3) reduces to the baryonic Tully–Fisher relation (8.1), and the latter relation yields a_0 directly from a single measured value of V_∞ and M_{gal}, without the need to specify $v(y)$. How, then, does the RAR provide an *independent* estimate of a_0?

---

[5] For example, in modified inertia theories the interpolation function is necessarily different for different orbits (Milgrom, 2006).

Only in the following sense. Recall that the prediction (8.3) follows from Milgrom's third postulate, while the prediction that the asymptotic rotation velocity is a function only of a galaxy's total mass, Equation (8.1), requires only the second postulate. Two galaxies with the same V_∞ can, and often do, have very different rotation curves. If Milgrom's third postulate was not correct, the predicted rotation curves might not be as observed, even if the asymptotic velocities are correctly predicted. Stated another way: The presence of Milgrom's constant in the BTFR is independent of its appearance as a 'transition acceleration' in the RAR.

Wu and Kroupa (2015) were the first to publish estimates of Milgrom's constant based on the RAR. Those authors used values of $a(R)$ and $g_N(R)$ that had previously been computed and tabulated by McGaugh (2004) and Sanders and McGaugh (2002) for a sample of 74 disk galaxies. The observational data used by the latter authors were taken from a number of still earlier studies and were rather heterogeneous: stellar and neutral hydrogen rotation curves (for $V(R)$ and $a(R)$), optical photometry in a variety of pass bands (for g_N). Restricting the data to the low-acceleration regime, Wu and Kroupa found a best-fit value of $a_0 \approx 1.24 \times 10^{-10}$ m s^{-2}, consistent with estimates based on the BTFR. In analyzing the full data set (including the high-acceleration regime), Wu and Kroupa tried two forms of the transition function in Equation (8.3):

$$\nu_n(y) = \left[\frac{1 + \left(1 + 4y^{-n}\right)^{1/2}}{2} \right]^{1/n} \tag{8.4}$$

with $n = 1$ and $n = 2$. The results, as given in Table 8.2, were similar but significantly different: $a_0 = (0.94, 1.21) \pm 0.03 \times 10^{-10}$ m s^{-2}.

Most recently, McGaugh et al. (2016) and Lelli et al. (2017) presented re-analyses of the RAR using 153 disk galaxies from the SPARC sample described above. In the second of these papers, the authors included, in addition, data from 87 early-type galaxies: 25 elliptical and lenticular (S0) galaxies for which accelerations could be inferred from neutral hydrogen or X-ray data, and 62 dwarf spheroidals (dSphs) having individual-star spectroscopy. In a precisely spherical system, circular motion in any fixed plane would be predicted by T_0 to obey a relation similar to (8.3). However, as discussed in detail in Chapter 5, non-disk galaxies are generically nonspherical and their internal motions are noncircular, and in fact Lelli et al. did not include the data from early-type galaxies when making quantitative estimates of a_0. They did, however, note the consistency of the latter data with the disk galaxy data, and concluded by stating:

The observed scatter [in the RAR] is very small ($\lesssim 0.13$ dex) and is largely driven by observational uncertainties: the radial acceleration relation has little (if any) intrinsic scatter. The tiny residuals show no correlation with either local or global galaxy properties.

The transition function assumed by both McGaugh et al. (2016) and Lelli et al. (2017) in their analyses of the disk-galaxy data was given as Equation (4.12):

$$\nu(y) = \left(1 - e^{-\sqrt{y}}\right)^{-1},$$

a form that had previously (McGaugh, 2008) been shown to accurately reproduce the Milky Way rotation curve. Figure 4.3 shows the RAR as constructed by these authors.

An important feature of the McGaugh et al. (2016) and Lelli et al. (2017) analyses was consideration of the *systematic* uncertainty in the inferred value of a_0 due to uncertainties in the mass-to-light ratio Υ. Those authors derived a systematic error that was an order of magnitude greater than the random error: $a_0 = 1.20 \pm 0.02$ (random)± 0.24 (systematic) $\times 10^{-10}$ m s^{-2}. It is reasonable to assume that the systematic error associated with Wu and Kroupa's estimate of a_0 is at least as large as this, since those authors did not use near-infrared photometry in their estimates of the stellar mass. A similar uncertainty presumably attaches to the other estimates in Table 8.2 based on optical data.

§§

Taking into account their (likely) systematic uncertainties, all of the a_0 estimates in Table 8.2 are consistent with each other and with a value that lies somewhere in the narrow range

$$1.2 \times 10^{-10} \text{ m s}^{-2} \leq a_0 \leq 1.3 \times 10^{-10} \text{ m s}^{-2}. \tag{8.5}$$

I conclude that a strong case can be made that the values of Milgrom's constant as determined, in various ways, from observational data are convergent. We are therefore justified in considering a_0 as a new constant of nature, like Avogadro's number or Planck's constant. And if Losee's reasoning is correct, the convergence of measured values of Milgrom's constant points to the necessity of a Lakatosian problemshift – in this case, to a theory of cosmology that is different from the standard cosmological model and that incorporates a universal acceleration scale equal to a_0.

§§

As noted above, Milgrom's constant appears in other predictions from his theory. For instance, a_0 appears in the predicted relation between mass and rms velocity of spherical isothermal systems (Equation 5.33). But as discussed in Chapter 5, few astrophysical systems appear to be accurately isothermal and that fact makes it difficult to use the prediction as a basis for estimating the value of a_0.

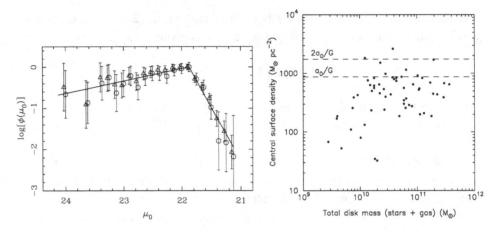

Figure 8.1 *Left panel*: The distribution of central surface brightnesses of disk galaxies. The horizontal axis is the disk central surface brightness in *B* magnitudes per square arcsecond; surface brightness increases to the right. This quantity is a measure of the light due to stars, with no contribution from the gas in the disk. Circles are based on data selected by diameter, triangles are based on magnitude-limited data There is a sharp falloff above a surface brightness of $\mu_0 \approx 21.9$ mag arcsec^{-2} (from McGaugh (1996), who used photometric measurements of Davies (1990)). *Right panel*: Each point is the central, surface *mass* density of one disk galaxy, including both stars and gas, as tabulated by Zavala et al. (2003). A sharp falloff is again apparent, above a value of $\sim a_0/G$. Figure (left panel) reprinted with permission from Stacy S. McGaugh, "The number, luminosity and mass density of spiral galaxies as a function of surface brightness," *Monthly Notices of the Royal Astronomical Society*, 280, p. 337–354, 1996. Reproduced by permission of Oxford University Press on behalf of the Royal Astronomical Society.

There is another prediction of Milgrom's theory that is similarly difficult to relate uniquely to a_0. But the observational data relating to this prediction are so striking – and seem so clearly to indicate the existence of an acceleration scale with magnitude close to a_0 – that the prediction is worth discussing here.

The relevant *data* are shown in Figure 8.1. The panel on the left is reproduced from McGaugh (1996), who used photometric data of Davies (1990) to compute the distribution of central surface brightnesses of galactic disks. The horizontal axis is μ_0, defined as the disk central surface brightness in (the rather bizarre, but quite standard) units of magnitudes in the *B* band per square arcsecond. There is an obvious break, or 'knee,' at $\mu_0 \approx 21.9$ mag arcsec^{-2}. Evidently, nature is reluctant to manufacture stellar disks with surface brightnesses that lie above a certain, well-defined value.[6]

---

[6] The existence of a characteristic surface brightness of galactic disks was noted already in 1970 by Kenneth Freeman, a result that is sometimes called 'Freeman's law.' Observational techniques since 1970 have improved to the extent that disks of very low surface brightness – even much lower than the background of the night sky – can now be detected and measured. What Freeman saw as, essentially, a single value of μ_0 is now understood to be the upper envelope of the surface brightness distribution shown in Figure 8.1.

Most of the mass of high surface brightness disks resides in the stars, and so an upper limit in surface brightness (due to star light) implies an upper limit in mass density. The panel on the right in Figure 8.1 plots data of Zavala et al. (2003), who computed and tabulated values of the central *mass* densities of galactic disks. The quantity plotted along the vertical axis is the total central surface density, $\Sigma(0)$, including contributions from both stars and gas; the horizontal axis is the total mass of the disk. There is indeed an approximate upper limit to $\Sigma(0)$ given by

$$\Sigma(0) \lesssim 2 \times 10^3 \, M_\odot \, \mathrm{pc}^{-2}. \tag{8.6}$$

This is close to twice the ratio a_0/G:

$$\frac{a_0}{G} \approx 1.80 \left(\frac{a_0}{1.2 \times 10^{-10} \, \mathrm{m \, s^{-2}}} \right) \mathrm{kg \, m^{-2}} \tag{8.7a}$$

$$\approx 0.87 \times 10^3 \left(\frac{a_0}{1.2 \times 10^{-10} \, \mathrm{m \, s^{-2}}} \right) M_\odot \, \mathrm{pc}^{-2}. \tag{8.7b}$$

If one chose to plot (say) the *average* surface densities of the disks within the inner few disk scale lengths, rather than the central values, the observed upper limit would be essentially equal to a_0/G.

Why should nature care about the existence of an acceleration scale when building galactic disks? One is tempted to invoke an argument based on the results of Chapter 5. Recall that under T_1 an effectively maximum value of the projected density was predicted to occur in two cases: the central projected density of the 'phantom dark matter' halo around a galaxy; and the mean projected density of the normal matter in an 'isothermal sphere.' However nothing in T_1 necessarily forbids the existence of thin, rotating configurations of arbitrarily high central density.

But high-density disks are likely to be in the Newtonian regime, and Newtonian stellar disks are notoriously unstable. 'Cold' disks – disks in which the stellar trajectories are close to circular – are unstable primarily to non-axisymmetric modes that convert an initially circular disk into a slowly rotating, oval distortion, or 'bar' (Toomre, 1981). Thin disks – particularly those that are 'hot,' i.e. having substantially noncircular motions – are unstable to bending modes, which convert the central parts of the disk into a spheroidal component, or 'bulge' (Merritt and Sellwood, 1994). The names 'bar' and 'bulge' were applied to disk galaxies long before the calculations that demonstrated unstable behavior: in fact, both sorts of feature are explicitly invoked as classificatory elements in the famous 'Hubble sequence' of galaxy types.

If we imagine starting from a high-density, unstable disk and gradually decreasing its density, there will come a point when all of the stars are in the low-acceleration regime, $a < a_0$. It is easy to see that, once this happens, the disk becomes 'more stable' than its Newtonian counterpart (Milgrom, 1989b; Brada and Milgrom, 1999). In the simplest analysis, the acceleration is related to the mass as $a \propto \sqrt{M}$ in the low-acceleration regime (e.g. Equation 3.1) while in the Newtonian regime $a \propto M$. This means that a change in the mass in some specified region – that is, a change in the density – implies a change in the acceleration of $\delta a/a = (1/2)\delta M/M$ as opposed to the Newtonian $\delta a/a = \delta M/M$; in other words, the disk is 'less responsive' to perturbations. One expects the transition to increased stability to occur when the inner parts of the disk – which experience the greatest a – all have $a \lesssim a_0$. This crude argument is a plausible explanation for the cutoff in disk surface densities displayed in Figure 8.1, if one assumes that the effect of the instability(ies) is to destroy the inner disk and replace it by a bulge or bar.

$$\S\S$$

In his remarks quoted near the start of this chapter, Jean Perrin seemed to take for granted that the convergence of measured values of Avogadro's number was extremely significant. "Our wonder is aroused," he wrote, "at the very remarkable agreement." Max Planck, also, seemed to feel no need to justify his claim that the agreement of different measured values of h was an extraordinary fact ("there is no other decision left for a critic who does not intend to resist the facts. . .") Both scientists accepted that the convergence of diverse experimental determinations of a new constant of nature was a strong argument in favor of theory change. With the benefit of hindsight, we can see that they were correct. But it would be nice to find an objective justification for their judgments.

Here is one possible way to look at it. Suppose one observes the Brownian displacement of a particle and uses those data to estimate the value of Avogadro's number, N, as it appears in Einstein's predicted relation. Given this – now established – value of N, one is in a position to make a *new* prediction about (for instance) black-body radiation: namely that the parameter N that appears in the argument of the exponent, $Nh\nu/RT$, has the same value that was measured previously in the Brownian motion experiment. (The original black-body prediction, since it did not specify the value of N, was a prediction only about the dependence of the spectrum on $h\nu/RT$.) In a similar way, one can generate one new prediction corresponding to each of the phenomena listed in Table 8.1, since in each case, the kinetic theory makes a prediction that contains the parameter N. In effect, the number of novel predictions has been increased by a number equal to itself, minus one (the prediction that was chosen to make the initial determination of N).

That argument is convincing enough, but it does not seem to justify the strong language used by Perrin and Planck. What is missing is some reason to think that the new set of predictions is *qualitatively* more remarkable than the original set.

Here is another way to look at it. Consider Milgrom's constant a_0 as it appears in two predictions: the baryonic Tully–Fisher relation, Equation (8.1), and the central surface density relation, Equation (5.22). The latter relation depends on a_0 through the quantity $\Sigma_0 \equiv a_0/(2\pi G)$. If we assume that the value of a_0 is the same in both relations, we are free to combine them mathematically and obtain a new prediction; for instance,

$$\Sigma_0 = \frac{1}{2\pi G^2} \left[\frac{V_\infty^4}{M_{\text{gal}}} \right],$$ (8.8)

where the quantity inside the brackets can in principle be determined from observations of a single disk galaxy. Equation (8.8) makes the surprising claim that we can predict the maximum surface density of 'dark matter haloes' around galaxies by measuring V_∞ and M_{gal} for *any single* disk galaxy. We can repeat this exercise for the radial acceleration relation, yielding (for instance) an expression that predicts the rotation curve of any disk galaxy, given a determination of the maximum observed surface density of 'dark matter haloes,' and so on.

This, perhaps, is a basis for the intuitive judgments of Perrin and Planck: namely that the convergence of the measured value of a 'constant of nature' implies a tight connection between facts that would otherwise not have been considered related.

§§

Of course, the word "otherwise" in the preceding sentence begs the question of what we *would have* expected, prior to the Milgromian prediction. And so once again it is important to deal with the question of novelty: not the novelty of the predicted facts themselves – that was dealt with, case by case, in the earlier chapters – but the novelty of relations such as Equation (8.8), which connect two seemingly unrelated quantities through a single constant with units of acceleration. Stated differently: How likely is it that nature should exhibit a universal acceleration scale with magnitude a_0?

At first blush, the answer is 'not very likely.' Certainly no one predicted the existence of such a scale prior to 1983, or claimed to see it in the data. But twenty years after the publication by Milgrom of his foundational postulates, two standard-model cosmologists, Manoj Kaplinghat and Michael Turner, argued that an acceleration scale, with magnitude comparable to a_0, *should* be imprinted on galaxies.

Kaplinghat and Turner (2002) begin by invoking four postulates of the standard cosmological model. Two of these were given in Chapter 1:

DM-2: Beginning at some early time, the universe contained a (nearly) uniform dark component, which subsequently evolved, as a collisionless fluid, in response to gravity.

DM-3: The dark matter that is postulated in DM-2 consists of elementary particles.

The other two standard-model postulates invoked by Kaplinghat and Turner were given in Chapter 6:

DM-4: The dark matter is 'cold,' that is, the velocity dispersion of the dark particles, at a time before the formation of gravitationally bound structures like galaxies, was small enough that it did not impede the formation of those structures.

DM-5: The spectrum of fluctuations in the dark matter at early times was 'scale invariant,' that is, perturbations of all sizes had the same average density contrast.

After a sufficiently long time, argue Kaplinghat and Turner, initially small perturbations in the dark matter would evolve into 'dark matter haloes,' regions in which the dark matter particles have reached a nearly steady state within their self-consistent potential wells. Kaplinghat and Turner argue that dark matter haloes that form from scale-invariant density perturbations are self-similar, in the sense that their density profiles – the dependence of density on distance from the center – can be written

$$\rho_{\text{DM}}(r) = \text{constant} \times (Gl)^{-1} \times S\,(l/l_0) \times \hat{\rho}_{\text{DM}}\left(\frac{r}{l}\right) \quad (8.9)$$

with $\hat{\rho}_{\text{DM}}(x)$ a universal (but unspecified) function. In Equation (8.9), l is a length that defines the size of the halo (e.g. its half-mass radius) and l_0 is the value of l for haloes that are just now forming, i.e. decoupling from the cosmic expansion and collapsing; the normalizing constant is a function of the collapse time of haloes of scale length l. Kaplinghat and Turner argue that the function $S(x)$ depends only weakly on halo size l and so can be treated effectively as a constant for all values of l corresponding to present-day galaxies. In such a dark matter halo, the gravitational acceleration varies with radius as

$$a_{\text{DM}}(r) \equiv r^{-2}GM_{\text{DM}}(<r)$$

$$= 4\pi \times \text{constant} \times S\,(l/l_0) \times \hat{a}_{\text{DM}}\left(\frac{r}{l}\right), \quad (8.10)$$

where $\hat{a}(y) \equiv y^{-2}\int_0^y \hat{\rho}(z)z^2dz$. Thus, galaxy-sized dark matter haloes have a nearly universal acceleration profile that differs from halo to halo only by a rescaling of the length.

The expression (8.10) still has no 'acceleration scale.' Kaplinghat and Turner next invoke a simple model of galaxy formation: they assume that the normal (i.e. non-dark) matter which makes up the visible part of a galaxy had initially the same distribution as the dark matter, but that after the dark matter had formed a fixed potential well, the normal matter underwent 'dissipation,' converting some fraction of its initial potential energy into heat and other radiation. In the process the normal matter would contract within the (essentially fixed) potential of the dark matter. Kaplinghat and Turner assume that these dissipative processes would cause the scale length of the normal matter to decrease by a universal factor of $\alpha \approx 10$ (a value that is consistent with the known sizes of galaxies), after which the matter would be cool enough that dissipation essentially stops. At this point, the system would have two components: a dark matter halo, and a (rotationally supported, or disk) galaxy at the center having a mean density $\sim 10^3$ times that of the dark matter. It follows that there should be a 'transition radius,' $r_t \approx 1/\alpha$, inside of which the gravitational acceleration is determined by the normal matter and outside of which the dark matter dominates. That radius is the solution to the equation

$$\frac{GM_{\text{gal}}(r \leq r_t)}{r_t^2} = a_{\text{DM}}(r_t). \tag{8.11}$$

Kaplinghat and Turner use this formula to estimate $a_{\text{DM}}(r_t)$ and find a value equal to a constant times cH_0, with weak dependence on galaxy size. This, they point out, is of the same order as Milgrom's a_0:

$$cH_0 = 7.3 \times 10^{-10} \left(\frac{H_0}{75 \text{ km s}^{-2}} \right) \text{m s}^{-2} \tag{8.12}$$

$$\approx 6.1 a_0.$$

Kaplinghat and Turner call this rough agreement a 'numerical coincidence': "Furthermore, a_{DM} is a fixed number since galaxies are bound and well relaxed today, while cH decreases with time. Thus, the approximate equality of a_{DM} with cH only holds today" (Kaplinghat and Turner, 2002, p. L21).

§

How convincing is this argument? Perhaps a better, first question to ask is: What facts do Kaplinghat and Turner claim to have explained? To quote from their paper:

In 1983, Milgrom (1983a, 1983b) made a remarkable observation: the need for the gravitational effect of nonluminous (dark) matter in galaxies arises only when the Newtonian acceleration is less than about $a_0 = 2 \times 10^{-8}$ cm s^{-2} ... This fact, which we will refer to as Milgrom's law, is the foundation for his modified Newtonian dynamics (MOND) alternative to particle dark matter (Kaplinghat and Turner, 2002, p. L19).

What Kaplinghat and Turner call 'Milgrom's law' is not to be found among the three predictions listed in Table 8.2, nor any of the other predictions discussed in previous chapters of this book. What they seem to have in mind is something similar to the words set in boldface in the following statement, which is taken from a review article by Milgrom:

A tight correlation is predicted between the locally measured acceleration, g, and mass discrepancy, η: $\eta \approx 1$ (no discrepancy) for $g \gg a_0$, **beginning to depart from 1 around a_0**, and $\eta \sim a_0/g \gg 1$ for $g \ll a_0$ (Milgrom, 2015, p. 108).

Milgrom's statement describes some consequences of his three foundational postulates. But it is clear that Kaplinghat and Turner's result (even if accepted at face value) only explains a small part of the empirical content of those postulates. It does not explain (*i*) the asymptotic flatness of disk galaxy rotation curves, (*ii*) the baryonic Tully–Fisher relation, or (*iii*) the radial acceleration relation – all of which are confirmed predictions of T_0. Nor does their analysis explain the central surface density relation or any of the other predictions of theory variant T_1. This is not a trivial shortcoming: because the two Milgromian predictions that relate to the central surface densities of disks – the central surface density relation, and the maximum surface density for disk stability – both relate a_0 directly to the density of *normal* matter, i.e. $a_0 \sim G\Sigma$. Such a result has no basis in Kaplinghat and Turner's derivation. Kaplinghat and Turner do acknowledge this, in a roundabout way. But the limited scope of their analysis does not allow them to address the question of convergence. At best, they have demonstrated only that the standard cosmological model is able to reproduce, in an order-of-magnitude sense, the characteristic acceleration at which disk galaxy rotation curves begin to depart from the Newtonian prediction.

In fact, they have not demonstrated even that much. As Milgrom (2002) and others have pointed out, one consequence of the Kaplinghat and Turner result is that *all* galaxies (or none of them) should exhibit a transition radius that is a fixed fraction of the dark-matter-halo scale length. In other words, galaxies should always contain a region near the center where the normal matter dominates and where $a > a_0$. But, as discussed in Chapters 4 and 5, there is a class of galaxies – the low surface brightness galaxies – that are characterized by $a < a_0$ *everywhere*, even near the center. And the postulates of T_0 correctly predict the rotation curves of these galaxies.

§§

The aim of this book is to assess the progressivity of the Milgromian research program, and not to speculate about how that program may evolve in the future. But it is tempting to wonder what it might *mean* for nature to incorporate a fundamental constant with dimensions of acceleration. After all, both Perrin and Planck had pretty clear ideas about how to interpret the existence of their two new constants of nature: recall that Perrin interpreted the convergence of experimental determinations of Avogadro's number as compelling evidence for "the real existence of the molecule," and Planck argued that the convergence in measured values of h confirmed the reality of "the quantum of action." What might be the implications of the existence of an 'acceleration constant' equal to a_0?

Here is one possible clue. Kaplinghat and Turner noted the rough numerical coincidence between a_0 and cH_0. That observation was not original with them: already in his first paper from 1983, Milgrom had pointed out the coincidence, and as the observational determinations of both a_0 and H_0 became more precise over the years, Milgrom and others noted an even more striking coincidence: that a_0 is essentially *identical* to $cH_0/(2\pi)$:

$$\frac{cH_0}{2\pi} = 1.16 \times 10^{-10} \left(\frac{H_0}{75 \text{ km s}^{-2}} \right) \text{ m s}^{-2}. \tag{8.13}$$

(The formal uncertainty in Hubble's constant is often given as a few percent, although there are persistent indications that the systematic error might be larger.)

The Hubble constant is definable in terms of local properties of the universe, without recourse to a cosmological model; indeed Georges Lemaître[7] first measured this parameter around 1927, long before the standard cosmological model had come into existence. But in its current state, the concordance cosmological model contains a number of other parameters which (when expressed as an acceleration, through appropriate factors of G and c) are comparable in value to cH_0 and hence to a_0. For instance, the so-called 'dark energy density parameter' Ω_Λ is defined as

$$\Omega_\Lambda = \frac{8\pi}{3} \frac{G\epsilon_\Lambda}{c^2 H_0^2}, \tag{8.14}$$

where ϵ_Λ is the energy density associated with dark energy. The parameter Ω_Λ is said to be ~0.75 and so one can write cH_0 in terms of ϵ_Λ as

$$\frac{cH_0}{2\pi} \approx 0.53 \left(\frac{\Omega_\Lambda}{0.75}\right)^{-1/2} (G\epsilon_\Lambda)^{1/2}. \tag{8.15}$$

And since the energy densities in the standard cosmological model associated with dark matter, dark energy and normal matter are all of the same order (the so-called 'cosmic coincidence'), any of these could be chosen as the parameter to compare with a_0.[8]

Thus, there is a numerical coincidence between a_0 and the 'cosmological acceleration scale.' This is redolent of *Mach's principle*: an idea that arose from Ernst Mach's (and George Berkeley's) critique of Newton's conceptions of 'absolute space' and 'absolute velocity,' and that Einstein cited as a motivation for his theory of relativity. Mach's principle (perhaps 'Mach's postulate' would be a better name) has been expressed in many different ways; a recent expression is "local physical laws are determined by the large-scale structure of the universe" (Ellis and Hawking, 1999, p. 1). In a similar way, Milgrom has suggested that the coincidence of $2\pi a_0$ and cH_0 may suggest that "The state of the Universe at large strongly enters, and affects, local dynamics of small systems" (Milgrom, 2015, p. 109).

Of course, Mach did not know of, nor was he trying to explain, the existence of an acceleration scale; his concern was with the origin of inertia itself and he took for granted that no such scale existed. Milgromian dynamics, on the other hand, suggests a relation between force and acceleration that is different from

---

[7] It is now generally agreed that Georges Lemaître, and not Edwin Hubble, deserves credit for discovering the law of universal expansion. The discovery paper is Lemaître (1927) where this important result is presented, remarkably, in a footnote. Equally remarkably, the footnote, and an associated equation in the text, were omitted from the English translation of Lemaître's paper that was published in 1931, two years after the publication of Hubble's paper. See van den Bergh (2011) and Livio (2011).

[8] Of course, to the extent that theory variants T_0–T_2 are correct, the density of dark matter, at least, should be essentially zero, or at least much lower than the current 'concordance' value of ρ_{DM}. One can only speculate what would be the consequences for dark energy.

the Newtonian relation that Mach was attempting to justify. But Milgrom (1983a, p. 370) has suggested:

If, for example, the inertia force is due to the interaction of the accelerated particle with an inertia field produced by [the] totality of mass in the universe... such that the inertia force is not proportional to the accelerations any more, the introduction of an acceleration constant into the local equations of dynamics is implied and CH_0 ... is the natural cosmological acceleration parameter which can play this role.

If we take seriously the numerical coincidence between $2\pi a_0$ and cH_0, then one final speculation comes naturally to mind. In the standard cosmological model, Hubble's parameter varies with time; so in order to preserve, for all times, the equality that currently exists, one would need to postulate a similar time variation of a_0. In particular, a_0 would need to have been larger in the past, and hence the departures from Newtonian dynamics that Milgrom's postulates imply would have been present in a wider range of systems, in the past, than in the current universe. And there our speculations must end.

9

Summary / Final Thoughts

The answer to the question: Why is one interpretation accepted over another?
can only be: because it seems more fruitful, more promising.

(Madison (1988))

We are now in a position to assess the progressivity of the Milgromian research
program, taken as a whole.

Let us begin by recalling the three main characteristics that Imre Lakatos
identified with theory change in progressive, i.e. historically successful, research
programs. As discussed in Chapter 2, these are:

(1) Each new theory in the research program should account for the successes of
the previous theory(ies) in the same program, but it should also make some
new, testable predictions: that is, it should have excess content.

(2) At least some of the novel predictions should be validated experimentally. Ideally, none would be refuted, but some failures of prediction are to be expected;
nor is the theory expected to make testable predictions in every physical regime
or domain to which it might be applied.

(3) Theory change should conform to the heuristic of the research program. One
consequence is that changes should not be made purely in response to chance
discoveries; ideally, those discoveries should be anticipated by the theory in the
course of its development.

Conditions (1) and (2) are sometimes described collectively as 'incorporation
with corroborated excess content'; condition (3) was called by Lakatos "heuristic
progress." Lakatos's original definition of 'novel prediction' was a fact (typically,
a relation between measurable quantities) that is "anticipated" by a theory: a fact
that is discovered in the process of subjecting the theory to potentially falsifying
tests. That definition has a basis in Karl Popper's ideas about theory corroboration.
As we have seen, a number of other definitions of 'novel prediction' have been
promoted by philosophers of science, but most are based on the idea (endorsed, as
we have seen, by both scientists and philosophers of science) that a theory should

make *surprising* predictions: it should (correctly) predict facts that would have been considered unlikely prior to the theory's construction. One such definition, due to Alan Musgrave, judges a fact to be novel on the basis of whether the leading, rival theory has difficulty explaining it. In our case, of course, the 'rival theory' is the standard, or ΛCDM, cosmological model.

§§

Table 7.1 summarizes the novel facts predicted by Milgrom's theory in the four variants (T_0–T_3) considered in this book. The research program including theory variants T_0 and T_1 is clearly progressive. Theory variant T_0 consists simply of Milgrom's three postulates from 1983, and so the requirement of 'heuristic progress' does not apply; but theory variant T_1, which contains a non-relativistic Lagrangian, clearly does satisfy this condition: a Lagrangian description was known, from the start, to be necessary if the theory was to avoid certain theoretical problems like non-conservation of momentum. Furthermore both T_0 and T_1 make a number of confirmed, novel predictions: at least three or four each, and (modulo one's judgments about degrees of corroboration) as many as nine or ten altogether.

Given the background knowledge that existed in 1983, these Milgromian predictions are surprising. Theory variants T_0 and T_1 allow one to make detailed, essentially exact (i.e. functional) predictions about the kinematics of a galaxy based on its observed distribution of normal matter alone. Confirmed Milgromian predictions include the baryonic Tully–Fisher relation (BTFR; Figure 4.1), the radial acceleration relation (RAR; Figure 4.3), and the central surface density relation (CSDR; Figure 5.2). Even under T_0, the rotation curve of a disk galaxy like the Milky Way is predictable, and under T_1 the testable content of the theory expands to include statements about the motion of stars perpendicular to the galactic plane as well.

It was, of course, the *failure* of the standard cosmological model to predict galaxy rotation curves that led, around 1980, to the first of the dark matter postulates (DM-1; Chapter 1). But even now, no algorithm exists, under the standard model, that can correctly predict the rotation curve of a galaxy based on its observed distribution of matter. One would be justified, on that basis alone, to conclude that the Milgromian predictions entailed by T_0 and T_1 satisfy Musgrave's condition (M) for novelty: the standard model can not, in principle, make any such predictions. The best the standard model can hope to do is to provide *statistical* explanations of the relations predicted by Milgrom, with the help of large-scale *simulations* of galaxy formation and evolution. But even standard-model cosmologists will acknowledge that these simulations have so far failed to accommodate many of the known properties of galaxies, including the properties that were successfully predicted by theory variants T_0 and T_1 in advance of their observational corroboration.

Perhaps the simplest way to justify that statement is to recall some of the anomalies that have achieved the status of 'named' problems under the standard model.[1] These include:

- The core-cusp problem;
- The too big to fail problem;
- The missing satellites problem;
- The problem of the satellite planes.

None of these problems exists for a Milgromian researcher, but for interestingly different reasons. The first ('core-cusp') problem describes the systematic failure of standard-model galaxy formation algorithms to accommodate the rotation curves of certain types of galaxies. Not only does this problem not exist under T_0 or T_1: the observations that are the basis for the 'anomaly' are correctly predicted by the modified dynamics; they constitute a *success* of Milgromian theory. The last three problems all involve dwarf galaxies; they reflect the difficulties of relating the dark matter properties of dwarf galaxies to the properties of their normal matter under the standard model. Dark matter does not exist in T_0 or T_1, and, as a consequence, these three problems simply do not arise. Stated differently: the dwarf-galaxy data that are anomalous from the standpoint of a standard-model researcher present no obvious problems to a Milgromian scientist.

But the standard-model failures of accommodation that are most evidentiarily relevant to T_0 and T_1 are arguably those that involve relations such as the BTFR and the RAR, which are predicted by Milgrom's theory to be 'exact,' and which have been confirmed, based on observations of galaxies, to exhibit essentially zero intrinsic scatter. Relations of this sort are extremely difficult to understand under the standard model, since the only mass that appears in them is the normal (i.e. non-dark) mass. A standard-model cosmologist would expect the *dark mass* to be the controlling variable. For instance, Bullock and Boylan-Kolchin (2017, p. 27), in discussing the problems posed by the RAR for the standard cosmology, write: "The real challenge, as we see it, is to understand how galaxies can have so much diversity in their rotation-curve shapes compared to naive ΛCDM expectations while also yielding tight correlations with baryonic [i.e. normal matter] content."

Theory variants T_0 and T_1 are progressive in another way. As detailed in Chapter 8, the single dimensional constant that appears in these theories – Milgrom's constant, or a_0 – is found to have the *same value*, whichever testable prediction is used in its determination. This convergence, itself, constitutes a

---

[1] There are quite a few other observational facts that constitute anomalies from the standpoint of the standard cosmological model but which have not yet been assigned catchy names. Silk and Mamon (2012) list seventeen; Kroupa (2012) lists twenty two. While a few of these are bona-fide failures of prediction – for instance, the observations that motivated the dark matter and dark energy postulates – the majority are better described as failures of accommodation: that is: facts that standard-model theorists have been unable to convincingly explain, even in a post-hoc sense.

successful novel prediction of Milgrom's theory, since the only dimensional parameter in the theory is a_0, and a_0 is postulated to be a universal constant. And recall that philosopher John Losee argued that convergence of the measured value of a new physical constant, *by itself*, justifies treating the theory containing the constant as a 'progressive problemshift': it warrants a transition to a new type of theory.

This almost perfect record of progressivity is marred by one failure of prediction. Theory variant T_1 fails to correctly predict the temperature of the intracluster gas in clusters of galaxies, the largest, self-gravitating systems in the universe. The mass required to reproduce the temperature data under T_1 is somewhat greater than what is observed. This anomaly may indicate that there is undetected mass in galaxy clusters; although, as noted in Chapter 5, the cluster data are also anomalous from the standpoint of the standard cosmological model. But taken at face value, the failure of T_1 to explain the observed dynamics of galaxy clusters constitutes a problem that must be addressed in a later variant of the theory.

§

Theory variant T_2 is the relativistic version of the theory. It was argued in Chapter 6 that the development of T_2 has sometimes violated the condition of 'heuristic progress.' As discussed in that chapter, and in Chapter 3, the heuristic that guides a theorist in her development of a relativistic variant of Milgrom's theory has been well articulated; for instance, the set of desiderata due to Jakob Bekenstein that are cited in Chapter 6. But those guidelines are, apparently, not detailed enough, or (to use Lakatos's expression) 'powerful' enough, to guide theory development in a way that anticipates the observations. As we saw in Chapter 6, Milgromian researchers have sometimes used chance observational discoveries – for instance, of gravitational lensing – to choose between the many relativistic theories that are consistent with Bekenstein's heuristic guidelines. (Of course, a similar charge can be leveled against the standard cosmological model; but we are not engaged here in *comparative* research program evaluation.)

The lack of a unique relativistic theory complicates the generation of testable predictions, but two predictions were identified in Chapter 6 that are, arguably, only weakly dependent on the form of the relativistic Lagrangian: an early epoch of cosmic reionization, and a particular value for the amplitude ratio of the first and second peaks in the power spectrum of cosmic microwave background (CMB) temperature fluctuations. As of this writing, the first of these two predictions is 'possibly corroborated' at best. But the second prediction, published by Stacy McGaugh in 1999, was beautifully confirmed a few years later by measurements of the CMB spectrum. McGaugh's prediction is, to date, the *only* successful, quantitative prediction that anyone has made about the CMB spectrum, and it must count as a major success of Milgromian theory.

While standard-model cosmologists did not *predict* this (or any other) quantitative feature of the CMB spectrum, they did eventually *accommodate* those data by adjusting their theory. As detailed in Chapter 6, those adjustments fit Karl Popper's definition of a 'conventionalist stratagem': cosmologists simply doubled (approximately) the assumed value of the mean density of baryons (i.e. normal matter) in the universe. In the process, they generated two inconsistencies in their theory that were not present before 2002 and which have persisted until the present day: the 'lithium problem' and the 'missing baryons problem.' Neither problem exists from the standpoint of a Milgromian researcher; both problems result directly from the need of standard-model cosmologists to include dark matter in their theory. Thus (it was argued in Chapter 6) McGaugh's successful prediction of the CMB peak ratio satisfies Musgrave's condition for novelty: because explaining the fact under the standard model required cosmologists to introduce inconsistencies into their model.

In spite of this success, no relativistic version of Milgrom's theory has yet been found that successfully reproduces the full CMB spectrum, as it is now determined from observations, without the help of auxiliary hypotheses. Nor do any of the proposed relativistic variants solve the problem of the 'missing mass' in clusters of galaxies.

§

Finally we come to theory variant T_3, due primarily to Lasha Berezhiani and Justin Khoury. These theorists proposed a modification to the hard core of the Milgromian research program. Impressed by the ability of the standard cosmological model to accommodate data like the CMB power spectrum, they introduced dark matter into Milgrom's theory. Einstein's theory of gravity and motion (or, in the appropriate regimes, Newton's) is assumed to be correct: there is no longer an 'acceleration scale' built into the gravitational action. The composition of the dark matter under T_3 is (as in the standard cosmological model) left unspecified, but Berezhiani and Khoury assume that, on the largest spatial scales, the dark matter is cold and collisionless: the same properties that are postulated by standard-model cosmologists. As a consequence, theory variant T_3 accommodates the CMB (and some other large-scale) data in precisely the same way that the standard model does. However, under conditions of sufficiently high density, for instance near the center of a galaxy, the dark matter is postulated to behave like a superfluid condensate. Berezhiani and Khoury postulate that the interaction of the dark matter with normal matter – via Newtonian gravity, but also via coupling of the superfluid phonons – results in an acceleration that approximates Milgrom's modified dynamics in the low-acceleration regime.

It was argued in Chapter 7 that the Milgromian research program containing theory variant T_3 is *not* progressive. By virtue of its modified hard core, T_3 loses a great deal of the explanatory power of theory variants T_0–T_2: it fails to satisfy Lakatos's condition of incorporation (or, as some philosophers would say, it suffers from 'Kuhn losses.') The remarkable predictive successes of theory variants T_0 and T_1 were dependent on the gravitational acceleration being determined by the distribution of normal matter alone. Under T_3, all such predictions involve dark matter as well; even the prediction of a disk galaxy rotation curve depends, to a greater or lesser extent, on what is *assumed* about the dark matter surrounding the galaxy beyond the superfluid core. Berezhiani and Khoury deal with this indeterminacy in the same way that standard-model cosmologists do; for instance, they accept the results of standard-model calculations of dark matter clustering and galaxy formation (except insofar as they would be affected by the superfluid nature of dark matter) as well as aspects of standard-model methodology, such as 'abundance matching,' that are used to relate the results of those simulations to observed galaxies. Furthermore, unlike the single (and well constrained) parameter a_0 of T_0–T_2, theory variant T_3 contains a number of unspecified parameters which must be determined from data. It is not clear how, or whether, theory variant T_3 would predict the remarkable convergence of measured values of a_0 that is discussed in Chapter 8.

With regard to Lakatos's requirement of 'corroborated excess content,' theory variant T_3 displays, at best, only modest evidence of progressivity. It was argued in Chapter 7 that the only 'excess content' of T_2 that is evidentiarily relevant is content that differs from that of the standard cosmological model – that is: predictions that depend on the postulated superfluid nature of the dark matter. Three, potentially testable predictions that meet this requirement were identified; these were judged to be 'partially corroborated,' 'possibly corroborated,' and 'neither confirmed nor refuted.'

The non-progressive character of theory variant T_3 has an obvious source: it is due, in largest part, to the adoption of standard-model postulates and standard-model methodology. As we have seen many times in this book, the standard cosmological model, since the adoption of the dark matter postulates beginning around 1980, has rarely, if ever, satisfied Lakatos's conditions for progressivity. Virtually all of its 'successes' have been successes of accommodation, not anticipation.

§

Application of Lakatos's criteria to the Milgromian research program suggests, therefore, that the program is remarkably progressive in its early variants, T_0 and T_1; only moderately progressive in variant T_2, the relativistic version of the theory; and that variant T_3, which attempts to ameliorate the shortcomings of T_2 by introducing dark matter into the theory, has not managed to bring the research program back into a progressive state.

What conclusions might one draw from this record of progress? It is tempting to try to extrapolate the existing record into the future. I accept the view of most philosophers that this would be a waste of time.[2] On the one hand, it is not at all obvious what sort of trend has been established by theory variants T_0–T_3. And even if a clear trend existed, the history of science contains numerous examples of research programs that progressed before degenerating, or that degenerated for a time before being 'rescued' and returned to a progressive state. The example of the atomic-kinetic research program was mentioned briefly in Chapter 7. Another example, discussed by Lakatos (1970, p. 53–55), is William Prout's (1815) program to show that atoms are composed of multiples of hydrogen.

A possibly less fraught question is: Has the Milgromian research program exhibited enough evidence of progress that one should consider devoting one's time to it? – given that doing so would probably entail one's *not* continuing to work on the standard cosmological model. I am going to risk the ire of standard-model cosmologists (and probably also some philosophers) by answering 'yes,' but with one qualification:

The most striking successes of the Milgromian research program are associated with theory variants T_0 and T_1. Given the background knowledge that existed *c*. 1980 (e.g. the known, asymptotic flatness of galaxy rotation curves), the proposal that the kinematics of any disk galaxy could be predicted, with high accuracy, from the observed distribution of *normal* matter alone was amazingly bold: rather as if one had predicted that the gravitational field of a planet is determined by (say) its spin angular momentum, or its surface area. There was simply no basis, under the standard model, for believing any such thing, and yet it turned out to be correct. The prediction of a single, universal acceleration scale (a_0) was equally bold and its experimental confirmation equally impressive.

These successes of T_0 and T_1 (together with the other successes listed in Table 5.1) suggest to me that (in the language of Lakatos and of Losee) a 'problemshift' – a transition to the Milgromian research program – is warranted.

My qualification is this: I do not think that a comparably compelling claim can be made about theory variants T_2 and T_3. But here an important distinction should be made.

Theory variant T_2 *retains* the hard-core postulates of T_0 and T_1, the same postulates that were responsible for the successes of those two theories. What seems to be *lacking* in T_2 is a suitable *heuristic*. As we have seen, the guidelines proposed by Milgrom and Bekenstein for the development of a relativistic theory are consistent with a wide range of possible theories, none of which has (so far) been successful at increasing the testable content of the theory much beyond that of T_1. By contrast, theory variant T_3 *discards* the elements of the Milgromian hard core that led to the

---

[2] E.g. Urbach (1978, p. 99): "The orthodox view among modern philosophers of science is that any attempt to appraise the future performance of scientific research is bound to be futile."

research program's early string of successes and replaces them with a new set of postulates that (as we saw in Chapter 7) seem to contain less testable content than the postulates they replace.

Based on these considerations, I think it would be reasonable to advise a young theorist who is so inclined to 'enter' the Milgromian research program at T_2: that is, to continue to assume the correctness of the foundational postulates, but to find a way to modify the heuristic so that it does more than lay down certain, very general, *desiderata* that a relativistic theory should satisfy. What is lacking, it would seem, is "some simple, new, and powerful, unifying idea" (Popper, 1963, p. 241) that can point the theorist toward a more successful relativistic version of the theory. Perhaps the rudiments of such an idea can be found in the remarks near the end of Chapter 8 about Mach's principle. Or perhaps the needed idea will turn out to be quite different from anything so far considered.

§§

As discussed briefly in Chapter 3, there are a number of similarities between Milgrom's research program and the research program initiated by Niels Bohr, in 1913, into what is now called the 'old' quantum mechanics. Like Milgrom's, Bohr's program was initially progressive. It correctly predicted the wavelengths of the hydrogen line emission spectrum: both the lines that had been measured prior to 1913, and the lines of the Lyman, Brackett and Pfund series when they were measured in 1914, 1922 and 1924, respectively. But beginning around 1922, seemingly intractable problems began to appear: the spectrum of helium, the anomalous Zeeman effect, the band spectra of molecules (Margenau, 1950, p. 311–313; Slater, 1960, p. 31–35). The Bohr–Kramers–Slater theory of 1924 made no successful predictions at all. In Lakatos's (1970, p. 67) words, "even this great programme came to a point where its heuristic power petered out. *Ad hoc* hypotheses multiplied and could not be replaced by content-increasing explanations."

In spite of the eventual degeneration of Bohr's research program, the set of foundational postulates in his 1913 papers are universally acknowledged as brilliant conjectures. That statement is true not simply in regard to *retrospective* judgments. James Jeans, already in 1913, referred to Bohr's new theory as "a most ingenious and suggestive, and I think we must add convincing, explanation of the laws of spectral series" (Jeans, 1914, p. 380). Einstein, also in 1913, declared, "There must be something behind it. I do not believe that the derivation of the absolute value of the Rydberg constant is purely fortuitous."[3] And even now, a century later, textbooks on quantum mechanics typically begin by deriving the energy levels of

---

[3] Quoted by Jammer (1966, p. 86). Jammer gives the original German in his note 107 as "da muß etwas dahinter sein; ich glaube nicht, daß die Rydbergkonstante durch Zufall in absoluten Werten ausgedrückt richtig herauskommt."

the hydrogen atom in precisely the same way that Bohr did in 1913. The eventual *failures* of Bohr's research program are, nowadays, typically not mentioned at all.

§

The situation is very different for Milgrom's theory. As documented elsewhere (Merritt, 2017), Milgrom's theory, and the observations that corroborate it, are almost universally ignored by contemporary textbook writers. And that fact constitutes a genuine puzzle: because the historical record shows that scientists do tend to be impressed by theories that make successful, novel predictions: even theories (like Bohr's) that are manifestly inconsistent, or that are deemed only partially successful. Whereas the majority of cosmologists working today – one could probably say, the vast majority – seem indifferent to Milgrom's theory, if not contemptuous of it.

I am not going to claim to understand the reasons for this attitude of malign neglect. But in reading the standard-model literature, one notices certain recurring patterns of exposition and argument that may shed light on the puzzle.

§

First: When writing about facts or relations that were first predicted by Milgromian researchers, standard-model cosmologists often fail to cite the relevant papers, or to acknowledge that what they are discussing is a successful, novel prediction of a competing theory. One comes away, after reading some of these papers, with the impression that the relation under discussion simply appeared one day out of thin air; the historical context is missing. Examples of this widespread, and troubling, phenomenon were documented elsewhere in this book: with regard to the baryonic Tully–Fisher relation (Chapter 4), the central surface density relation (Chapter 5), and the Milgromian prediction of an early epoch of cosmic reionization (Chapter 6). Although it was not specifically noted in Chapter 6, research articles by standard-model cosmologists on the CMB never mention McGaugh's successful, Milgromian prediction of 1999, the first and only successful prediction of its kind. And I am not aware of any article by standard-model researchers that credits Milgrom's theory with the successful prediction of the vertical force law in the Milky Way (Chapter 5), a fact that the standard model has only been able to accommodate by invoking ad hoc adjustments to the local dark matter distribution.

One wonders whether this systemic 'failure to cite' is deliberate, or whether it reflects a general ignorance on the part of standard-model cosmologists of the published record. If the latter, this book may serve as a useful corrective. But whatever the explanation, the result is a serious misrepresentation of the degree to which discoveries deriving from Milgrom's theory have shaped, and continue to shape, the research directions of astrophysicists and cosmologists, *including* standard-model cosmologists.

§

Second: Articles in the standard-model literature that *do* mention Milgrom's theory rarely credit the theory with more than a single successful prediction. Rarely, if ever, does one find an acknowledgement that Milgrom's theory has produced a *string* of confirmed novel predictions – at least a half dozen, and perhaps twice that number (Table 7.1). One striking case was discussed in Chapter 8: Kaplinghat and Turner's (2002) paper on (what they call) "Milgrom's law" attributes to Milgrom only the vague prediction that "the need for the gravitational effect of nonluminous (dark) matter in galaxies arises only when the Newtonian acceleration is less than about $a_0 = 2 \times 10^{-8}$ cm s^{-2}." Similarly, the review articles by Silk and Mamon (2012) and Peebles (2015) credit Milgrom's theory only with the prediction of the baryonic Tully–Fisher relation. And the standard-model literature is full of statements implying that Milgrom's postulates are the modern-day equivalent of Ptolemy's epicycles and equants: that they do nothing more than reproduce facts that were already known in 1983. An example is the statement by Freese (2017, p. 2) that "While these [Milgromian] models have been shown to fail ... they may provide an interesting phenomenological fit on small scales." Lisanti (2017, p. 401), after describing the asymptotic flatness of galaxy rotation curves, writes: "MOdified Newtonian Dynamics (MOND) is a class of phenomenological models that seek to address this point." And van den Bosch and Dalcanton (2000) write: "both MOND and DM were constructed to fit the rotation curves of disk galaxies." Of course, the *prediction* of rotation curves is one of the most remarkable successes of Milgrom's theory; it is only the standard cosmological model that is "constructed to fit rotation curves" through the postulate DM-1 (Chapter 1).

An 'exception that proves the rule' is Bullock and Boylan-Kolchin's (2017) review article on "Small-scale challenges to the ΛCDM paradigm." These authors mention *two* successful predictions of the Milgromian research program: the baryonic Tully–Fisher relation and the radial acceleration relation. But then they imply that one relation is entailed by the other (italics added):

Among the more puzzling aspects of galaxy phenomenology in the context of ΛCDM are the tight scaling relations between dynamical properties and baryonic properties, even within systems that are dark matter dominated. One well-known example of this is the baryonic Tully–Fisher relation ... *A generalization of the baryonic Tully–Fisher relation known as the radial acceleration relation (RAR)* was recently introduced by McGaugh, Lelli & Schombert (2016) (Bullock and Boylan-Kolchin, 2017, p. 26).[4]

---

[4] Note the careless attribution here. As documented in Chapter 4, the RAR was discussed by Milgromian researchers already in the 1990s.

It is possible to imagine a theory that predicts the same dependence of asymptotic rotation speed on galaxy mass as Milgrom's theory, but which fails to correctly predict galaxy rotation curves in regions where $a \approx a_0$.[5] In Milgrom's (2002, p. L82) words: "the appearance of a_0 in the [baryonic] Tully–Fisher (TF) relation is independent of its role as a transition acceleration." The fact that Milgrom's theory correctly does *both* things counts twice in its favor when it comes to reckoning the theory's empirical success.

From an epistemic point of view, this is important. Recall the reasoning that led Popper, Lakatos, Zahar and many others to find evidential support solely in *novel* predictions. Those authors argued that it is easy to "cleverly engineer" a theory to explain known facts, but that it would be unlikely for an incorrect theory to correctly predict a new, and unexpected, fact. Of course, it would be foolish to assign numerical values to such probabilities, and as far as I know, no one has ever tried. But: the laws of probability state that the joint probability of two independent events is the product of the individual probabilities. If it is unlikely that an incorrect theory would correctly predict one novel fact, it is *very* unlikely that the same theory would correctly predict *two* novel facts; *extremely* unlikely that it would correctly predict *three* novel facts; and so on.

Here again, the problem may simply be a weak grasp of Milgrom's theory on the part of standard-model cosmologists. But whatever the explanation, the result is an extreme misrepresentation, in the standard-model literature, of the degree to which Milgrom's theory has been empirically successful.

§

Third: As documented throughout this book, standard-model cosmologists routinely ignore the distinction between prediction and accommodation. That is: they assume that a post hoc *explanation* of a fact has the same evidentiary status, with regard to the standard cosmological model, that successful *prediction* of the fact has with regard to Milgrom's theory. For instance, Peebles (2015, p. 12247), in a review article on dark matter, describes the baryonic Tully–Fisher relation as follows: "MOND made a strikingly successful prediction. This success is not an argument against ΛCDM, of course, unless it can be shown that the relation is improbable within ΛCDM." One finds no acknowledgement, here, that the successful Milgromian prediction of the BTFR will always count for more, evidentiarily, than any post hoc, standard model accommodation of that relation possibly can. Furthermore, standard-model authors rarely express a need to explain more than a single Milgromian prediction; as in the article by Peebles, it is taken for granted, apparently, that the successful accommodation of just one of the novel predictions listed in Table 7.1 is sufficient to justify a neglect of Milgrom's theory *in toto*.

---

[5] For instance, the standard cosmological model. After adjusting their galaxy formation codes to accommodate – approximately – the BTFR, theorists are still left with the 'core-cusp problem.'

We might want to give standard-model cosmologists the benefit of the doubt here. Perhaps their 'one-and-done' approach is simply a reflection of their limited knowledge of Milgrom's theory; perhaps (as in the examples cited above) they believe that the single Milgromian prediction that they are targeting for explanation (whichever one it happens to be) is the theory's only successful prediction, and so, by accommodating it, they have accomplished all they need to.

But surely this is too generous. One need not be a card-carrying Popperian to appreciate that the successful prediction of a fact, in advance of its experimental confirmation, counts for more than a post hoc accommodation of that fact. Here is how Leibniz expressed it, some two and a half centuries before Popper: "Those hypotheses deserve the highest praise ... by whose aid predictions can be made, even about phenomena or observations which have not been tested before" (Leibniz, 1678). (In the sentence that precedes this one, Leibniz accords lesser praise to hypotheses that do nothing more than explain known facts.) Why, then, do standard-model cosmologists seem to assume the epistemic equivalence of prediction and accommodation? Stated differently: can we identify some theory of knowledge (valid or otherwise) on the basis of which the standard-model approach to theory corroboration makes logical sense?

I believe we can. Karl Popper, in his *Realism and the Aim of Science* (1983, p. 234), identified two main "attitudes" with respect to corroboration of scientific theories:

(a) The uncritical or verificationist attitude: one looks out for 'verification' or 'confirmation' or 'instantiation', and one finds it, as a rule. Every observed 'instance' of the theory is thought to 'confirm' the theory.

(b) The critical attitude, or falsificationist attitude: one looks for falsification, or for counter-instances. Only if the most conscientious search for counter-instances does not succeed may we speak of a corroboration of the theory.

Attitude (b), which is what Popper called 'critical rationalism,' is, of course, the attitude that Popper endorsed, and that Lakatos incorporated into his *Methodology*. A critical rationalist demands that a theory make testable predictions, and he is alert to possible refutations. Verificationism, Popper argued, is a species of inductivism, and induction is a fallacy; it does not exist. But it would be natural for a *verificationist*, one with standard-model commitments, to argue as follows: "By accommodating the standard cosmological model to a Milgromian prediction, I have verified the standard cosmological model; and since two competing theories can not both be correct, then, by verifying the standard model, I have refuted Milgrom's theory."[6]

---

[6] This attitude is captured perfectly in the original title of Keller and Wadsley's 2017 paper on the radial acceleration relation: "La Fin du MOND? Λ CDM is Fully Consistent with SPARC Acceleration Law." (This is "version 1" as posted at arXiv.org.) To their credit, Keller and Wadsley changed the title for the published version of their paper.

The attitude of *Milgromian* researchers toward theory corroboration has been consistently critical-rationalist.[7] Testable predictions, often very surprising ones, have been generated from the theory; each predicted fact has been tested by comparison with observations; and again and again (and again, and again, and again) the Milgromian prediction has been confirmed: the theory has survived the attempt to refute it. And when (as has happened once or twice) the theory has *failed* to pass a test, Milgromian researchers have taken the failure seriously.

Where might such a difference in attitudes come from? Standard-model cosmologists, I am sure, are just as impressed as other scientists by instances in which a theory anticipates an observation. But they are committed to their theory; and their theory is simply not very good at making successful, novel predictions. And so it is understandable, perhaps, that standard-model cosmologists would gravitate toward a theory of knowledge, and a methodology, that favor verificationism over critical rationalism.

§§

Here is a final thought. It is currently fashionable in some scientific circles to pooh-pooh Popper's criterion of demarcation. In the words of cosmologist Sean Carroll (2014): "The falsifiability criterion gestures toward something true and important about science, but it is a blunt instrument."

Exhibit A in the case against falsifiability is the standard cosmological model itself. As we have seen (and as most standard-model cosmologists will admit), that theory is simply not well suited to making refutable predictions. There are too many causal steps between the theory's postulates – about dark matter, for instance – and the observable properties of a galaxy; too many physical mechanisms involved in galaxy formation and evolution; too much uncertainty in the initial conditions. Standard-model cosmologists have resigned themselves to the post hoc accommodation of new data, typically via large-scale computer simulations, and typically only in a statistical sense.

It would be reasonable, given this state of affairs, to suppose that the science of cosmology has evolved beyond the class of theories to which the ideas of Popper or Lakatos can be applied. It would be reasonable to conclude that falsifiability is too strict a standard to hold scientific theories to.

It *would* be reasonable, that is, except for one thing. Milgrom's theory *is* eminently testable. That, by itself, is unremarkable, since one can always construct testable theories if one is content for the tests to fail. But again and again, Milgrom's predictions have been shown to be correct. Thus: there exists a theory of cosmology that is both testable, *and* viable. And that is something that anyone who would sound the death knell of falsifiability must come to terms with.

---

[7] At least with regard to theory variants T_0–T_2.

References

Aaronson, M. 1983. Accurate radial velocities for carbon stars in Draco and Ursa Minor: The first hint of a dwarf spheroidal mass-to-light ratio. *The Astrophysical Journal Letters*, **266**, L11–L15.

Aaronson, M., Huchra, J., and Mould, J. 1979. The infrared luminosity/H I velocity-width relation and its application to the distance scale. *The Astrophysical Journal*, **229**, 1–13.

Abbott, B. P., Abbott, R., Abbott, T. D., Acernese, F., Ackley, K., Adams, C., Adams, T., Addesso, P., Adhikari, R. X., Adya, V. B., et al. 2017. GW170817: observation of gravitational waves from a binary neutron star inspiral. *Physical Review Letters*, **119**(16), 161101.

Adams, T. F. 1976. The detectability of deuterium Lyman alpha in QSOs. *Astronomy and Astrophysics*, **50**, 461.

Aguirre, A., Schaye, J., and Quataert, E. 2001. Problems for modified Newtonian dynamics in clusters and the Lyα forest? *The Astrophysical Journal*, **561**, 550–558.

Albornoz Vásquez, D., Belikov, A., Coc, A., Silk, J., and Vangioni, E. 2012. Neutron injection during primordial nucleosynthesis alleviates the primordial Li7 problem. *Physical Review D*, **86**, 063501.

Alder, H. L., and Roessler, E. B. 1968. *Introduction to Probability and Statistics*. Freeman.

Alonso, A., Arribas, S., and Martínez-Roger, C. 1999. The effective temperature scale of giant stars (F0-K5). I. The effective temperature determination by means of the IRFM. *Astronomy and Astrophysics Supplement*, **139**, 335–358.

Anderson, J. D., Laing, P. A., Lau, E. L., Liu, A. S., Nieto, M. M., and Turyshev, S. G. 1998. Indication, from Pioneer 10/11, Galileo, and Ulysses data, of an apparent anomalous, weak, long-range acceleration. *Physical Review Letters*, **81**, 2858–2861.

Ando, S., Cyburt, R. H., Hong, S. W., and Hyun, C. H. 2006. Radiative neutron capture on a proton at big-bang nucleosynthesis energies. *Physical Review C*, **74**, 025809.

Andreon, S. 2010. The stellar mass fraction and baryon content of galaxy clusters and groups. *Monthly Notices of the Royal Astronomical Society*, **407**, 263–276.

Angulo, C., Casarejos, E., Couder, M., Demaret, P., Leleux, P., Vanderbist, F., Coc, A., Kiener, J., Tatischeff, V., Davinson, T., Murphy, A. S., Achouri, N. L., Orr, N. A., Cortina-Gil, D., Figuera, P., Fulton, B. R., Mukha, I., and Vangioni, E. 2005. The ^{7}Be(d,p)2α cross section at big bang energies and the primordial ^{7}Li abundance. *The Astrophysical Journal Letters*, **630**, L105–L108.

Angus, G. W. 2009. Is an 11eV sterile neutrino consistent with clusters, the cosmic microwave background and modified Newtonian dynamics? *Monthly Notices of the Royal Astronomical Society*, **394**, 527–532.

Angus, G. W., Famaey, B., and Buote, D. A. 2008. X-ray group and cluster mass profiles in MOND: Unexplained mass on the group scale. *Monthly Notices of the Royal Astronomical Society*, **387**, 1470–1480.

Aoki, W., Barklem, P. S., Beers, T. C., Christlieb, N., Inoue, S., García Pérez, A. E., Norris, J. E., and Carollo, D. 2009. Lithium abundances of extremely metal-poor turnoff stars. *The Astrophysical Journal*, **698**, 1803–1812.

Arraut, I. 2014. Can a nonlocal model of gravity reproduce dark matter effects in agreement with MOND? *International Journal of Modern Physics D*, **23**, 1450008.

Asplund, M., Lambert, D. L., Nissen, P. E., Primas, F., and Smith, V. V. 2006. Lithium isotopic abundances in metal-poor halo stars. *The Astrophysical Journal*, **644**, 229–259.

Audouze, J., and Tinsley, B. M. 1976. Chemical evolution of galaxies. *Annual Reviews of Astronomy and Astrophysics*, **14**, 43–79.

Ayer, A. J. 1946. *Language, Truth and Logic*. Victor Gollantz.

Bacon, F. 1621/1863. *Novum organum. The Works of Francis Bacon, Volume VIII*. Taggard and Thompson. Edited by J. Spedding, R. L. Ellis and D. D. Heath.

Bahcall, J. N. 1984a. K giants and the total amount of matter near the sun. *The Astrophysical Journal*, **287**, 926–944.

Bahcall, J. N. 1984b. Self-consistent determinations of the total amount of matter near the sun. *The Astrophysical Journal*, **276**, 169–181.

Bahcall, J. N. 1987. Dark matter in the galactic disk. Pages 17–27 of: Kormendy, J., and Knapp, G. R. (eds), *Dark Matter in the Universe*. IAU Symposium, vol. 117.

Bahcall, J. N., and Soneira, R. M. 1980. The universe at faint magnitudes. I. Models for the galaxy and the predicted star counts. *The Astrophysical Journal Supplement*, **44**, 73–110.

Bahcall, J. N., Flynn, C., and Gould, A. 1992. Local dark matter from a carefully selected sample. *The Astrophysical Journal*, **389**, 234–250.

Balashev, S. A., Zavarygin, E. O., Ivanchik, A. V., Telikova, K. N., and Varshalovich, D. A. 2016. The primordial deuterium abundance: subDLA system at $z_{abs} = 2.437$ towards the QSO J1444+2919. *Monthly Notices of the Royal Astronomical Society*, **458**, 2188–2198.

Barkana, R., and Loeb, A. 2001. In the beginning: the first sources of light and the reionization of the universe. *Physics Reports*, **349**, 125–238.

Barvinsky, A. O. 2003. Nonlocal action for long-distance modifications of gravity theory. *Physics Letters B*, **572**, 109–116.

Begeman, K. G., Broeils, A. H., and Sanders, R. H. 1991. Extended rotation curves of spiral galaxies: Dark haloes and modified dynamics. *Monthly Notices of the Royal Astronomical Society*, **249**, 523–537.

Begum, A., Chengalur, J. N., Karachentsev, I. D., and Sharina, M. E. 2008. Baryonic Tully–Fisher relation for extremely low mass galaxies. *Monthly Notices of the Royal Astronomical Society*, **386**, 138–144.

Bekenstein, J. D. 1992. New gravitational theories as alternatives to dark matter. Pages 905–924 of: Satō, F., and Nakamura, T. (eds), *Marcel Grossmann Meeting on General Relativity*.

Bekenstein, J. D. 1993. Relation between physical and gravitational geometry. *Physical Review D*, **48**, 3641–3647.

Bekenstein, J. D. 2004. Relativistic gravitation theory for the modified Newtonian dynamics paradigm. *Physical Review D*, **70**, 083509.

Bekenstein, J. D., and Milgrom, M. 1984. Does the missing mass problem signal the breakdown of Newtonian gravity? *The Astrophysical Journal*, **286**, 7–14.

Bekenstein, J. D., and Sagi, E. 2008. Do Newton's G and Milgrom's a_0 vary with cosmological epoch? *Physical Review D*, **77**, 103512.

Bennett, C. L., Halpern, M., Hinshaw, G., Jarosik, N., Kogut, A., Limon, M., Meyer, S. S., Page, L., Spergel, D. N., Tucker, G. S., Wollack, E., Wright, E. L., Barnes, C., Greason, M. R., Hill, R. S., Komatsu, E., Nolta, M. R., Odegard, N., Peiris, H. V., Verde, L., and Weiland, J. L. 2003. First-year *Wilkinson Microwave Anisotropy Probe*

(*WMAP*) observations: preliminary maps and basic results. *The Astrophysical Journal Supplement*, **148**, 1–27.

Berezhiani, L., and Khoury, J. 2015. Theory of dark matter superfluidity. *Physical Review D*, **92**, 103510.

Berezhiani, L., and Khoury, J. 2016. Dark matter superfluidity and galactic dynamics. *Physics Letters B*, **753**, 639–643.

Berezhiani, L., Famaey, B., and Khoury, J. 2017. Phenomenological consequences of superfluid dark matter with baryon-phonon coupling. *ArXiv e-prints*, 1711.05748.

Bernstein, G. M., Guhathakurta, P., Raychaudhury, S., Giovanelli, R., Haynes, M. P., Herter, T., and Vogt, N. P. 1994. Tests of the Tully–Fisher relation. 1: Scatter in infrared magnitude versus 21 cm width. *The Astronomical Journal*, **107**, 1962–1976.

Bertulani, C. A., and Kajino, T. 2016. Frontiers in nuclear astrophysics. *Progress in Particle and Nuclear Physics*, **89**, 56–100.

Bienaymé, O. 2009. Potential-density pairs and vertical tilt of the stellar velocity ellipsoid. *Astronomy and Astrophysics*, **500**, 781–784.

Bienaymé, O., Robin, A. C., and Creze, M. 1987. The mass density in our Galaxy. *Astronomy and Astrophysics*, **180**, 94–110.

Bienaymé, O., Soubiran, C., Mishenina, T. V., Kovtyukh, V. V., and Siebert, A. 2006. Vertical distribution of Galactic disk stars. *Astronomy and Astrophysics*, **446**, 933–942.

Bienaymé, O., Famaey, B., Wu, X., Zhao, H. S., and Aubert, D. 2009. Galactic kinematics with modified Newtonian dynamics. *Astronomy and Astrophysics*, **500**, 801–805.

Bienaymé, O., Famaey, B., Siebert, A., Freeman, K. C., Gibson, B. K., Gilmore, G., Grebel, E. K., Bland-Hawthorn, J., Kordopatis, G., Munari, U., Navarro, J. F., Parker, Q., Reid, W., Seabroke, G. M., Siviero, A., Steinmetz, M., Watson, F., Wyse, R. F. G., and Zwitter, T. 2014. Weighing the local dark matter with RAVE red clump stars. *Astronomy and Astrophysics*, **571**, A92.

Blackwell, D. E., Petford, A. D., and Shallis, M. J. 1980. Use of the infra-red flux method for determining stellar effective temperatures and angular diameters – The stellar temperature scale. *Astronomy and Astrophysics*, **82**, 249–252.

Blackwell, D. E., Petford, A. D., Arribas, S., Haddock, D. J., and Selby, M. J. 1990. Determination of temperatures and angular diameters of 114 F-M stars using the infrared flux method (IRFM). *Astronomy and Astrophysics*, **232**, 396–410.

Blome, H.-J., Chicone, C., Hehl, F. W., and Mashhoon, B. 2010. Nonlocal modification of Newtonian gravity. *Physical Review D*, **81**, 065020.

Blumenthal, G. R., Faber, S. M., Flores, R., and Primack, J. R. 1986. Contraction of dark matter galactic halos due to baryonic infall. *The Astrophysical Journal*, **301**, 27.

Bode, P., Ostriker, J. P., and Vikhlinin, A. 2009. Exploring the energetics of intracluster gas with a simple and accurate model. *The Astrophysical Journal*, **700**, 989–999.

Bodenheimer, P. 1965. Studies in stellar evolution. II. Lithium depletion during the pre-main contraction. *The Astrophysical Journal*, **142**, 451–461.

Boesgaard, A. M., and Steigman, G. 1985. Big bang nucleosynthesis: Theories and observations. *Annual Reviews of Astronomy and Astrophysics*, **23**, 319–378.

Bohr, N. 1913. On the constitution of atoms and molecules. Part III. Systems containing several nuclei. *The London, Edinburgh, and Dublin Philosophical Magazine and Journal of Science*, **26**, 857–875.

Bosma, A. 1983. HI velocity fields and rotation curves. Pages 11–20 of: Athanassoula, E. (ed), *Internal Kinematics and Dynamics of Galaxies*. IAU Symposium, vol. 100.

Bovy, J., and Rix, H.-W. 2013. A direct dynamical measurement of the Milky Way's disk surface density profile, disk scale length, and dark matter profile at 4 kpc $\lesssim$ R $\lesssim$ 9 kpc. *The Astrophysical Journal*, **779**, 115.

Bovy, J., and Trelayne, S. 2012. On the local dark matter density. *The Astrophysical Journal*, **756**, 89.

Boylan-Kolchin, M., Bullock, J. S., and Kaplinghat, M. 2011. Too big to fail? The puzzling darkness of massive Milky Way subhaloes. *Monthly Notices of the Royal Astronomical Society*, **415**, L40–L44.

Brada, R., and Milgrom, M. 1995. Exact solutions and approximations of MOND fields of disc galaxies. *Monthly Notices of the Royal Astronomical Society*, **276**, 453–459.

Brada, R., and Milgrom, M. 1999. The modified Newtonian dynamics predicts an absolute maximum to the acceleration produced by "dark halos". *The Astrophysical Journal Letters*, **512**, L17–L18.

Bradford, J. D., Geha, M. C., and Blanton, M. R. 2015. A study in blue: The baryon content of isolated low-mass galaxies. *The Astrophysical Journal*, **809**, 146.

Bradford, J. D., Geha, M. C., and van den Bosch, F. C. 2016. A slippery slope: Systematic uncertainties in the line width baryonic Tully–Fisher relation. *The Astrophysical Journal*, **832**(1), 11.

Bridle, S. L., Zehavi, I., Dekel, A., Lahav, O., Hobson, M. P., and Lasenby, A. N. 2001. Cosmological parameters from velocities, cosmic microwave background and supernovae. *Monthly Notices of the Royal Astronomical Society*, **321**, 333–340.

Bristow, P. D., and Phillipps, S. 1994. On the baryon content of the universe. *Monthly Notices of the Royal Astronomical Sociey*, **267**, 13.

Broeils, A. H. 1992. The mass distribution of the dwarf spiral NGC 1560. *Astronomy and Astrophysics*, **256**, 19–32.

Broggini, C., Canton, L., Fiorentini, G., and Villante, F. L. 2012. The cosmological ^{7}Li problem from a nuclear physics perspective. *Journal of Cosmology and Astroparticle Physics*, **6**, 030.

Bromm, V., and Larson, R. B. 2004. The first stars. *Annual Review of Astronomy and Astrophysics*, **42**, 79–118.

Brown, L., and Schramm, D. N. 1988. The lithium isotope ratio in Population II halo dwarfs: A proposed test of the late decaying massive particle nucleosynthesis scenario. *The Astrophysical Journal Letters*, **329**, L103–L106.

Bruneton, J.-P., and Esposito-Farèse, G. 2007. Field-theoretical formulations of MOND-like gravity. *Physical Review D*, **76**, 124012.

Bull, P., Akrami, Y., Adamek, J., Baker, T., Bellini, E., Beltrán Jiménez, J., Bentivegna, E., Camera, S., Clesse, S., and Davis, J. H. 2016. Beyond Λ CDM: Problems, solutions, and the road ahead. *Physics of the Dark Universe*, **12**, 56–99.

Bullock, J. S. 2010. Notes on the missing satellites problem. *ArXiv e-prints*, arXiv:1009.4505.

Bullock, J. S., and Boylan-Kolchin, M. 2017. Small-scale challenges to the ΛCDM paradigm. *Annual Review of Astronomy and Astrophysics*, **55**, 343–387.

Bullock, J. S., Kolatt, T. S., Sigad, Y., Somerville, R. S., Kravtsov, A. V., Klypin, A. A., Primack, J. R., and Dekel, A. 2001. Profiles of dark haloes: Evolution, scatter and environment. *Monthly Notices of the Royal Astronomical Society*, **321**, 559–575.

Burles, S., and Tytler, D. 1998a. The deuterium abundance toward Q1937-1009. *The Astrophysical Journal*, **499**, 699–712.

Burles, S., and Tytler, D. 1998b. The deuterium abundance toward QSO 1009+2956. *The Astrophysical Journal*, **507**, 732–744.

Campbell, N. 1921. *What is Science?* Methuen.

Cappellari, M., Bacon, R., Bureau, M., Damen, M. C., Davies, R. L., de Zeeuw, P. T., Emsellem, E., Falcón-Barroso, J., Krajnović, D., Kuntschner, H., McDermid, R. M., Peletier, R. F., Sarzi, M., van den Bosch, R. C. E., and van de Ven, G. 2006. The SAURON project–IV. The mass-to-light ratio, the virial mass estimator and the Fundamental Plane of elliptical and lenticular galaxies. *Monthly Notices of the Royal Astronomical Society*, **366**, 1126–1150.

Cardone, V. F., Angus, G., Diaferio, A., Tortora, C., and Molinaro, R. 2011. The modified Newtonian dynamics fundamental plane. *Monthly Notices of the Royal Astronomical Society*, **412**, 2617–2630.

Carrier, M. 1988. On novel facts: A discussion of criteria for non-ad-hoc-ness in the methodology of scientific research programmes. *Zeitschrift für allgemeine Wissenschaftstherorie*, **19**, 205–231.

Carrier, M. 2002. Explaining scientific progress: Lakatos' methodological account of Kuhnian patterns of theory change. In Kampis, G., Kvasz, L., and Stöltzner, M. (eds), *Appraising Lakatos: Mathematics, Methodology and the Man*. Kluwer.

Carroll, S. 2014. Falsifiability. In *2014: What Scientific Idea is Ready for Retirement?* https://www.edge.org/response-detail/25322/.

Casagrande, L., Ramírez, I., Meléndez, J., Bessell, M., and Asplund, M. 2010. An absolutely calibrated T_{eff} scale from the infrared flux method. Dwarfs and subgiants. *Astronomy and Astrophysics*, **512**, A54.

Catinella, B., Kauffmann, G., Schiminovich, D., Lemonias, J., Scannapieco, C., Wang, J., Fabello, S., Hummels, C., Moran, S. M., Wu, R., Cooper, A. P., Giovanelli, R., Haynes, M. P., Heckman, T. M., and Saintonge, A. 2012. The GALEX Arecibo SDSS survey–IV. Baryonic mass–velocity–size relations of massive galaxies. *Monthly Notices of the Royal Astronomical Society*, **420**, 1959–1976.

Cattaneo, A., Tollet, E., Kucukbas, M., Mamon, G. A., Guiderdoni, B., Blaizot, J., Devriendt, J. E. G., Dekel, A., and Thob, A. C. R. 2017. The new semi-analytic code GalICS 2.0: Reproducing the galaxy stellar mass function and the Tully–Fisher relation simultaneously. *Monthly Notices of the Royal Astronomical Society*, **471**, 1401–1427.

Chakraborty, N., Fields, B. D., and Olive, K. A. 2011. Resonant destruction as a possible solution to the cosmological lithium problem. *Physical Review D*, **83**, 063006.

Chamcham, K., Silk, J., Barrow, J. D., and Saunders, S. 2017. *The Philosophy of Cosmology*. Cambridge University Press.

Chandrasekhar, S. 1967. *An Introduction to the Study of Stellar Structure*. Dover.

Chang, C.-K., Ko, C.-M., and Peng, T.-H. 2011. Information on the Milky Way from the Two Micron All Sky Survey whole sky star count: The structure parameters. *The Astrophysical Journal*, **740**, 34.

Chiu, M.-C., Ko, C.-M., and Shu, C. 2017. Origin of the fundamental plane of elliptical galaxies in the Coma cluster without fine-tuning. *Physical Review D*, **95**, 063020.

Ciotti, L., Lanzoni, B., and Renzini, A. 1996. The tilt of the fundamental plane of elliptical galaxies – I. Exploring dynamical and structural effects. *Monthly Notices of the Royal Astronomical Society*, **282**, 1–12.

Civitarese, O., and Mosquera, M. E. 2013. Nuclear structure constrains on resonant energies: A solution of the cosmological ^{7}Li problem? *Nuclear Physics A*, **898**, 1–13.

Clark, P. 1976. Atomism versus thermodynamics. Pages 41–106 of: Howson, C. (ed), *Method and Appraisal in the Physical Sciences*. Cambridge University Press.

Clifton, T., Ferreira, P. G., Padilla, A., and Skordis, C. 2012. Modified gravity and cosmology. *Physics Reports*, **513**, 1–189.

Coc, A. 2013. Primordial nucleosynthesis. *Acta Physica Polonica B*, **44**, 521.

Coc, A., and Vangioni, E. 2010. Big-Bang nucleosynthesis with updated nuclear data. *Journal of Physics Conference Series*, **202**, 012001.

Coc, A., and Vangioni, E. 2017. Primordial nucleosynthesis. *International Journal of Modern Physics E*, **26**, 1741002.

Coc, A., Vangioni-Flam, E., Descouvemont, P., Adahchour, A., and Angulo, C. 2004. Updated big bang nucleosynthesis compared with *Wilkinson Microwave Anisotropy*

Probe observations and the abundance of light elements. *The Astrophysical Journal*, **600**, 544–552.

Coc, A., Goriely, S., Xu, Y., Saimpert, M., and Vangioni, E. 2012. Standard big bang nucleosynthesis up to CNO with an improved extended nuclear network. *The Astrophysical Journal*, **744**, 158.

Coc, A., Uzan, J.-P., and Vangioni, E. 2013. Standard big-bang nucleosynthesis after Planck. *ArXiv e-prints*, 1307.6955.

Coc, A., Pospelov, M., Uzan, J.-P., and Vangioni, E. 2014. Modified big bang nucleosynthesis with nonstandard neutron sources. *Physical Review D*, **90**, 085018.

Coc, A., Petitjean, P., Uzan, J.-P., Vangioni, E., Descouvemont, P., Iliadis, C., and Longland, R. 2015. New reaction rates for improved primordial D/H calculation and the cosmic evolution of deuterium. *Physical Review D*, **92**, 123526.

Coleman, P. H., and Pietronero, L. 1992. The fractal structure of the universe. *Physics Reports*, **213**, 311–389.

Collins, M. L. M., Chapman, S. C., Rich, R. M., Ibata, R. A., Martin, N. F., Irwin, M. J., Bate, N. F., Lewis, G. F., Peñarrubia, J., Arimoto, N., Casey, C. M., Ferguson, A. M. N., Koch, A., McConnachie, A. W., and Tanvir, N. 2013. A kinematic study of the Andromeda dwarf spheroidal system. *The Astrophysical Journal*, **768**, 172.

Conn, A. R., Ibata, R. A., Lewis, G. F., Parker, Q. A., Zucker, D. B., Martin, N. F., McConnachie, A. W., Irwin, M. J., Tanvir, N., and Fardal, M. A. 2012. A Bayesian approach to locating the red giant branch tip magnitude. II. Distances to the satellites of M31. *The Astrophysical Journal*, **758**, 11.

Contaldi, C. R., Wiseman, T., and Withers, B. 2008. TeVeS gets caught on caustics. *Physical Review D*, **78**, 044034.

Cooke, R., Pettini, M., Jorgenson, R. A., Murphy, M. T., and Steidel, C. C. 2014. Precision measures of the primordial abundance of deuterium. *The Astrophysical Journal*, **781**.

Copi, C. J., Schramm, D. N., and Turner, M. S. 1995. Big-bang nucleosynthesis and the baryon density of the universe. *Science*, **267**, 192–199.

Courteau, S., McDonald, M., Widrow, L. M., and Holtzman, J. 2007. The bulge–halo connection in galaxies: A physical interpretation of the V_c–σ_0 relation. *The Astrophysical Journal Letters*, **655**(1), L21–L24.

Creasey, P., Sameie, O., Sales, L. V., Yu, H.-B., Vogelsberger, M., and Zavala, J. 2017. Spreading out and staying sharp: Creating diverse rotation curves via baryonic and self-interaction effects. *Monthly Notices of the Royal Astronomical Society*, **468**, 2283–2295.

Crézé, M., Chereul, E., Bienayme, O., and Pichon, C. 1998. The distribution of nearby stars in phase space mapped by Hipparcos. I. The potential well and local dynamical mass. *Astronomy and Astrophysics*, **329**, 920–936.

Crighton, N. H. M., Webb, J. K., Ortiz-Gil, A., and Fernández-Soto, A. 2004. Deuterium/hydrogen in a new Lyman limit absorption system at $z = 3.256$ towards PKS1937-1009. *Monthly Notices of the Royal Astronomical Society*, **355**, 1042–1052.

Crupi, V., Festa, R., and Buttasi, C. 2010. Towards a grammar of Bayesian confirmation. Pages 73–93 of: Suárez, M., Dorato, M., and Rédei, M. (eds), *EPSA Epistemology and Methodology of Science: Launch of the European Philosophy of Science Association*. Springer.

Cuddeford, P., and Amendt, P. 1991. Extended stellar hydrodynamics for galactic discs. II. *Monthly Notices of the Royal Astronomical Society*, **253**, 427–444.

Cyburt, R. H., Fields, B. D., and Olive, K. A. 2008. An update on the big bang nucleosynthesis prediction for ^{7}Li: The problem worsens. *Journal of Cosmology and Astroparticle Physics*, **11**, 012.

Daly, R. A., and Djorgovski, S. G. 2005. Direct determinations of the redshift behavior of the pressure, energy density, and equation of state of the dark energy and the acceleration of the universe. *International Journal of Modern Physics A*, **20**, 1113–1120.

Davies, J. I. 1990. Visibility and the selection of galaxies. *Monthly Notices of the Royal Astronomical Society*, **244**, 8–24.

de Bernardis, P., Ade, P. A. R., Bock, J. J., Bond, J. R., Borrill, J., Boscaleri, A., Coble, K., Crill, B. P., De Gasperis, G., Farese, P. C., Ferreira, P. G., Ganga, K., Giacometti, M., Hivon, E., Hristov, V. V., Iacoangeli, A., Jaffe, A. H., Lange, A. E., Martinis, L., Masi, S., Mason, P. V., Mauskopf, P. D., Melchiorri, A., Miglio, L., Montroy, T., Netterfield, C. B., Pascale, E., Piacentini, F., Pogosyan, D., Prunet, S., Rao, S., Romeo, G., Ruhl, J. E., Scaramuzzi, F., Sforna, D., and Vittorio, N. 2000. A flat universe from high-resolution maps of the cosmic microwave background radiation. *Nature*, **404**, 955–959.

de Bernardis, P., Ade, P. A. R., Bock, J. J., Bond, J. R., Borrill, J., Boscaleri, A., Coble, K., Contaldi, C. R., Crill, B. P., De Troia, G., Farese, P., Ganga, K., Giacometti, M., Hivon, E., Hristov, V. V., Iacoangeli, A., Jaffe, A. H., Jones, W. C., Lange, A. E., Martinis, L., Masi, S., Mason, P., Mauskopf, P. D., Melchiorri, A., Montroy, T., Netterfield, C. B., Pascale, E., Piacentini, F., Pogosyan, D., Polenta, G., Pongetti, F., Prunet, S., Romeo, G., Ruhl, J. E., and Scaramuzzi, F. 2002. Multiple peaks in the angular power spectrum of the cosmic microwave background: Significance and consequences for cosmology. *The Astrophysical Journal*, **564**, 559–566.

de Blok, W. J. G. 2010. The core-cusp problem. *Advances in Astronomy*, **2010**, 789293.

de Blok, W. J. G., and McGaugh, S. S. 1997. The dark and visible matter content of low surface brightness disc galaxies. *Monthly Notices of the Royal Astronomical Society*, **290**, 533–552.

de Blok, W. J. G., McGaugh, S. S., and Rubin, V. C. 2001. High-resolution rotation curves of low surface brightness galaxies. II. Mass models. *The Astronomical Journal*, **122**, 2396–2427.

Deffayet, C., Esposito-Farèse, G., and Woodard, R. P. 2011. Nonlocal metric formulations of modified Newtonian dynamics with sufficient lensing. *Physical Review D*, **84**, 124054.

Deffayet, C., Esposito-Farèse, G., and Woodard, R. P. 2014. Field equations and cosmology for a class of nonlocal metric models of MOND. *Physical Review D*, **90**, 064038.

Del Popolo, A., and Le Delliou, M. 2017. Small scale problems of the ΛCDM model: A short review. *Galaxies*, **5**, 17–46.

Deliyannis, C. P., Demarque, P., and Kawaler, S. D. 1990. Lithium in halo stars from standard stellar evolution. *The Astrophysical Journal Supplement*, **73**, 21–65.

Descouvemont, P., Adahchour, A., Angulo, C., Coc, A., and Vangioni-Flam, E. 2004. Compilation and R-matrix analysis of big bang nuclear reaction rates. *Atomic Data and Nuclear Data Tables*, **88**, 203–236.

Deser, S., and Woodard, R. P. 2007. Nonlocal cosmology. *Physical Review Letters*, **99**, 111301.

Desmond, H. 2017. The scatter, residual correlations and curvature of the SPARC baryonic Tully–Fisher relation. *Monthly Notices of the Royal Astronomical Society*, **472**, L35–L39.

Desmond, H., and Wechsler, R. H. 2015. The Tully–Fisher and mass–size relations from halo abundance matching. *Monthly Notices of the Royal Astronomical Society*, **454**, 322–343.

Desmond, H., and Wechsler, R. H. 2017. The Faber–Jackson relation and Fundamental Plane from halo abundance matching. *Monthly Notices of the Royal Astronomical Society*, **465**, 820–833.

Di Leva, A., Gialanella, L., and Strieder, F. 2016. Experimental status of ^{7}Be production and destruction at astrophysical relevant energies. *Journal of Physics: Conference Series*, **665**, 012002.

Dodelson, S. 2011. The real problem with MOND. *International Journal of Modern Physics D*, **20**, 2749–2753.

Dodelson, S., and Widrow, L. M. 1994. Sterile neutrinos as dark matter. *Physical Review Letters*, **72**, 17–20.

Donato, F., Gentile, G., Salucci, P., Frigerio Martins, C., Wilkinson, M. I., Gilmore, G., Grebel, E. K., Koch, A., and Wyse, R. 2009. A constant dark matter halo surface density in galaxies. *Monthly Notices of the Royal Astronomical Society*, **397**, 1169–1176.

Drewes, M. 2013. The phenomenology of right handed neutrinos. *International Journal of Modern Physics E*, **22**, 1330019–593.

Duncan, D. K. 1981. Lithium abundances, K line emission and ages of nearby solar type stars. *The Astrophysical Journal*, **248**, 651–669.

Dutton, A. A., Conroy, C., van den Bosch, F. C., Simard, L., Mendel, J. T., Courteau, S., Dekel, A., More, S., and Prada, F. 2011. Dark halo response and the stellar initial mass function in early-type and late-type galaxies. *Monthly Notices of the Royal Astronomical Society*, **416**, 322–345.

Dutton, A. A., Macciò, A. V., Mendel, J. T., and Simard, L. 2013. Universal IMF versus dark halo response in early-type galaxies: Breaking the degeneracy with the Fundamental Plane. *Monthly Notices of the Royal Astronomical Society*, **432**(3), 2496–2511.

Earman, J. 1996. *Bayes or Bust? A Critical Examination of Bayesian Confirmation Theory*. MIT Press.

Einasto, J. 2005. Dark matter: Early considerations. Page 241 of: Blanchard, A., and Signore., M. (eds), *Frontiers of Cosmology*, vol. 187.

Ellis, G. F. R., and Hawking, S. W. 1999. *The Large Scale Structure of Space-Time*. Cambridge University Press.

Epstein, R. I., Lattimer, J. M., and Schramm, D. N. 1976. The origin of deuterium. *Nature*, **263**, 198–202.

Faber, S. M., and Gallagher, J. S. 1979. Masses and mass-to-light ratios of galaxies. *Annual Review of Astronomy and Astrophysics*, **17**, 135–187.

Faber, S. M., and Jackson, R. E. 1976. Velocity dispersions and mass-to-light ratios for elliptical galaxies. *The Astrophysical Journal*, **204**, 668–683.

Fabian, A. C. 1991. On the baryon content of the Shapley Supercluster. *Monthly Notices of the Royal Astronomical Society*, **253**, 29P.

Fabjan, D., Borgani, S., Tornatore, L., Saro, A., Murante, G., and Dolag, K. 2010. Simulating the effect of active galactic nuclei feedback on the metal enrichment of galaxy clusters. *Monthly Notices of the Royal Astronomical Society*, **401**(3), 1670–1690.

Famaey, B., and McGaugh, S. 2012. Modified Newtonian dynamics (MOND): Observational phenomenology and relativistic extensions. *Living Reviews in Relativity*, **15**.

Famaey, B., Gentile, G., Bruneton, J.-P., and Zhao, H. 2007. Insight into the baryon–gravity relation in galaxies. *Physical Review D*, **75**, 063002.

Felten, J. E. 1984. Milgrom's revision of Newton's laws: Dynamical and cosmological consequences. *The Astrophysical Journal*, **286**, 3–6.

Feyerabend, P. 1970. Consolations for the specialist. Pages 197–230 of: Lakatos, I., and Musgrave, A. (eds), *Criticism and the Growth of Knowledge*. Cambridge University Press.

Feyerabend, P. 1975. *Against Method: Outline of an Anarchistic Theory of Knowledge*. Verso.

Fields, B. D. 2011. The primordial lithium problem. *Annual Review of Nuclear and Particle Science*, **61**, 47–68.

Fields, B. D., Olive, K. A., and Vangioni-Flam, E. 2005. Implications of a new temperature scale for halo dwarfs on LiBeB and chemical evolution. *The Astrophysical Journal*, **623**, 1083–1091.

Flores, R. A., and Primack, J. R. 1994. Observational and theoretical constraints on singular dark matter halos. *The Astrophysical Journal Letters*, **427**, L1.

Flynn, C., Holmberg, J., Portinari, L., Fuchs, B., and Jahreiß, H. 2006. On the mass-to-light ratio of the local Galactic disc and the optical luminosity of the Galaxy. *Monthly Notices of the Royal Astronomical Society*, **372**, 1149–1160.

Forbes, D. A., and Kroupa, P. 2011. What is a galaxy? Cast your vote here. *Publications of the Astronomical Society of Australia*, **28**, 77–82.

Frebel, A., and Norris, J. E. 2013. Metal-poor stars and the chemical enrichment of the universe. Page 55 of: Oswalt, T. D., and Gilmore, G. (eds), *Planets, Stars and Stellar Systems. Volume 5: Galactic Structure and Stellar Populations*.

Freeman, K. C. 1970. On the disks of spiral and S0 galaxies. *The Astrophysical Journal*, **160**, 811.

Freeman, K. C. 1999. Historical introduction. Pages 3–8 of: Davies, J. I., Impey, C., and Phillips, S. (eds), *The Low Surface Brightness Universe*. Astronomical Society of the Pacific Conference Series, vol. 170.

Freese, K. 2017. Status of dark matter in the universe. *International Journal of Modern Physics D*, **26**, 1730012.

Fukugita, M., and Peebles, P. J. E. 2004. The cosmic energy inventory. *The Astrophysical Journal*, **616**, 643–668.

Fukugita, M., Hogan, C. J., and Peebles, P. J. E. 1998. The cosmic baryon budget. *The Astrophysical Journal*, **503**, 518–530.

Garbari, S., Liu, C., Read, J. I., and Lake, G. 2012. A new determination of the local dark matter density from the kinematics of K dwarfs. *Monthly Notices of the Royal Astronomical Society*, **425**, 1445–1458.

Gardner, M. 1982. Predicting novel facts. *The British Journal for the Philosophy of Science*, **33**, 1–15.

Garrison-Kimmel, S., Boylan-Kolchin, M., Bullock, J. S., and Kirby, E. N. 2014. Too big to fail in the Local Group. *Monthly Notices of the Royal Astronomical Society*, **444**, 222–236.

Geha, M., Blanton, M. R., Masjedi, M., and West, A. A. 2006. The baryon content of extremely low mass dwarf galaxies. *The Astrophysical Journal*, **653**, 240–254.

Gehren, T. 1981. The temperature scale of solar-type stars. *Astronomy and Astrophysics*, **100**, 97–106.

Gentile, G., Salucci, P., Klein, U., Vergani, D., and Kalberla, P. 2004. The cored distribution of dark matter in spiral galaxies. *Monthly Notices of the Royal Astronomical Society*, **351**, 903–922.

Gentile, G., Salucci, P., Klein, U., and Granato, G. L. 2007. NGC 3741: The dark halo profile from the most extended rotation curve. *Monthly Notices of the Royal Astronomical Society*, **375**, 199–212.

Gerbal, D., Durret, F., Lachieze-Rey, M., and Lima-Neto, G. 1992. Analysis of X-ray galaxy clusters in the framework of modified Newtonian dynamics. *Astronomy and Astrophysics*, **262**, 395–400.

Gilmore, G., Wilkinson, M. I., Wyse, R. F. G., Kleyna, J. T., Koch, A., Wyn Evans, N., and Grebel, E. K. 2007. The observed properties of dark matter on small spatial scales. *The Astrophysical Journal*, **663**, 948–959.

Giunti, C., and Laveder, M. 2008. ν_e disappearance in MiniBooNE. *Physical Review D*, **77**, 093002.

Glass, J. C., and Johnson, W. 1989. *Economics: Progression, Stagnation or Degeneration?* Iowa State University Press.

Glymour, C. 1980. *Theory and Evidence*. Princeton University Press.

Gnedin, N. I., and Ostriker, J. P. 1992. Light element nucleosynthesis: A false clue? *The Astrophysical Journal*, **400**, 1–20.

Goldstein, A., Veres, P., Burns, E., Briggs, M. S., Hamburg, R., Kocevski, D., Wilson-Hodge, C. A., Preece, R. D., Poolakkil, S., Roberts, O. J., Hui, C. M., Connaughton, V., Racusin, J., von Kienlin, A., Canton, T. Dal, Christensen, N., Littenberg, T., Siellez, K., Blackburn, L., Broida, J., Bissaldi, E., Cleveland, W. H., Gibby, M. H., Giles, M. M., Kippen, R. M., McBreen, S., McEnery, J., Meegan, C. A., Paciesas, W. S., and Stanbro, M. 2017. An ordinary short gamma-ray burst with extraordinary implications: Fermi-GBM detection of GRB 170817A. *The Astrophysical Journal*, **848**, L14.

Gonzalez, A. H., Zaritsky, D., and Zabludoff, A. I. 2007. A census of baryons in galaxy clusters and groups. *The Astrophysical Journal*, **666**, 147–155.

González Hernández, J. I., and Bonifacio, P. 2009. A new implementation of the infrared flux method using the 2MASS catalogue. *Astronomy and Astrophysics*, **497**, 497–509.

Governato, F., Brook, C., Mayer, L., Brooks, A., Rhee, G., Wadsley, J., Jonsson, P., Willman, B., Stinson, G., Quinn, T., and Madau, P. 2010. Bulgeless dwarf galaxies and dark matter cores from supernova-driven outflows. *Nature*, **463**, 203–206.

Greene, G. L., and Geltenbort, P. 2016. The neutron enigma. *Scientific American*, **314**, 36–41.

Greiter, M., Wilczek, F., and Witten, E. 1989. Hydrodynamic relations in superconductivity. *Modern Physics Letters B*, **3**, 903–918.

Grünbaum, A. 1976. Is falsifiability the touchstone of scientific rationality? Karl Popper versus inductivism. Pages 213–252 of: Cohen, R. S., Feyerabend, P. K., and Wartofsky, M. W. (eds), *Essays in Memory of Imre Lakatos*. D. Reidel.

Grünbaum, A. 1989. The degeneration of Popper's theory of demarcation. Pages 141–161 of: D'Agostino, F., and Jarvie, I. C. (eds), *Freedom and Rationality. Essays in Honor of John Watkins*. Kluwer.

Gruyters, P., Korn, A. J., Richard, O., Grundahl, F., Collet, R., Mashonkina, L. I., Osorio, Y., and Barklem, P. S. 2013. Atomic diffusion and mixing in old stars. IV. Weak abundance trends in the globular cluster NGC 6752. *Astronomy and Astrophysics*, **555**, A31.

Gurovich, S., Freeman, K., Jerjen, H., Staveley-Smith, L., and Puerari, I. 2010. The slope of the baryonic Tully–Fisher relation. *The Astronomical Journal*, **140**, 663–676.

Hammache, F., Coc, A., de Séréville, N., Stefan, I., Roussel, P., Ancelin, S., Assié, M., Audouin, L., Beaumel, D., Franchoo, S., Fernandez-Dominguez, B., Fox, S., Hamadache, C., Kiener, J., Laird, A., Le Crom, B., Lefebvre-Schuhl, A., Lefebvre, L., Matea, I., Matta, A., Mavilla, G., Mrazek, J., Morfouace, P., de Oliveira Santos, F., Parikh, A., Perrot, L., Sanchez-Benitez, A. M., Suzuki, D., Tatischeff, V., Ujic, P., and Vandebrouck, M. 2013. Search for new resonant states in ^{10}C and ^{11}C and their impact on the cosmological lithium problem. *Physical Review C*, **88**, 062802.

Hehl, F. W., and Mashhoon, B. 2009a. Formal framework for a nonlocal generalization of Einstein's theory of gravitation. *Physical Review D*, **79**, 064028.

Hehl, F. W., and Mashhoon, B. 2009b. Nonlocal gravity simulates dark matter. *Physics Letters B*, **673**, 279–282.

Hempel, C. G. 1937. Le problème de la vérité. *Theoria*, **3**, 206–246.

Hempel, C. G. 1945. Studies in the logic of confirmation. *Mind*, **54**, 1–26, 97–121.

Hempel, C. G. 1965. *Aspects of Scientific Explanation*. The Free Press.

Hempel, C. G. 1973. The meaning of theoretical terms: A critique of the standard empiricist construal. Pages 351–378 of: Suppes, P., Henkin, L., Joya, A., and Moisil, G. C. (eds), *Logic, Methodology and Philosophy of Science*, vol. 4. North Holland Publishers.

Herschel, J. 1842. *Preliminary Discourse on the Study of Natural Philosophy*. London.

Hill, E. R. 1960. The component of the galactic gravitational field perpendicular to the galactic plane, K_z. *Bulletin of the Astronomical Institutes of the Netherlands*, **15**, 1.

Hinshaw, G., Nolta, M. R., Bennett, C. L., Bean, R., Doré, O., Greason, M. R., Halpern, M., Hill, R. S., Jarosik, N., Kogut, A., Komatsu, E., Limon, M., Odegard, N., Meyer, S. S., Page, L., Peiris, H. V., Spergel, D. N., Tucker, G. S., Verde, L., Weiland, J. L.,

Wollack, E., and Wright, E. L. 2007. Three-year Wilkinson microwave anisotropy probe (WMAP) observations: Temperature analysis. *The Astrophysical Journal Supplement Series*, **170**, 288–334.

Hinshaw, G., Larson, D., Komatsu, E., Spergel, D. N., Bennett, C. L., Dunkley, J., Nolta, M. R., Halpern, M., Hill, R. S., Odegard, N., Page, L., Smith, K. M., Weiland, J. L., Gold, B., Jarosik, N., Kogut, A., Limon, M., Meyer, S. S., Tucker, G. S., Wollack, E., and Wright, E. L. 2013. Nine-year Wilkinson Microwave Anisotropy Probe (WMAP) observations: Cosmological parameter results. *The Astrophysical Journal Supplement*, **208**, 19.

Hodson, A. O., Zhao, H., Khoury, J., and Famaey, B. 2017. Galaxy clusters in the context of superfluid dark matter. *Astronomy and Astrophysics*, **607**, A108.

Holmberg, J., and Flynn, C. 2000. The local density of matter mapped by Hipparcos. *Monthly Notices of the Royal Astronomical Society*, **313**, 209–216.

Holmberg, J., and Flynn, C. 2004. The local surface density of disc matter mapped by Hipparcos. *Monthly Notices of the Royal Astronomical Society*, **352**, 440–446.

Hosford, A., Ryan, S. G., García Pérez, A. E., Norris, J. E., and Olive, K. A. 2009. Lithium abundances of halo dwarfs based on excitation temperature. I. Local thermodynamic equilibrium. *Astronomy and Astrophysics*, **493**, 601–612.

Hosford, A., García Pérez, A. E., Collet, R., Ryan, S. G., Norris, J. E., and Olive, K. A. 2010. Lithium abundances of halo dwarfs based on excitation temperatures. II. Non-local thermodynamic equilibrium. *Astronomy and Astrophysics*, **511**, A47.

Hosiasson-Lindenbaum, J. 1940. On confirmation. *Journal of Symbolic Logic*, **5**, 133–148.

Howson, C. (ed). 1976. *Method and Appraisal in the Physical Sciences*. Cambridge University Press.

Howson, C., and Urbach, P. (eds). 2006. *Scientific Reasoning: The Bayesian Approach*, 3rd edn. Open Court Publishing Company.

Hu, W., and Dodelson, S. 2002. Cosmic microwave background anisotropies. *Annual Review of Astronomy and Astrophysics*, **40**, 171–216.

Hu, W., Sugiyama, N., and Silk, J. 1997. The physics of microwave background anisotropies. *Nature*, **386**, 37–43.

Hull, D. L. 2010. *Science as a Process: An Evolutionary Account of the Social and Conceptual Development of Science*. Science and Its Conceptual Foundations series. University of Chicago Press.

Hume, D. 1739–40/1978. *A Treatise of Human Nature*. L. A. Selby-Bigge and P. H. Nidditch (eds.), 2nd edn. Clarendon Press, 1978.

Hume, D. 1748/1975. *An Enquiry concerning Human Understanding*. Cited from *Enquiries concerning Human Understanding and concerning the Principles of Morals*, L. A. Selby-Bigge and P. H. Nidditch (eds.), 3rd edn. Clarendon Press, 1975.

Ivanchik, A. V., Petitjean, P., Balashev, S. A., Srianand, R., Varshalovich, D. A., Ledoux, C., and Noterdaeme, P. 2010. HD molecules at high redshift: The absorption system at $z = 2.3377$ towards Q 1232 + 082. *Monthly Notices of the Royal Astronomical Society*, **404**, 1583–1590.

Jammer, M. 1966. *The Conceptual Development of Quantum Mechanics*. McGraw-Hill.

Jeans, J. H. 1914. Discussion on radiation. Pages 376–386 of: *Report of the 83rd Meeting of the British Association for the Advancement of Science*.

Jeans, J. H. 1922. The motions of stars in a Kapteyn universe. *Monthly Notices of the Royal Astronomical Society*, **82**, 122–132.

Jedamzik, K. 2004. Did something decay, evaporate, or annihilate during big bang nucleosynthesis? *Physical Review D*, **70**, 063524.

Jurieu, P. 1687. *Traité de la Nature et de la Grâce*. F. Halma.

Kadvany, J. 2001. *Imre Lakatos and the Guises of Reason*. Duke University Press.

Kahn, F. D., and Woltjer, L. 1959. Intergalactic matter and the galaxy. *The Astrophysical Journal*, **130**, 705.

Kahya, E. O. 2008. A decisive test to confirm or rule out the existence of dark matter emulators using gravitational wave observations. *Classical and Quantum Gravity*, **25**, 184008.

Kahya, E. O., and Woodard, R. P. 2007. A generic test of modified gravity models which emulate dark matter. *Physics Letters B*, **652**, 213–216.

Kaplinghat, M., and Turner, M. 2002. How cold dark matter theory explains Milgrom's law. *The Astrophysical Journal Letters*, **569**, L19–L22.

Kaplinghat, M., Chu, M., Haiman, Z., Holder, G. P., Knox, L., and Skordis, C. 2003. Probing the reionization history of the universe using the cosmic microwave background polarization. *The Astrophysical Journal*, **583**, 24–32.

Kapteyn, J. C. 1922. First attempt at a theory of the arrangement and motion of the sidereal system. *The Astrophysical Journal*, **55**, 302.

Karachentsev, I. D., Kaisina, E. I., and Kashibadze Nasonova, O. G. 2017. The local Tully–Fisher relation for dwarf galaxies. *The Astronomical Journal*, **153**(1), 6.

Keller, B. W., and Wadsley, J. W. 2017. ΛCDM is consistent with SPARC radial acceleration relation. *The Astrophysical Journal Letters*, **835**, L17.

Kent, S. M. 1987. Dark matter in spiral galaxies. II. Galaxies with H I rotation curves. *The Astronomical Journal*, **93**, 816.

Kernan, P. J., and Krauss, L. M. 1994. Refined big bang nucleosynthesis constraints on Ω_B and N_ν. *Physical Review Letters*, **72**, 3309–3312.

Kim, M., Rahat, M. H., Sayeb, M., Tan, L., Woodard, R. P., and Xu, B. 2016. Determining cosmology for a nonlocal realization of MOND. *Physical Review D*, **94**, 104009.

Kirkman, D., Tytler, D., Suzuki, N., O'Meara, J. M., and Lubin, D. 2003. The cosmological baryon density from the deuterium-to-hydrogen ratio in QSO absorption systems: D/H toward Q1243+3047. *The Astrophysical Journal Supplement*, **149**, 1–28.

Kirsebom, O. S., and Davids, B. 2011. One fewer solution to the cosmological lithium problem. *Physical Review C*, **84**, 058801.

Kitcher, P. 1995. Author's response. *Philosophy and Phenomenological Research*, **55**, 653–673.

Klypin, A., Gottlöber, S., Kravtsov, A. V., and Khokhlov, A. M. 1999. Galaxies in N-body simulations: Overcoming the overmerging problem. *The Astrophysical Journal*, **516**, 530–551.

Kolb, E. W., and Turner, M. S. 1994. *The Early Universe*. Addison-Wesley.

Kormendy, J., Drory, N., Bender, R., and Cornell, M. E. 2010. Bulgeless giant galaxies challenge our picture of galaxy formation by hierarchical clustering. *The Astrophysical Journal*, **723**, 54–80.

Korn, A. J. 2012. Shedding light on lithium evolution: The globular cluster perspective. *Memorie della Societa Astronomica Italiana Supplementi*, **22**, 64.

Korn, A. J., Grundahl, F., Richard, O., Barklem, P. S., Mashonkina, L., Collet, R., Piskunov, N., and Gustafsson, B. 2006. A probable stellar solution to the cosmological lithium discrepancy. *Nature*, **442**, 657–659.

Korn, A. J., Grundahl, F., Richard, O., Mashonkina, L., Barklem, P. S., Collet, R., Gustafsson, B., and Piskunov, N. 2007. Atomic diffusion and mixing in old stars. I. Very Large Telescope FLAMES-UVES observations of stars in NGC 6397. *The Astrophysical Journal*, **671**, 402–419.

Kourkchi, E., Khosroshahi, H. G., Carter, D., and Mobasher, B. 2012. Dwarf galaxies in the Coma cluster – II. Spectroscopic and photometric fundamental planes. *Monthly Notices of the Royal Astronomical Society*, **420**, 2835–2850.

Kowalski, M., Rubin, D., Aldering, G., Agostinho, R. J., Amadon, A., Amanullah, R., Balland, C., Barbary, K., Blanc, G., Challis, P. J., Conley, A., Connolly, N. V., Covarrubias, R., Dawson, K. S., Deustua, S. E., Ellis, R., Fabbro, S., Fadeyev, V., Fan, X., Farris, B., Folatelli, G., Frye, B. L., Garavini, G., Gates, E. L., Germany, L., Goldhaber, G., Goldman, B., Goobar, A., Groom, D. E., Haissinski, J., Hardin, D.,

Hook, I., Kent, S., Kim, A. G., Knop, R. A., Lidman, C., Linder, E. V., Mendez, J., Meyers, J., Miller, G. J., Moniez, M., Mourão, A. M., Newberg, H., Nobili, S., Nugent, P. E., Pain, R., Perdereau, O., Perlmutter, S., Phillips, M. M., Prasad, V., Quimby, R., Regnault, N., Rich, J., Rubenstein, E. P., Ruiz-Lapuente, P., Santos, F. D., Schaefer, B. E., Schommer, R. A., Smith, R. C., Soderberg, A. M., Spadafora, A. L., Strolger, L. G., Strovink, M., Suntzeff, N. B., Suzuki, N., Thomas, R. C., Walton, N. A., Wang, L., Wood-Vasey, W. M., and Yun, J. L. 2008. Improved cosmological constraints from new, old, and combined supernova data sets. *The Astrophysical Journal*, **686**, 749–778.

Kragh, H. 2012. "The most philosophically of all the sciences": Karl Popper and physical cosmology. *Perspectives on Science*, **21**, 325–357.

Kroupa, P. 2012. The dark matter crisis: falsification of the current standard model of cosmology. *Publications of the Astronomical Society of Australia*, **29**, 395–433.

Kroupa, P. 2014. The planar satellite distributions around Andromeda, the Milky Way and other galaxies, and their implications for fundamental physics. Page 183 of: Iodice, E., and Corsini, E. M. (eds), *Multi-Spin Galaxies*. Astronomical Society of the Pacific Conference Series, vol. 486.

Kroupa, P. 2015a. Galaxies as simple dynamical systems: Observational data disfavor dark matter and stochastic star formation. *Canadian Journal of Physics*, **93**, 169–202.

Kroupa, P. 2015b. Lessons from the Local Group (and beyond) on dark matter. Pages 337–352 of: Freeman, K., Elmegreen, B., Block, D., and Woolway, M. (eds), *Lessons from the Local Group*. Springer International Publishing.

Kroupa, P., Famaey, B., de Boer, K. S., Dabringhausen, J., Pawlowski, M. S., Boily, C. M., Jerjen, H., Forbes, D., Hensler, G., and Metz, M. 2010. Local-Group tests of dark-matter concordance cosmology: Towards a new paradigm for structure formation. *Astronomy and Astrophysics*, **523**, A32.

Kroupa, P., Pawlowski, M., and Milgrom, M. 2012. The failures of the standard model of cosmology require a new paradigm. *International Journal of Modern Physics D*, **21**, 1230003.

Kuhn, T. S. 1962. *The Structure of Scientific Revolutions*. The University of Chicago Press.

Kuhn, T. S. 1970. Logic of discovery or psychology of research? Pages 1–24 of: Lakatos, I., and Musgrave, A. (eds), *Criticism and the Growth of Knowledge*. Cambridge University Press.

Kuijken, K., and Gilmore, G. 1989a. The mass distribution in the galactic disc – II. Determination of the surface mass density of the galactic disc near the Sun. *Monthly Notices of the Royal Astronomical Society*, **239**, 605–649.

Kuijken, K., and Gilmore, G. 1989b. The mass distribution in the galactic disc – III. The local volume mass density. *Monthly Notices of the Royal Astronomical Society*, **239**, 651–664.

Kuijken, K., and Gilmore, G. 1989c. The mass distribution in the galactic disc – I. A technique to determine the integral surface mass density of the disc near the sun. *Monthly Notices of the Royal Astronomical Society*, **239**, 571–603.

Kuijken, K., and Gilmore, G. 1991. The galactic disk surface mass density and the galactic force $K(z)$ at $z = 1.1$ kiloparsecs. *The Astrophysical Journal Letters*, **367**, L9–L13.

Kusakabe, Motohiko, Cheoun, Myung-Ki, and Kim, K. S. 2014. General limit on the relation between abundances of D and ^{7}Li in big bang nucleosynthesis with nucleon injections. *Physical Review D*, **90**, 045009.

Kyburg, H. E., Jr. 1970. *Probability and Inductive Logic*. Macmillan.

Lakatos, I. 1970. Falsification and the methodology of scientific research programmes. In Lakatos (1978), 8–101.

Lakatos, I. 1971. History of science and its rational reconstructions. In Lakatos (1978), 102–138.

Lakatos, I. 1973. Introduction: Science and pseudoscience. In Lakatos (1978), 1–7.

Lakatos, I. 1974. Popper on demarcation and induction. In Lakatos (1978), 139–167.

Lakatos, I. 1978. *The Methodology of Scientific Research Programmes (Philosophical Papers Volume I)*. Worrall, J. and Currie, G. (eds). Cambridge University Press.

Lakatos, I., and Zahar, E. 1976. Why did Copernicus's research programme supersede Ptolemy's? In Lakatos (1978), 168–192.

Lakatos, I., Worrall, J., and Zahar, E. 1976. *Proofs and Refutations: The Logic of Mathematical Discovery*. Philosophical Papers, vol. 1. Cambridge University Press.

Lake, G. 1989. Testing modifications of gravity. *The Astrophysical Journal Letters*, **345**, L17.

Lambert, D. L. 2004. Lithium in very metal-poor dwarf stars: Problems for standard big bang nucleosynthesis? Pages 206–223 of: Allen, R. E., Nanopoulos, D. V., and Pope, C. N. (eds), *The New Cosmology: Conference on Strings and Cosmology*. American Institute of Physics Conference Series, vol. 743.

Landau, L. D. 1941. The theory of superfluidity of helium-II. *Journal of Physics - USSR*, **5**, 71.

Landau, L. D. 1947. On the theory of superfluidity of helium-II. *Journal of Physics - USSR*, **11**, 91.

Lange, A. E., Ade, P. A., Bock, J. J., Bond, J. R., Borrill, J., Boscaleri, A., Coble, K., Crill, B. P., de Bernardis, P., Farese, P., Ferreira, P., Ganga, K., Giacometti, M., Hivon, E., Hristov, V. V., Iacoangeli, A., Jaffe, A. H., Martinis, L., Masi, S., Mauskopf, P. D., Melchiorri, A., Montroy, T., Netterfield, C. B., Pascale, E., Piacentini, F., Pogosyan, D., Prunet, S., Rao, S., Romeo, G., Ruhl, J. E., Scaramuzzi, F., and Sforna, D. 2001. Cosmological parameters from the first results of Boomerang. *Physical Review D*, **63**, 042001.

Laudan, L. 1983. The demise of the demarcation problem. Pages 111–128 of: Cohen, R. S., and Laudan, L. (eds), *Physics, Philosophy, and Psychoanalysis: Essays in Honor of Adolf Grünbaum*. D. Reidel.

Laudan, L. 1984. *Science and Values*. University of California Press.

Laudan, R., Laudan, L., and Donovan, A. 1988. Testing theories of scientific change. Pages 3–46 of: Donovan, A., Laudan, L., and Laudan, R. (eds), *Scrutinizing Science*, vol. 13. Kluwer.

Lazutkina, A. 2017. Theoretical terms of contemporary cosmology as intellectual artifacts. *ArXiv e-prints*, arXiv:1707.05235.

Leibniz, G. W. 1678. Letter to Herman Conring, 19 March. Pages 186–191 of: Loemker, L. (ed), *Gottfried Wilhelm Leibniz: Philosophical Papers and Letters*, 2nd edn. Reidel, 1970.

Lelli, F., McGaugh, S. S., and Schombert, J. M. 2016a. SPARC: Mass models for 175 disk galaxies with Spitzer photometry and accurate rotation curves. *The Astronomical Journal*, **152**, 157.

Lelli, F., McGaugh, S. S., Schombert, J. M., and Pawlowski, M. S. 2016b. The relation between stellar and dynamical surface densities in the central regions of disk galaxies. *The Astrophysical Journal Letters*, **827**, L19.

Lelli, F., McGaugh, S. S., and Schombert, J. M. 2016c. The small scatter of the baryonic Tully–Fisher relation. *The Astrophysical Journal Letters*, **816**, L14.

Lelli, F., McGaugh, S. S., Schombert, J. M., and Pawlowski, M. S. 2017. One law to rule them all: The radial acceleration relation of galaxies. *The Astrophysical Journal*, **836**, 152.

Lem, S. 1970. Robots in science fiction. *The Journal of Omphalistic Epistemology*, Jan., 8–20.

Lemaître, G. 1927. Un Univers homogène de masse constante et de rayon croissant rendant compte de la vitesse radiale des nébuleuses extra-galactiques. *Annales de la Société Scientifique de Bruxelles*, **47**, 49–59.

Lind, K., Melendez, J., Asplund, M., Collet, R., and Magic, Z. 2013. The lithium isotopic ratio in very metal-poor stars. *Astronomy and Astrophysics*, **554**, A96.

Linsky, J. L. 2003. Atomic deuterium/hydrogen in the Galaxy. *Space Science Reviews*, **106**, 49–60.

Lisanti, M. 2017. Lectures on dark matter physics. Pages 399–446 of: *New Frontiers in Fields and Strings (TASI 2015)*.

Liu, J., Chen, X., and Ji, X. 2017. Current status of direct dark matter detection experiments. *Nature Physics*, **13**(3), 212–216.

Livio, M. 2011. Mystery of the missing text solved. *Nature*, **479**, 171–173.

Llinares, C., Knebe, A., and Zhao, H. 2008. Cosmological structure formation under MOND: A new numerical solver for Poisson's equation. *Monthly Notices of the Royal Astronomical Society*, **391**, 1778–1790.

Losee, J. 2004. *Theories of Scientific Progress*. Routledge.

Losee, J. 2005. *Theories on the Scrap Heap*. University of Pittsburgh Press.

Ludlow, A. D., Benítez-Llambay, A., Schaller, M., Theuns, T., Frenk, C. S., Bower, R., Schaye, J., Crain, R. A., Navarro, J. F., and Fattahi, A. 2017. Mass-discrepancy acceleration relation: A natural outcome of galaxy formation in cold dark matter halos. *Physical Review Letters*, **118**, 161103.

Mackey, A. D., and Gilmore, G. F. 2003. Surface brightness profiles and structural parameters for globular clusters in the Fornax and Sagittarius dwarf spheroidal galaxies. *Monthly Notices of the Royal Astronomical Society*, **340**, 175–190.

Madison, G. B. 1988. A critique of Hirsch's *Validity*. Chapter 1, pages 3–24 of: *The Hermeneutics of Postmodernity*. Indiana University Press.

Magee, B. 1997. *Confessions of a Philosopher: A Journey Through Western Philosophy*. Random House.

Mahoney, M. J., and DeMonbreun, B. G. 1977. Psychology of the scientist: An analysis of problem-solving bias. *Cognitive Therapy and Research*, **1**, 229–238.

Margenau, H. 1950. *The Nature of Physical Reality: A Philosophy of Modern Physics*. McGraw-Hill.

Mashchenko, S., Couchman, H. M. P., and Wadsley, J. 2006. The removal of cusps from galaxy centres by stellar feedback in the early Universe. *Nature*, **442**, 539–542.

Mashian, N., Oesch, P. A., and Loeb, A. 2016. An empirical model for the galaxy luminosity and star formation rate function at high redshift. *Monthly Notices of the Royal Astronomical Society*, **455**, 2101–2109.

Masterman, M. 1970. The nature of a paradigm. Pages 59–90 of: Lakatos, I., and Musgrave, A. (eds), *Criticism and the Growth of Knowledge*. Cambridge University Press.

Mathews, G. J., Kajino, T., and Shima, T. 2005. Big bang nucleosynthesis with a new neutron lifetime. *Physical Review D*, **71**, 021302.

Matteucci, F. 2003. What determines galactic evolution? *Astrophysics and Space Science*, **284**, 539–548.

Mayer, L., and Moore, B. 2004. The baryonic mass–velocity relation: Clues to feedback processes during structure formation and the cosmic baryon inventory. *Monthly Notices of the Royal Astronomical Sociey*, **354**, 477–484.

McCrea, W. H., and Milne, E. A. 1934. Newtonian universes and the curvature of space. *The Quarterly Journal of Mathematics*, **5**.

McCulloch, M. E. 2007. Modelling the Pioneer anomaly as modified inertia. *Monthly Notices of the Royal Astronomical Society*, **376**, 338–342.

McGaugh, S. S. 1996. The number, luminosity and mass density of spiral galaxies as a function of surface brightness. *Monthly Notices of the Royal Astronomical Society*, **280**, 337–354.

McGaugh, S. S. 1999a. Distinguishing between cold dark matter and modified Newtonian dynamics: Predictions for the microwave background. *The Astrophysical Journal Letters*, **523**, L99–L102.

McGaugh, S. S. 1999b. How galaxies don't form: The effective force law in disk galaxies. In Merritt, D. R., Valluri, M., and Sellwood, J. A. (eds), *Galaxy Dynamics: A Rutgers Symposium*. Astronomical Society of the Pacific Conference Series, vol. 182.

McGaugh, S. S. 2004. The mass discrepancy–acceleration relation: Disk mass and the dark matter distribution. *The Astrophysical Journal*, **609**, 652–666.

McGaugh, S. S. 2005. The baryonic Tully–Fisher relation of galaxies with extended rotation curves and the stellar mass of rotating galaxies. *The Astrophysical Journal*, **632**, 859–871.

McGaugh, S. S. 2008. Milky Way mass models and MOND. *The Astrophysical Journal*, **683**, 137–148.

McGaugh, S. S. 2011. Novel test of modified Newtonian dynamics with gas rich galaxies. *Physical Review Letters*, **106**, 121303.

McGaugh, S. S. 2012. The baryonic Tully–Fisher relation of gas-rich galaxies as a test of ΛCDM and MOND. *The Astronomical Journal*, **143**, 40.

McGaugh, S. S. 2015. A tale of two paradigms: The mutual incommensurability of ΛCDM and MOND. *Canadian Journal of Physics*, **93**, 250–259.

McGaugh, S. S., and de Blok, W. J. G. 1998a. Testing the dark matter hypothesis with low surface brightness galaxies and other evidence. *The Astrophysical Journal*, **499**, 41–65.

McGaugh, S. S., and de Blok, W. J. G. 1998b. Testing the hypothesis of modified dynamics with low surface brightness galaxies and other evidence. *The Astrophysical Journal*, **499**, 66–81.

McGaugh, S. S., and Milgrom, M. 2013a. Andromeda dwarfs in light of modified Newtonian dynamics. *The Astrophysical Journal*, **766**, 22.

McGaugh, S. S., and Milgrom, M. 2013b. Andromeda dwarfs in light of MOND. II. Testing prior predictions. *The Astrophysical Journal*, **775**, 139.

McGaugh, S. S., Schombert, J. M., Bothun, G. D., and de Blok, W. J. G. 2000. The baryonic Tully–Fisher relation. *The Astrophysical Journal Letters*, **533**, L99–L102.

McGaugh, S. S., Lelli, F., and Schombert, J. M. 2016. Radial acceleration relation in rotationally supported galaxies. *Physical Review Letters*, **117**, 201101.

Meléndez, J., and Ramírez, I. 2004. Reappraising the Spite lithium plateau: Extremely thin and marginally consistent with WMAP data. *The Astrophysical Journal Letters*, **615**, L33–L36.

Merritt, D. 1987. The distribution of dark matter in the coma cluster. *The Astrophysical Journal*, **313**, 121–135.

Merritt, D. 2001. Brownian motion of a massive binary. *The Astrophysical Journal*, **556**, 245–264.

Merritt, D. 2013. *Dynamics and Evolution of Galactic Nuclei*. Princeton: Princeton University Press.

Merritt, D. 2017. Cosmology and convention. *Studies in History and Philosophy of Modern Physics*, **57**, 41–52.

Merritt, D., and Milosavljević, M. 2002. Dynamics of dark-matter cusps. Pages 79–89 of: Klapdor-Kleingrothaus, H. V., and Viollier, R. D. (eds), *Dark Matter in Astro- and Particle Physics*. Springer.

Merritt, D., and Sellwood, J. A. 1994. Bending instabilities in stellar systems. *The Astrophysical Journal*, **425**, 551–567.

Michaud, G., Fontaine, G., and Beaudet, G. 1984. The lithium abundance: Constraints on stellar evolution. *The Astrophysical Journal*, **282**, 206–213.

Milgrom, M. 1983a. A modification of the Newtonian dynamics as a possible alternative to the hidden mass hypothesis. *The Astrophysical Journal*, **270**, 365–370.

Milgrom, M. 1983b. A modification of the Newtonian dynamics: Implications for galaxies. *The Astrophysical Journal*, **270**, 371–383.

Milgrom, M. 1983c. A modification of the Newtonian dynamics: Implications for galaxy systems. *The Astrophysical Journal*, **270**, 384–389.

Milgrom, M. 1984. Isothermal spheres in the modified dynamics. *The Astrophysical Journal*, **287**, 571–576.

Milgrom, M. 1989a. Alternatives to dark matter. *Comments on Astrophysics*, **13**, 215–230.

Milgrom, M. 1989b. On stability of galactic disks in the modified dynamics and the distribution of their mean surface brightness. *The Astrophysical Journal*, **338**, 121–127.

Milgrom, M. 1994a. Dynamics with a nonstandard inertia–acceleration relation: An alternative to dark matter in galactic systems. *Annals of Physics*, **229**, 384–415.

Milgrom, M. 1994b. Modified dynamics predictions agree with observations of the HI kinematics in faint dwarf galaxies contrary to the conclusions of Lo, Sargent, and Young. *The Astrophysical Journal*, **429**, 540–544.

Milgrom, M. 1997. Nonlinear conformally invariant generalization of the Poisson equation to $D > 2$ dimensions. *Physical Review E*, **56**, 1148–1159.

Milgrom, M. 1999. The modified dynamics as a vacuum effect. *Physics Letters A*, **253**, 273–279.

Milgrom, M. 2001a. MOND–A pedagogical review. *Acta Physica Polonica B*, **32**, 3613–3627.

Milgrom, M. 2001b. The shape of 'dark matter' haloes of disc galaxies according to MOND. *Monthly Notices of the Royal Astronomical Society*, **326**, 1261–1264.

Milgrom, M. 2002. Do modified Newtonian dynamics follow from the cold dark matter paradigm? *The Astrophysical Journal Letters*, **571**, L81–L83.

Milgrom, M. 2006. MOND as modified inertia. *EAS Publications Series*, **20**, 217–224.

Milgrom, M. 2008. The MOND paradigm. *ArXiv e-prints*, 0801.3133.

Milgrom, M. 2009. Bimetric MOND gravity. *Physical Review D*, **80**, 123536.

Milgrom, M. 2009a. The MOND limit from spacetime invariance. *The Astrophysical Journal*, **698**, 1630–1638.

Milgrom, M. 2009b. The central surface density of 'dark haloes' predicted by MOND. *Monthly Notices of the Royal Astronomical Society*, **398**, 1023–1026.

Milgrom, M. 2010. Quasi-linear formulation of MOND. *Monthly Notices of the Royal Astronomical Society*, **403**, 886–895.

Milgrom, M. 2011a. MD or DM? Modified dynamics at low accelerations vs dark matter. *ArXiv e-prints*, 1101.5122.

Milgrom, M. 2011b. MOND–particularly as modified inertia. *Acta Physica Polonica B*, **42**, 2175–2184.

Milgrom, M. 2014. MOND laws of galactic dynamics. *Monthly Notices of the Royal Astronomical Society*, **437**, 2531–2541.

Milgrom, M. 2015. MOND theory. *Canadian Journal of Physics*, **93**, 107–118.

Milgrom, M. 2016a. MOND impact on and of the recently updated mass-discrepancy–acceleration relation. *ArXiv e-prints*, arXiv:1609.06642.

Milgrom, M. 2016b. The ΛCDM simulations of Keller and Wadsley do not account for the MOND mass-discrepancy–acceleration relation. *ArXiv e-prints*, 1610.07538.

Milgrom, M. 2016c. Universal modified Newtonian dynamics relation between the baryonic and "dynamical" central surface densities of disc galaxies. *Physical Review Letters*, **117**, 141101.

Milgrom, M., and Braun, E. 1988. The rotation curve of DDO 154: A particularly acute test of the modified dynamics. *The Astrophysical Journal*, **334**, 130–133.

Milgrom, M., and Sanders, R. H. 2005. MOND predictions of 'halo' phenomenology in disc galaxies. *Monthly Notices of the Royal Astronomical Society*, **357**, 45–48.

Milgrom, M., and Sanders, R. H. 2008. Rings and shells of "dark matter" as MOND artifacts. *The Astrophysical Journal*, **678**, 131–143.

Miller, D. 2014a. *Critical Rationalism: A Restatement and Defence*. Open Court.

Miller, D. 2014b. Some hard questions for critical rationalism. *Discusiones Filosóficas*, **15**(24), 15–40.

Moffat, J. W. 2006. Scalar–tensor–vector gravity theory. *Journal of Cosmology and Astroparticle Physics*, **2006**, 004.

Moni Bidin, C., Carraro, G., Méndez, R. A., and Smith, R. 2012. Kinematical and chemical vertical structure of the galactic thick disk. II. A lack of dark matter in the solar neighborhood. *The Astrophysical Journal*, **751**, 30.

Moore, B. 1994. Evidence against dissipation-less dark matter from observations of galaxy haloes. *Nature*, **370**, 629–631.

Moore, B., Ghigna, S., Governato, F., Lake, G., Quinn, T., Stadel, J., and Tozzi, P. 1999. Dark matter substructure within galactic halos. *The Astrophysical Journal*, **524**, L19–L22.

Mott, A., Steffen, M., Caffau, E., Spada, F., and Strassmeier, K. G. 2017. Lithium abundance and $^6Li/^7Li$ ratio in the active giant HD 123351. I. A comparative analysis of 3D and 1D NLTE line-profile fits. *Astronomy and Astrophysics*, **604**, A44.

Motterlini, M. (ed). 1999. *For and Against Method: Including Lakatos's Lectures on Scientific Method and the Lakatos–Feyerabend Correspondence.* The University of Chicago Press.

Mucciarelli, A., Salaris, M., Lovisi, L., Ferraro, F. R., Lanzoni, B., Lucatello, S., and Gratton, R. G. 2011. Lithium abundance in the globular cluster M4: From the turn-off to the red giant branch bump. *Monthly Notices of the Royal Astronomical Society*, **412**, 81–94.

Musgrave, A. 1971. Kuhn's second thoughts. *The British Journal for the Philosophy of Science*, **22**, 287–297.

Musgrave, A. 1974. Logical versus historical theories of confirmation. *The British Journal for the Philosophy of Science*, **25**, 1–23.

Musgrave, A. 1978. Evidential support, falsification, heuristics, and anarchism. Pages 181–202 of: Radnitzky, G., and Andersson, G. (eds), *Progress and Rationality in Science*. Boston Studies in the Philosophy of Science, vol. 58. Dordrecht.

Musil, R. 1961. *Notebooks*. Quoted in the Foreword to *The Man Without Qualities, I*, p. xii. Secker and Warburg, 1961.

Naab, T., and Ostriker, J. P. 2017. Theoretical challenges in galaxy formation. *Annual Review of Astronomy and Astrophysics*, **55**, 59–109.

Nagai, D., Kravtsov, A. V., and Vikhlinin, A. 2007. Effects of galaxy formation on thermodynamics of the intracluster medium. *The Astrophysical Journal*, **668**(1), 1–14.

Natoli, J., and Hutcheon, L. (eds). 1993. *A Postmodern Reader*. State University of New York Press.

Netterfield, C. B., Ade, P. A. R., Bock, J. J., Bond, J. R., Borrill, J., Boscaleri, A., Coble, K., Contaldi, C. R., Crill, B. P., de Bernardis, P., Farese, P., Ganga, K., Giacometti, M., Hivon, E., Hristov, V. V., Iacoangeli, A., Jaffe, A. H., Jones, W. C., Lange, A. E., Martinis, L., Masi, S., Mason, P., Mauskopf, P. D., Melchiorri, A., Montroy, T., Pascale, E., Piacentini, F., Pogosyan, D., Pongetti, F., Prunet, S., Romeo, G., Ruhl, J. E., and Scaramuzzi, F. 2002. A measurement by BOOMERANG of multiple peaks in the angular power spectrum of the cosmic microwave background. *The Astrophysical Journal*, **571**, 604–614.

Newton-Smith, W. H. 1981. *The Rationality of Science*. Routledge and Kegan Paul.

Nicastro, F., Krongold, Y., Mathur, S., and Elvis, M. 2017. A decade of warm hot inter-galactic medium searches: Where do we stand and where do we go? *Astronomische Nachrichten*, **338**, 281–286.

Nicolis, A. 2011. Low-energy effective field theory for finite-temperature relativistic superfluids. *ArXiv e-prints*, 1108.2513.

Nipoti, C., Londrillo, P., and Ciotti, L. 2007a. Galaxy merging in modified Newtonian dynamics. *Monthly Notices of the Royal Astronomical Sociey*, **381**, L104–L108.

Nipoti, C., Londrillo, P., Zhao, H., and Ciotti, L. 2007b. Vertical dynamics of disc galaxies in modified Newtonian dynamics. *Monthly Notices of the Royal Astronomical Society*, **379**, 597–604.

Nordlander, T., Korn, A. J., Richard, O., and Lind, K. 2012. Lithium in globular clusters: Significant systematics. Atomic diffusion, the temperature scale, and pollution in NGC 6397. *Memorie della Societa Astronomica Italiana Supplementi*, **22**, 110.

Nusser, A. 2002. Modified Newtonian dynamics of large-scale structure. *Monthly Notices of the Royal Astronomical Society*, **331**, 909–916.

Oh, K. S., Lin, D. N. C., and Richer, H. B. 2000. Globular clusters in the Fornax dwarf spheroidal galaxy. *The Astrophysical Journal*, **531**, 727–738.

O'Hear, A. 1980. *Karl Popper*. Routledge and Kegan Paul.

Olive, K. A. 2004. Big bang nucleosynthesis in the post-WMAP era. Pages 190–205 of: Allen, R. E., Nanopoulos, D. V., and Pope, C. N. (eds), *The New Cosmology: Conference on Strings and Cosmology*. American Institute of Physics Conference Series, vol. 743.

Olive, K. A., Schramm, D. N., Turner, M. S., Yang, J., and Steigman, G. 1981. Big-bang nucleosynthesis as a probe of cosmology and particle physics. *The Astrophysical Journal*, **246**, 557–568.

Olive, K. A., et al. 2014. Review of particle physics (Particle Data Group). *Chinese Physics C*, **38**, 090001.

O'Malley, P. D., Bardayan, D. W., Adekola, A. S., Ahn, S., Chae, K. Y., Cizewski, J. A., Graves, S., Howard, M. E., Jones, K. L., Kozub, R. L., Lindhardt, L., Matos, M., Moazen, B. M., Nesaraja, C. D., Pain, S. D., Peters, W. A., Pittman, S. T., Schmitt, K. T., Shriner, Jr., J. F., Smith, M. S., Spassova, I., Strauss, S. Y., and Wheeler, J. L. 2011. Search for a resonant enhancement of the ^{7}Be + d reaction and primordial ^{7}Li abundances. *Physical Review C*, **84**, 042801.

Oman, K. A., Navarro, J. F., Fattahi, A., Frenk, C. S., Sawala, T., White, S. D. M., Bower, R., Crain, R. A., Furlong, M., and Schaller, M. 2015. The unexpected diversity of dwarf galaxy rotation curves. *Monthly Notices of the Royal Astronomical Society*, **452**, 3650–3665.

Oort, J. H. 1960. Note on the determination of K_z and on the mass density near the Sun. *Bulletin of the Astronomical Institutes of the Netherlands*, **15**, 45–53.

Ostrogradski, M. V. 1850. Mémoires sur les equations differentielles relatives au problème des isopérimètres. *Mémoires présentés à l'Académie impériale des sciences de Saint-Pétersbourg*, 385. Series VI (4).

Page, L., Nolta, M. R., Barnes, C., Bennett, C. L., Halpern, M., Hinshaw, G., Jarosik, N., Kogut, A., Limon, M., Meyer, S. S., Peiris, H. V., Spergel, D. N., Tucker, G. S., Wollack, E., and Wright, E. L. 2003. First-year Wilkinson Microwave Anisotropy Probe (WMAP) observations: Interpretation of the TT and TE angular power spectrum peaks. *The Astrophysical Journal Supplement Series*, **148**, 233–241.

Pagel, B. E. J. 1986. Nucleosynthesis. *Philosophical Transactions of the Royal Society of London Series A*, **320**, 557–564.

Papastergis, E., Giovanelli, R., Haynes, M. P., and Shankar, F. 2015. Is there a "too big to fail" problem in the field? *Astronomy and Astrophysics*, **574**, A113.

Patton, D. R., Carlberg, R. G., Marzke, R. O., Pritchet, C. J., da Costa, L. N., and Pellegrini, P. S. 2000. New techniques for relating dynamically close galaxy pairs to merger and accretion rates: Application to the second Southern Sky Redshift Survey. *The Astrophysical Journal*, **536**, 153–172.

Pawlowski, M. S., Famaey, B., Jerjen, H., Merritt, D., Kroupa, P., Dabringhausen, J., Lüghausen, F., Forbes, D. A., Hensler, G., and Hammer, F. 2014. Co-orbiting satellite galaxy structures are still in conflict with the distribution of primordial dwarf galaxies. *Monthly Notices of the Royal Astronomical Society*, **442**, 2362–2380.

Peacock, J. A. 1999. *Cosmological Physics*. Cambridge University Press.

Peebles, P. J. E. 2015. Dark matter. *Proceedings of the National Academy of Science*, **112**, 12246–12248.

Peletier, R. F., and Willner, S. P. 1991. Infrared images, Virgo spirals, and the Tully–Fisher law. *The Astrophysical Journal*, **382**, 382–395.

Peñarrubia, J., Pontzen, A., Walker, M. G., and Koposov, S. E. 2012. The coupling between the core/cusp and missing satellite problems. *The Astrophysical Journal*, **759**, L42.

Perrin, J. 1916. *Atoms*. Van Nostrand. trans. D. Ll. Hammick.

Peterson, R. C., and Carney, B. W. 1979. Abundance analyses of metal-poor stars. II. Yellow spectra of five dwarfs. *The Astrophysical Journal*, **231**, 762–780.

Pettini, M., and Bowen, D. V. 2001. A new measurement of the primordial abundance of deuterium: Toward convergence with the baryon density from the cosmic microwave background? *The Astrophysical Journal*, **560**, 41–48.

Pettini, M., and Cooke, R. 2012. A new, precise measurement of the primordial abundance of deuterium. *Monthly Notices of the Royal Astronomical Society*, **425**, 2477–2486.

Piffl, T., Blimey, J., McMillan, P. J., Steinmetz, M., Helmi, A., Wyse, R. F. G., Bienaymé, O., Bland-Hawthorn, J., Freeman, K., Gibson, B., Gilmore, G., Grebel, E. K., Kordopatis, G., Navarro, J. F., Parker, Q., Reid, W. A., Seabroke, G., Siebert, A., Watson, F., and Zwitter, T. 2014. Constraining the Galaxy's dark halo with RAVE stars. *Monthly Notices of the Royal Astronomical Society*, **445**, 3133–3151.

Pizzone, R. G., Spartá, R., Bertulani, C. A., Spitaleri, C., La Cognata, M., Lalmansingh, J., Lamia, L., Mukhamedzhanov, A., and Tumino, A. 2014. Big bang nucleosynthesis revisited via Trojan horse method measurements. *The Astrophysical Journal*, **786**, 112–120.

Planck, M. 1922. *The origin and development of the quantum theory. Nobel Prize in Physics Award Address, 1920*. Clarendon Press. Trans. H. T. Clarke and L. Silberstein. Reprinted in *The World of the Atom*, eds. H. A. Boorse and L. Motz (New York: Basic Books, 1966), p. 496–500.

Planck Collaboration XIII. 2016. *Planck* 2015 results. XIII. Cosmological parameters. *Astronomy and Astrophysics*, **594**, A13.

Planck Collaboration XLVI. 2016. *Planck* intermediate results. XLVI. Reduction of large-scale systematic effects in HFI polarization maps and estimation of the reionization optical depth. *Astronomy and Astrophysics*, **596**, A107.

Planck Collaboration XVI. 2014. *Planck* 2013 results. XVI. Cosmological parameters. *Astronomy and Astrophysics*, **571**, A16.

Polido, P., Jablonski, F., and Lépine, J. R. D. 2013. A galaxy model from Two Micron All Sky Survey star counts in the whole sky, including the plane. *The Astrophysical Journal*, **778**, 32–49.

Ponomareva, A. A., Verheijen, M. A. W., Peletier, R. F., and Bosma, A. 2017. The multiwavelength Tully–Fisher relation with spatially resolved H I kinematics. *Monthly Notices of the Royal Astronomical Society*, **469**, 2387–2400.

Popkin, R. H. 2003. *The History of Scepticism: From Savonarola to Bayle*. Oxford University Press.

Popper, K. 1945. *The Open Society and Its Enemies. Vol. 2. The High Tide of Prophecy: Hegel, Marx, and the Aftermath*. Routledge.

Popper, K. 1959. *The Logic of Scientific Discovery*. Basic Books.

Popper, K. 1963. *Conjectures and Refutations: The Growth of Scientific Knowledge*. Routledge & Kegan Paul.

Popper, K. 1972. *Objective Knowledge: An Evolutionary Approach*. Oxford University Press.

Popper, K. 1974. Intellectual autobiography. Pages 3–181 of: Schilpp, P. A. (ed), *The Philosophy of Karl Popper*. Open Court.

Popper, K. 1983. *Realism and the Aim of Science*. Rowman and Littlefield.

Prout, William. 1815. On the relation between the specific gravities of bodies in their gaseous state and the weights of their atoms. *Annals of Philosophy*, **6**, 321–330.

Read, J. I. 2014. The local dark matter density. *Journal of Physics G Nuclear Physics*, **41**, 063101.

Read, J. I., Goerdt, T., Moore, B., Pontzen, A. P., Stadel, J., and Lake, G. 2006. Dynamical friction in constant density cores: A failure of the Chandrasekhar formula. *Monthly Notices of the Royal Astronomical Society*, **373**, 1451–1460.

Reeves, H. 1994. On the origin of the light elements ($Z < 6$). *Reviews of Modern Physics*, **66**, 193–216.

Reno, M. H., and Seckel, D. 1988. Primordial nucleosynthesis: The effects of injecting hadrons. *Physical Review D*, **37**, 3441–3462.

Rich, J. 2010. *Fundamentals of Cosmology*. 2nd edn. Springer-Verlag.

Richard, O., Michaud, G., and Richer, J. 2005. Implications of *WMAP* observations on Li abundance and stellar evolution models. *The Astrophysical Journal*, **619**, 538–548.

Riemer-Sørensen, S., Webb, J. K., Crighton, N., Dumont, V., Ali, K., Kotuš, S., Bainbridge, M., Murphy, M. T., and Carswell, R. 2015. A robust deuterium abundance; re-measurement of the $z = 3.256$ absorption system towards the quasar PKS 1937-101. *Monthly Notices of the Royal Astronomical Society*, **447**, 2925–2936.

Riess, A. G., Macri, L. M., Hoffmann, S. L., Scolnic, D., Casertano, S., Filippenko, A. V., Tucker, B. E., Reid, M. J., Jones, D. O., Silverman, J. M., Chornock, R., Challis, P., Yuan, W., Brown, P. J., and Foley, R. J. 2016. A 2.4% determination of the local value of the Hubble constant. *The Astrophysical Journal*, **826**, 56.

Robertson, B. E., Ellis, R S., Furlanetto, S. R., and Dunlop, J. S. 2015. Cosmic reionization and early star-forming galaxies: A joint analysis of new constraints from *PLANCK* and the *Hubble Space Telescope*. *The Astrophysical Journal*, **802**, L19.

Romatka, R. 1992. *Alternativen zur "dunklen Materie"*. Ph.D. thesis, Max-Planck-Institut für Physik, Munich.

Rubin, V. C. 1983. Systematics of H II rotation curves. Pages 3–8 of: Athanassoula, E. (ed), *Internal Kinematics and Dynamics of Galaxies*. IAU Symposium, vol. 100.

Rubin, V. C., Ford, Jr., W. K., and Thonnard, N. 1980. Rotational properties of 21 SC galaxies with a large range of luminosities and radii, from NGC 4605 ($R = 4$ kpc) to UGC 2885 ($R = 122$ kpc). *The Astrophysical Journal*, **238**, 471–487.

Ryan, S. G., Beers, T. C., Olive, K. A., Fields, B. D., and Norris, J. E. 2000. Primordial lithium and big bang nucleosynthesis. *The Astrophysical Journal Letters*, **530**, L57–L60.

Ryle, G. 1954. *Dilemmas*. Cambridge University Press.

Sagi, E. 2009. Preferred frame parameters in the tensor-vector-scalar theory of gravity and its generalization. *Physical Review D*, **80**, 044032.

Sales, L. V., Navarro, J. F., Oman, K., Fattahi, A., Ferrero, I., Abadi, M., Bower, R., Crain, R. A., Frenk, C. S., Sawala, T., Schaller, M., Schaye, J., Theuns, T., and White, S. D. M. 2017. The low-mass end of the baryonic Tully–Fisher relation. *Monthly Notices of the Royal Astronomical Society*, **464**, 2419–2428.

Salucci, P. 2001. The constant-density region of the dark haloes of spiral galaxies. *Monthly Notices of the Royal Astronomical Society*, **320**, L1–L5.

Sancisi, R. 2004. The visible matter–dark matter coupling. Page 233 of: Ryder, S., Pisano, D., Walker, M., and Freeman, K. (eds), *Dark Matter in Galaxies*. IAU Symposium, vol. 220.

Sanders, R. H. 1990. Mass discrepancies in galaxies: Dark matter and alternatives. *Astronomy and Astrophysics Review*, **2**, 1–28.

Sanders, R. H. 1994. A Faber–Jackson relation for clusters of galaxies: Implications for modified dynamics. *Astronomy and Astrophysics*, **284**, L31–L34.

Sanders, R. H. 1997. A stratified framework for scalar-tensor theories of modified dynamics. *The Astrophysical Journal*, **480**, 492–502.

Sanders, R. H. 1998. Cosmology with modified Newtonian dynamics (MOND). *Monthly Notices of the Royal Astronomical Society*, **296**, 1009–1018.

Sanders, R. H. 1999. The virial discrepancy in clusters of galaxies in the context of modified Newtonian dynamics. *The Astrophysical Journal Letters*, **512**, L23–L26.

Sanders, R. H. 2000. The fundamental plane of elliptical galaxies with modified Newtonian dynamics. *Monthly Notices of the Royal Astronomical Society*, **313**, 767–774.

Sanders, R. H. 2001. The formation of cosmic structure with modified Newtonian dynamics. *The Astrophysical Journal*, **560**, 1–6.

Sanders, R. H. 2003. Clusters of galaxies with modified Newtonian dynamics. *Monthly Notices of the Royal Astronomical Society*, **342**, 901–908.

Sanders, R. H. 2007. Neutrinos as cluster dark matter. *Monthly Notices of the Royal Astronomical Society*, **380**, 331–338.

Sanders, R. H. 2008. Forming galaxies with MOND. *Monthly Notices of the Royal Astronomical Society*, **386**, 1588–1596.

Sanders, R. H. 2010. The universal Faber–Jackson relation. *Monthly Notices of the Royal Astronomical Society*, **407**, 1128–1134.

Sanders, R. H. 2015. A historical perspective on modified Newtonian dynamics. *Canadian Journal of Physics*, **93**, 126–138.

Sanders, R. H., and Land, D. D. 2008. MOND and the lensing fundamental plane: No need for dark matter on galaxy scales. *Monthly Notices of the Royal Astronomical Society*, **389**, 701–705.

Sanders, R. H., and McGaugh, S. S. 2002. Modified Newtonian dynamics as an alternative to dark matter. *Annual Reviews of Astronomy and Astrophysics*, **40**, 263–317.

Sanders, R. H., and Noordermeer, E. 2007. Confrontation of modified Newtonian dynamics with the rotation curves of early-type disc galaxies. *Monthly Notices of the Royal Astronomical Society*, **379**, 702–710.

Santos-Santos, I. M., Brook, C. B., Stinson, G., Di Cintio, A., Wadsley, J., Domínguez-Tenreiro, R., Gottlöber, S., and Yepes, G. 2016. The distribution of mass components in simulated disc galaxies. *Monthly Notices of the Royal Astronomical Society*, **455**, 476–483.

Sarazin, C. L. 1988. *X-Ray Emission from Clusters of Galaxies*. Cambridge Astrophysics Series. Cambridge University Press.

Sarkar, S. 1996. Big bang nucleosynthesis and physics beyond the standard model. *Reports on Progress in Physics*, **59**, 1493–1609.

Savchenko, V., Ferrigno, C., Kuulkers, E., Bazzano, A., Bozzo, E., Brandt, S., Chenevez, J., Courvoisier, T. J.-L., Diehl, R., Domingo, A., Hanlon, L., Jourdain, E., von Kienlin, A., Laurent, P., Lebrun, F., Lutovinov, A., Martin-Carrillo, A., Mereghetti, S., Natalucci, L., Rodi, J., Roques, J.-P., Sunyaev, R., and Ubertini, P. 2017. INTEGRAL detection of the first prompt gamma-ray signal coincident with the gravitational-wave event GW170817. *The Astrophysical Journal*, **848**, L15.

Sbordone, L., Bonifacio, P., Caffau, E., Ludwig, H.-G., Behara, N. T., González Hernández, J. I., Steffen, M., Cayrel, R., Freytag, B., van't Veer, C., Molaro, P., Plez, B., Sivarani, T., Spite, M., Spite, F., Beers, T. C., Christlieb, N., François, P., and Hill, V. 2010. The metal-poor end of the Spite plateau. I. Stellar parameters, metallicities, and lithium abundances. *Astronomy and Astrophysics*, **522**, A26–A47.

Schaffner, K. 1969. Correspondence rules. *Philosophy of Science*, **36**, 280–290.

Schaye, J., Crain, R. A., Bower, R. G., Furlong, M., Schaller, M., Theuns, T., Dalla Vecchia, C., Frenk, C. S., McCarthy, I. G., Helly, J. C., Jenkins, A., Rosas-Guevara, Y. M., White, S. D. M., Baes, M., Booth, C. M., Camps, P., Navarro, J. F., Qu, Y., Rahmati, A.,

Sawala, T., Thomas, P. A., and Trayford, J. 2015. The EAGLE project: Simulating the evolution and assembly of galaxies and their environments. *Monthly Notices of the Royal Astronomical Society*, **446**, 521–554.

Schilpp, P. A. 1974. *The Philosophy of Karl Popper, Books 1 and 2*. The Library of Living Philosophers. Open Court.

Schmalzing, J., Sommer-Larsen, J., and Goetz, M. 2000. Constraints on the redshift of reionization from CMB data. *arXiv e-prints*, astro–ph/0010063.

Schneider, P. 2015. *Extragalactic Astronomy and Cosmology*. 2nd edn. Springer.

Scholl, C., Fujita, Y., Adachi, T., von Brentano, P., Fujita, H., Górska, M., Hashimoto, H., Hatanaka, K., Matsubara, H., Nakanishi, K., Ohta, T., Sakemi, Y., Shimbara, Y., Shimizu, Y., Tameshige, Y., Tamii, A., Yosoi, M., and Zegers, R. G. T. 2011. High-resolution study of the $^9\mathrm{Be}(^3\mathrm{He}, t)^9\mathrm{B}$ reaction up to the $^9\mathrm{B}$ triton threshold. *Physical Review C*, **84**, 014308.

Schombert, J. M., McGaugh, S. S., and Eder, J. A. 2001. Gas mass fractions and the evolution of low surface brightness dwarf galaxies. *The Astronomical Journal*, **121**, 2420–2430.

Schramm, D. N. 1982. Constraints on the density of baryons in the universe. *Philosophical Transactions of the Royal Society of London Series A*, **307**, 43–53.

Schramm, D. N. 1991. Big bang nucleosynthesis: The standard model and alternatives. *Physica Scripta Volume T*, **36**, 22–29.

Schramm, D. N. 1998. Primordial nucleosynthesis. *Proceedings of the National Academy of Science*, **95**, 42–46.

Schramm, D. N., and Turner, M. S. 1998. Big-bang nucleosynthesis enters the precision era. *Reviews of Modern Physics*, **70**, 303–318.

Scully, S., Cassé, M., Olive, K. A., and Vangioni-Flam, E. 1997. The effects of an early galactic wind on the evolution of D, $^3\mathrm{He}$, and Z. *The Astrophysical Journal*, **476**, 521–533.

Seljak, U., and Zaldarriaga, M. 1996. A line-of-sight integration approach to cosmic microwave background anisotropies. *The Astrophysical Journal*, **469**, 437.

Sellwood, J. A., and McGaugh, Stacy S. 2005. The compression of dark matter halos by baryonic infall. *The Astrophysical Journal*, **634**, 70–76.

Serpico, P. D., Esposito, S., Iocco, F., Mangano, G., Miele, G., and Pisanti, O. 2004. Nuclear reaction network for primordial nucleosynthesis: A detailed analysis of rates, uncertainties and light nuclei yields. *Journal of Cosmology and Astroparticle Physics*, **12**, 010.

Shaposhnikov, M. 2010. Sterile neutrinos. Page 228 of: Bertone, G. (ed), *Particle Dark Matter: Observations, Models and Searches*. Cambridge University Press.

Sherman, P. W., Jarvis, J. U. M., and Alexander, R. D. (eds). 1991. *The Biology of the Naked Mole-Rat*. Princeton University Press.

Shull, J. M., Smith, B. D., and Danforth, C. W. 2012. The baryon census in a multiphase intergalactic medium: 30% of the baryons may still be missing. *The Astrophysical Journal*, **759**, 23.

Siebert, A., Bienaymé, O., and Soubiran, C. 2003. Vertical distribution of Galactic disk stars. II. The surface mass density in the Galactic plane. *Astronomy and Astrophysics*, **399**, 531–541.

Siebert, A., Bienaymé, O., Blimey, J., Bland-Hawthorn, J., Campbell, R., Freeman, K. C., Gibson, B. K., Gilmore, G., Grebel, E. K., Helmi, A., Munari, U., Navarro, J. F., Parker, Q. A., Seabroke, G., Siviero, A., Steinmetz, M., Williams, M., Wyse, R. F. G., and Zwitter, T. 2008. Estimation of the tilt of the stellar velocity ellipsoid from RAVE and implications for mass models. *Monthly Notices of the Royal Astronomical Society*, **391**, 793–801.

Siegel, M. H., Majewski, S. R., Reid, I. N., and Thompson, I. B. 2002. Star counts redivivus. IV. Density laws through photometric parallaxes. *The Astrophysical Journal*, **578**, 151–175.

Silk, J. 2004. Dark matter theory. Pages 67–77 of: Freeman, W. L. (ed), *Measuring and Modeling the Universe*. Carnegie Observatories Astrophysics Series, vol. 2.

Silk, J., and Mamon, G. A. 2012. The current status of galaxy formation. *Research in Astronomy and Astrophysics*, **12**, 917–946.

Skordis, C. 2008. Generalizing tensor-vector-scalar cosmology. *Physical Review D*, **77**, 123502.

Skordis, C., and Zlosnik, T. 2012. Geometry of modified Newtonian dynamics. *Physical Review D*, **85**, 044044.

Slater, C. T., Bell, E. F., and Martin, N. F. 2011. Andromeda XXVIII: A dwarf galaxy more than 350 kpc from Andromeda. *The Astrophysical Journal*, **742**, L14.

Slater, J. C. 1960. *Quantum Theory of Atomic Structure*. Vol. 1. McGraw-Hill.

Smith, M. C., Whiteoak, S. H., and Evans, N. W. 2012. Slicing and dicing the Milky Way disk in the Sloan digital sky survey. *The Astrophysical Journal*, **746**, 181.

Smith, M. S., Kawano, L. H., and Malaney, R. A. 1993. Experimental, computational, and observational analysis of primordial nucleosynthesis. *The Astrophysical Journal Supplement Series*, **85**, 219–247.

Son, D. T., and Wingate, M. 2006. General coordinate invariance and conformal invariance in nonrelativistic physics: Unitary Fermi gas. *Annals of Physics*, **321**, 197–224.

Sorce, J. G., and Guo, Q. 2016. The baryonic Tully–Fisher relation cares about the galaxy sample. *Monthly Notices of the Royal Astronomical Sociey*, **458**, 2667–2675.

Sorce, J. G., Courtois, H. M., Tully, R. B., Seibert, M., Scowcroft, V., Freedman, W. L., Madore, B. F., Persson, S. E., Monson, A., and Rigby, J. 2013. Calibration of the mid-infrared Tully–Fisher relation. *The Astrophysical Journal*, **765**, 94.

Soussa, M. E., and Woodard, R. P. 2004. A generic problem with purely metric formulations of MOND. *Physics Letters B*, **578**, 253–258.

Spergel, D. N., Verde, L., Peiris, H. V., Komatsu, E., Nolta, M. R., Bennett, C. L., Halpern, M., Hinshaw, G., Jarosik, N., Kogut, A., Limon, M., Meyer, S. S., Page, L., Tucker, G. S., Weiland, J. L., Wollack, E., and Wright, E. L. 2003. First-year Wilkinson Microwave Anisotropy Probe (*WMAP*) observations: Determination of cosmological parameters. *The Astrophysical Journal Supplement*, **148**, 175–194.

Spite, F. 1984. Intérêt cosmologique de l'étude du lithium dans l'univers. *L'Astronomie*, **98**, 371–380.

Spite, F., and Spite, M. 1982. Abundance of lithium in unevolved halo stars and old disk stars: Interpretation and consequences. *Astronomy and Astrophysics*, **115**, 357–366.

Spite, M., Maillard, J. P., and Spite, F. 1984. Abundance of lithium in another sample of halo dwarfs, and in the spectroscopic binary BD-0 deg 4234. *Astronomy and Astrophysics*, **141**, 56–60.

Spite, M., Spite, F., and Bonifacio, P. 2012. The cosmic lithium problem: An observer's perspective. *Memorie della Societa Astronomica Italiana Supplementi*, **22**, 9.

Srianand, R., Gupta, N., Petitjean, P., Noterdaeme, P., and Ledoux, C. 2010. Detection of 21-cm, H_2 and deuterium absorption at $z > 3$ along the line of sight to J1337+3152. *Monthly Notices of the Royal Astronomical Society*, **405**, 1888–1900.

Stachniewicz, S., and Kutschera, M. 2001. The first compact objects in the MOND model. *Acta Physica Polonica B*, **32**, 3629.

Stachniewicz, S., and Kutschera, M. 2003. The first compact objects in the Λ-dominated universe. *Monthly Notices of the Royal Astronomical Society*, **339**, 616–622.

Stachniewicz, S., and Kutschera, M. 2005. The end of the dark ages in modified Newtonian dynamics. *Monthly Notices of the Royal Astronomical Society*, **362**, 89–94.

Stark, D. V., McGaugh, S. S., and Swaters, R. A. 2009. A first attempt to calibrate the baryonic Tully–Fisher relation with gas-dominated galaxies. *The Astronomical Journal*, **138**, 392–401.

Steigman, G. 2007. Primordial nucleosynthesis in the precision cosmology era. *Annual Review of Nuclear and Particle Science*, **57**, 463–491.

Suppes, P. 1967. What is a scientific theory? Pages 55–67 of: Morgenbesser, S. (ed), *Philosophy of Science Today*. Basic Books.

Tegmark, M., and Zaldarriaga, M. 2000. New microwave background constraints on the cosmic matter budget: Trouble for nucleosynthesis? *Physical Review Letters*, **85**, 2240–2243.

Tenneti, A., Mao, Y.-Y., Croft, R. A. C., Di Matteo, T., Kosowsky, A., Zago, F., and Zentner, A. R. 2017. The radial acceleration relation in disc galaxies in the MassiveBlack-II simulation. *ArXiv e-prints*, 1703.05287.

Tollerud, E. J., Geha, M. C., Vargas, L. C., and Bullock, J. S. 2013. The outer limits of the M31 system: Kinematics of the dwarf galaxy satellites And XXVIII & And XXIX. *The Astrophysical Journal*, **768**, 50.

Toomre, A. 1963. On the distribution of matter within highly flattened galaxies. *The Astrophysical Journal*, **138**, 385.

Toomre, A. 1977. Mergers and some consequences. Page 401 of: Tinsley, B. M., and Larson, D. Campbell, R. B. G. (eds), *Evolution of Galaxies and Stellar Populations*.

Toomre, A. 1981. What amplifies the spirals? Pages 111–136 of: Fall, S. M., and Lynden-Bell, D. (eds), *Structure and Evolution of Normal Galaxies*.

Trachternach, C., de Blok, W. J. G., McGaugh, S. S., van der Hulst, J. M., and Dettmar, R.-J. 2009. The baryonic Tully–Fisher relation and its implication for dark matter halos. *Astronomy and Astrophysics*, **505**, 577–587.

Trimble, V. 1996. H_0: The incredible shrinking constant, 1925–1975. *Publications of the Astronomical Society of the Pacific*, **108**, 1073–1082.

Trujillo-Gomez, S., Klypin, A., Primack, J., and Romanowsky, A. J. 2011. Galaxies in ΛCDM with halo abundance matching: luminosity–velocity relation, baryonic mass–velocity relation, velocity function, and clustering. *The Astrophysical Journal*, **742**, 16.

Tully, R. B., and Courtois, H. M. 2012. Cosmicflows-2: I-band luminosity–H I linewidth calibration. *The Astrophysical Journal*, **749**, 78–94.

Tully, R. B., and Fisher, J. R. 1977. A new method of determining distances to galaxies. *Astronomy and Astrophysics*, **54**, 661–673.

Tully, R. B., and Pierce, M. J. 2000. Distances to galaxies from the correlation between luminosities and line widths. III. Cluster template and global measurement of H_0. *The Astrophysical Journal*, **533**, 744–780.

Turner, M. S. 1999. Cosmology solved? Quite possibly! *Publications of the Astronomical Society of the Pacific*, **111**, 264–273.

Turner, M. S. 2000. The dark side of the universe: From Zwicky to accelerated expansion. *Physics Reports*, **333**, 619–635.

Urbach, P. 1978. The objective promise of a research programme. Pages 99–116 of: Radnitzky, G., and Andersson, G. (eds), *Progress and Rationality in Science*. Boston Studies in the Philosophy of Science, vol. 58. D. Reidel.

van den Bergh, S. 2011. The curious case of Lemaître's equation no. 24. *Journal of the Royal Astronomical Society of Canada*, **105**, 151.

van den Bosch, F. C., and Dalcanton, J. J. 2000. Semianalytical models for the formation of disk galaxies. II. Dark matter versus Modified Newtonian Dynamics. *The Astrophysical Journal*, **534**, 146–164.

van den Bosch, F. C., Mo, H. J., and Yang, X. 2003. Towards cosmological concordance on galactic scales. *Monthly Notices of the Royal Astronomical Society*, **345**, 923–938.

Vandervoort, P. O. 1970. The equilibria of a galaxy which is a superposition of subsystems. *The Astrophysical Journal*, **162**, 453–462.

Verheijen, M. A. W. 2001. The Ursa Major cluster of galaxies. V. H I rotation curve shapes and the Tully–Fisher relations. *The Astrophysical Journal*, **563**, 694–715.

Vikhlinin, A., Kravtsov, A., Forman, W., Jones, C., Markevitch, M., Murray, S. S., and Van Speybroeck, L. 2006. Chandra sample of nearby relaxed galaxy clusters: Mass, gas fraction, and mass–temperature relation. *The Astrophysical Journal*, **640**, 691–709.

Vogelsberger, M., Genel, S., Springel, V., Torrey, P., Sijacki, D., Xu, D., Snyder, G., Bird, S., Nelson, D., and Hernquist, L. 2014. Properties of galaxies reproduced by a hydrodynamic simulation. *Nature*, **509**, 177–182.

Walch, S., and Naab, T. 2015. The energy and momentum input of supernova explosions in structured and ionized molecular clouds. *Monthly Notices of the Royal Astronomical Society*, **451**, 2757–2771.

Walker, M. A. 1999. Collisional baryonic dark matter haloes. *Monthly Notices of the Royal Astronomical Society*, **308**, 551–556.

Walker, T. P., Steigman, G., Schramm, D. N., Olive, K. A., and Kang, H.-S. 1991. Primordial nucleosynthesis redux. *The Astrophysical Journal*, **376**, 51–69.

Walker, T. P., Steigman, G., Schramm, D. N., Olive, K. A., and Kang, H.-S. 1996. Primordial nucleosynthesis redux. Pages 43–61 of: Schramm, D. N. (ed), *The Big Bang and Other Explosions in Nuclear and Particle Astrophysics*. World Scientific Publishing Co.

Wang, X., Tegmark, M., Jain, B., and Zaldarriaga, M. 2003. Last stand before WMAP: Cosmological parameters from lensing, CMB, and galaxy clustering. *Physical Review D*, **68**, 123001.

Weinberg, S. 2008. *Cosmology*. Oxford University Press.

Weinzirl, T., Jogee, S., Khochfar, S., Burkert, A., and Kormendy, J. 2009. Bulge n and B/T in high-mass galaxies: Constraints on the origin of bulges in hierarchical models. *The Astrophysical Journal*, **696**, 411–447.

Whewell, W. 1847. *The Philosophy of the Inductive Sciences: Founded Upon Their History*. John W. Parker.

White, S. D. M., Navarro, J. F., Evrard, A. E., and Frenk, C. S. 1993. The baryon content of galaxy clusters: A challenge to cosmological orthodoxy. *Nature*, **366**, 429–433.

Willman, B., and Strader, J. 2012. "'Galaxy," defined. *The Astrophysical Journal*, **144**, 76.

Wisdom, J. O. 1968. Refutation by observation and refutation by theory. Pages 65–67 of: Lakatos, I., and Musgrave, A. (eds), *Problems in the Philosophy of Science*. North-Holland.

Woolley, R., and Stewart, J. M. 1967. Motion of A stars perpendicular to the galactic plane-II. *Monthly Notices of the Royal Astronomical Society*, **136**, 329.

Worrall, J. 1978a. Research programmes, empirical support, and the Duhem problem: Replies to criticism. Pages 321–338 of: Radnitzky, G., and Andersson, G. (eds), *Progress and Rationality in Science*. Boston Studies in the Philosophy of Science, vol. 58. D. Reidel.

Worrall, J. 1978b. The ways in which the methodology of scientific research programmes improves on Popper's methodology. Pages 45–70 of: Radnitzky, G., and Andersson, G. (eds), *Progress and Rationality in Science*. Boston Studies in the Philosophy of Science, vol. 58. D. Reidel.

Worrall, J. 2007. Miracles and models: Why reports of the death of structural realism may be exaggerated. Pages 125–154 of: O'Hear, A. (ed), *Philosophy of Science*. Royal Institute of Philosophy Supplement Series, vol. 61. Cambridge University Press.

Wu, X., and Kroupa, P. 2015. Galactic rotation curves, the baryon-to-dark-halo-mass relation and space-time scale invariance. *Monthly Notices of the Royal Astronomical Society*, **446**, 330–344.

Yang, J., Turner, M. S., Steigman, G., Schramm, D. N., and Olive, K. A. 1984. Primordial nucleosynthesis: A critical comparison of theory and observation. *The Astrophysical Journal*, **281**, 493–511.

Zahar, E. 1973. Why did Einstein's research programme supersede Lorentz's? *The British Journal for the Philosophy of Science*, **24**, 95–123 and 223–262.

Zaritsky, D., Courtois, H., Muñoz-Mateos, J.-C., Sorce, J., Erroz-Ferrer, S., Comerón, S., Gadotti, D. A., Gil de Paz, A., Hinz, J. L., Laurikainen, E., Kim, T., Laine, J., Menéndez-Delmestre, K., Mizusawa, T., Regan, M. W., Salo, H., Seibert, M., Sheth, K., Athanassoula, E., Bosma, A., Cisternas, M., Ho, L. C., and Holwerda, B. 2014. The baryonic Tully–Fisher relationship for S^4G galaxies and the "condensed" baryon fraction of galaxies. *The Astronomical Journal*, **147**, 134.

Zavala, J., Avila-Reese, V., Hernández-Toledo, H., and Firmani, C. 2003. The luminous and dark matter content of disk galaxies. *Astronomy and Astrophysics*, **412**, 633–650.

Zhang, L., Rix, H.-W., van de Ven, G., Bovy, J., Liu, C., and Zhao, G. 2013. The gravitational potential near the Sun from SEGUE K-dwarf kinematics. *The Astrophysical Journal*, **772**, 108–121.

Zhao, H. 2008. Reinterpreting MOND: Coupling of Einsteinian gravity and spin of cosmic neutrinos? *ArXiv e-prints*, 0805.4046.

Zlosnik, T. G., Ferreira, P. G., and Starkman, G. D. 2007. Modifying gravity with the aether: An alternative to dark matter. *Physical Review D*, **75**, 044017.

Zwicky, F. 1937. On the masses of nebulae and of clusters of nebulae. *The Astrophysical Journal*, **86**, 217–245.

Index

Printed in the United States
by Baker & Taylor Publisher Services